Psychology Revivals

The Wave Theory of Difference and Similarity

Two experimental procedures prompted the empirical development of psychophysical models: those that measure response frequency, often referred to as response probability; and those that measure response time, sometimes referred to as reaction time. The history of psychophysics is filled with theories that predict one or the other of these two responses. Yet the persistent reappearance of empirical relationships between these two measures of performance makes clear the need for a theory that both predicts and relates these two measures. Most likely, both response measures are the result of a single process that generates empirical laws relating response time and response probability. It is this process — its theory, description, and application — that is the topic of *The Wave Theory of Difference and Similarity*. Originally published in 1992, the author of this book has set out to provide a theoretical foundation for formulating new theories that systematize earlier results and to stimulate new concepts and introduce new tools for exploring mental phenomena and improving mental measurement.

The Wave Theory of Difference and Similarity

Stephen W. Link

Routledge
Taylor & Francis Group
LONDON AND NEW YORK

First published in 1992
by Lawrence Erlbaum Associates, Inc.

This edition first published in 2020 by Routledge
2 Park Square, Milton Park, Abingdon, Oxon OX14 4RN

and by Routledge
52 Vanderbilt Avenue, New York, NY 10017

Routledge is an imprint of the Taylor & Francis Group, an informa business

Publisher's Note

The publisher has gone to great lengths to ensure the quality of this reprint but points out that some imperfections in the original copies may be apparent.

Disclaimer

The publisher has made every effort to trace copyright holders and welcomes correspondence from those they have been unable to contact.

A Library of Congress record exists under ISBN: 0805809260

ISBN: 978-0-367-15046-4 (hbk)
ISBN: 978-0-429-05470-9 (ebk)

THE WAVE THEORY OF DIFFERENCE AND SIMILARITY

SCIENTIFIC PSYCHOLOGY SERIES
Monographs

Stephen W. Link • The Wave Theory of Difference and Similarity

William R. Uttal • The Swimmer: An Integrated Computational Model of a Perceptual-Motor System

SCIENTIFIC PSYCHOLOGY SERIES
Edited Volumes

Hans-Georg Geissler, Stephen W. Link, and James T. Townsend • Cognition, Information Processing, and Psychophysics: Basic Issues

F. Gregory Ashby • Multidimensional Models of Perception and Cognition

THE WAVE THEORY OF DIFFERENCE AND SIMILARITY

Stephen W. Link

LEA LAWRENCE ERLBAUM ASSOCIATES, PUBLISHERS
1992 Hillsdale, New Jersey Hove and London

Lawrence Erlbaum Associates, Inc., Publishers
365 Broadway
Hillsdale, New Jersey 07642

Library of Congress Cataloging-in-Publication Data

Link, Stephen W.
Wave theory of difference and similarity / Stephen W. Link.
p. .cm.
Includes bibliographical references and indexes.
ISBN 0-8058-0926-0
1. Psychophysics. 2. Similarity judgment. 3. Difference (Psychology) 4. Wave-discrimination, Theory of. I. Title.
BF237.L46 1992
152.8—dc20 91-27650
CIP

To
Richard Atkinson and William Estes

Table of Contents

Preface

This volume presents and extends the classical psychological theory of judgment. The classical theory originated with Gustav Fechner's extension of Gauss' Theory of Error. Fechner's theory of the comparison of two sensations, and subsequent experimental studies of sensory discrimination and aesthetics, opened to scientific measurement previously unseen mental events. Part I of Wave Theory presents Fechner's theory, the basis for modern statistical theories of choice, and the many outstanding scientific achievements of later scientists who tested and advanced Fechner's work. Throughout these advances important roles are played by the three primary measures of discrimination: response frequency, response time, and response confidence.

Part I is a platform for the development of Part II. The many seemingly disparate results of Part I are rejoined through a new synthesis presented in Part II. Rather than attributing variability to error, as does the Gaussian Theory of Error, variability is created by transceivers that, in response to stimulation, generate a Poisson wave consisting of quantum releases of energy. The comparisons of such Poisson Waves provides the basis for bipolar judgments between stimuli (e.g., *more* or *less* and *larger* or *smaller*). A separate mechanism compares Poisson Waves for the judgments of equality, of *sameness* or *difference*.

The analysis of these processes reveals the origins of Weber's Law, Fechner's Law and relates Weber's constants to the exponents of the Power Law advocated by S. S. Stevens. Wave Discrimination Theory gives rise to a theory of sensation and a theory of feeling. Wave Similarity Theory provides a basis for aesthetic judgments and interrelates the various response measures used to investigate feelings of equality.

The division of this book into two parts is a consequence of the unusual development of the field of psychology. Scientific psychology's origin lay in Fechner's brilliant extension of Gauss's Theory of Error. The mathematical difficulties produced by this extension obliged Fechner to seek the aid of the great German topologist A. F. Möbius for confirmation of these new theoretical results. Yet, by the turn of the 19th century, mathematics, the *primum mobile* at the founding of scientific psychology, was not part of the psychologist's curriculum at all. Indeed, no less an eminent authority than E. B. Titchener, scion of American experimental psychology, proclaimed that "Knowledge of elementary mathematics is part of a man's general scientific outfit: but one may work a lifetime, and with success, in psychology, without needing the knowledge" (1901, p. xxiii).

Fortunately mathematics remained an important tool for developing psychological theory in spite of psychology's drift into behaviorism and serendipity. No one did more to advance the use of mathematics in psychology than Patrick Suppes, Director of Stanford University's Institute for Mathematical Studies in the Social Sciences. Suppes' genius brought brilliance to Stanford. Richard Atkinson, Gordon Bower, William Estes, Duncan Luce, William McGill, Frank Restle . . . the list of extraordinary scientists that peopled the Institute's Ventura Hall reads more like a Dean's faculty wish-list than a historical fact.

In this regard I thank the faculty and fellows of the Institute for their endless patience and brilliant lectures. Although the idea of using the difference between Poisson waves in conjunction with an accumulative process is my own, I owe much of my deeper knowledge about Poisson processes in psychophysics to William McGill's instruction. At one time my analyses depended on the mathematical methods common to electrical engineering; methods that I learned during my association with James C. Bliss at the Control Systems Laboratory of Stanford Research Institute. Later, I discovered how to relate the theory of stochastic processes to the analysis of waveforms through the techniques of sequential analysis. Here I owe much to the stimulating and often amazing lectures of Charles Stein.

Although theory sometimes advances on the horns of experimental data, mathematical theorizing and experimental investigation are complementary rather than opposing processes. Part I presents advances of theory and data in a way that profits from the historical methods found in Boring's (1950) "History of Experimental Psychology" and Kimble's (1961) "Hilgard and Marquis' Conditioning and Learning." To Howard Gruber and Albert Hastorf I owe much of my sense of the history of ideas in psychology, although the significance of the experiments in Part I gained much from discussions with Clyde Coombs and Frank Restle and from the helpful suggestions of Warren Torgerson, Leon Festinger, and Dorwin Cartwright.

In something of a departure from tradition, the citations in the text are supplemented in the reference section by articles that are meant to tweak the reader's interest. Due to the need for detailed discussion of applications of Wave Theory to classical discrimination experiments, much of this important additional material could not be discussed. In the case of sequence effects, for example, much previous analysis focussed only on correct responses. The interpretation of those results is in keeping with the theory presented here, but many experiments require additional data before a thorough analysis can occur. Like all research, this previous work guided the development of theory and posed questions critical to the development of Wave Theory. While the contribution may not be entirely evident, and may at times even seem at variance with the ideas of Wave Theory, these articles created the crucible in which ideas mixed to ignite new accounts of these outstanding scientific contributions. Not to reference them would be ingenuous. To reference them provides the curious reader with pathways for personal explorations of some of the most delightful of scientific works.

The manuscript was read and commented on in its entirety by James Townsend whose editorial eye spotted many possibilities for improvement. Part I was also thoroughly read by Donna Lamping and David Noreen. Each brought a highly specialized talent and insight to the critical evaluation of the presentation. The final copy is much improved by their efforts. In addition, the editorial staff at Lawrence Erlbaum Associates assisted by their generous advice and recommendations. To each and all I give my most sincere thanks.

No list of acknowledgments would be complete without expressing thanks to the taxpayers of the United States and Canada who became partners in the scientific enterprise through the distribution of grants and fellowships to psychological scientists. Among these the Natural Sciences and Engineering Research Council, the National Science Foundation, and the National Research Council all contributed directly toward the research described here.

Wave Theory applies directly to psychological processes, but it is also a general theory with immediate applications to judgments made in other contexts. Thus, the principles apply equally well to the analysis of opposing forces in sociological or econometric theory. I hope that the Wave Theory answers John von Neumann's (1955/1932) speculations about quantum theory and psychophysical parallelism, but goes even further to show how feeling is the ultimate outcome of measurement.

—Stephen Link

Our sensations and feelings are intensive magnitudes, which form a temporal series. The course of mental events has, therefore, at any rate two dimensions; and with this fact is given the general possibility of its representation in mathematical form. Otherwise, indeed, Herbart could hardly have lighted upon the idea of applying mathematics to psychology. And his attempt has the indisputable merit of proving once and for all the possibility of an application of mathematical methods to the sphere of mind.

—Wundt, 1904

Our words have no meaning beyond what they acquire through our learning of them, and all our learning of them goes back directly or indirectly to the association of utterances with concurrent sensory stimulation. The sentences of science, no matter how theoretical, acquire what meaning they have through a network of sentence-to-sentence links whose starting point is sensory stimulation.

—Quine, 1985

All depends then on our ability to express stimulus magnitude and sensation intensity as functions of one another.

—Fechner, 1860

I

PSYCHOPHYSICAL ORIGINS OF DIFFERENCE AND SIMILARITY

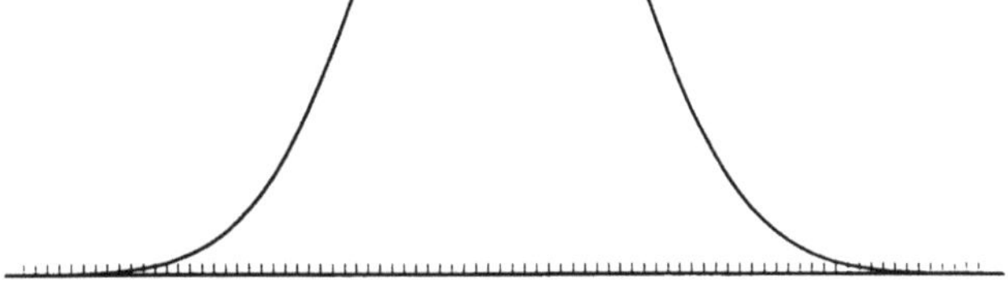

Introduction

PSYCHOPHYSICAL ANALYSIS

The first astonishing discovery that injury to the body caused sensation in the mind is now lost in prehistoric memory. Yet, such puzzling discoveries were likely precursors to later, more refined, philosophical analyses of relationships between the body and mind. Following an initial spate of philosophical curiosity, centuries passed before the scientific concepts needed to analyze our hidden feelings sprang into existence. Then, these secret elements of mental life became the subject matter of Psychophysics and of its offspring, Experimental Psychology.

Gustav Fechner created Psychophysics about 1850 in Leipzig, Germany. By 1860 his famous *Elemente der Psychophysik* proclaimed Psychophysics to be "an exact theory of the functionally dependent relations of body and soul or, more generally, of the material and the mental, of the physical and psychological worlds". From a solid mathematical base it became a powerful discipline devoted to the measurement of psychological phenomena: measurements of what many now consider to be the mind.

What exactly do Psychophysical methods measure? At first, the measurement of a sensory threshold was challenge enough. Then, having determined the physical magnitude that generated the first awareness of a stimulus, psychophysicists established the additional stimulation needed to produce a just noticeable increment in sensation. These measures of sensation depended upon determining, somehow, the amount of physical stimulation applied to an exterior body surface. Yet, only by measuring the amount of *internal* mental sensation, without reference to external physical events, could psychophysics create a tool to advance the scientific study of unseen psychological events.

That tool appeared in L. L. Thurstone's remarkable advance beyond the Fechnerian theory of sensory measurement. According to Thurstone's Law of Comparative Judgment (1927), mental events were compared between themselves rather than by reference to a hypothesized threshold. By creating a new theory of psychological judgment, the Thurstonian theory produced measurements of such psychological objects as attitudes, feelings, preferences, meanings, annoyances, social values, and the appeal of political candidates. Many mental measurements were made. The values obtained proved sufficiently precise to order the objects of inquiry and to measure differences between them. Now, however, the goal of psychophysical theory is the measurement of more elusive psychological entities and ultimately, the measurement of thought itself.

The profound influence of psychophysical tools and concepts is easily discernible in the development of certain older areas of psychology. In other areas, the benefit of psychophysical analysis becomes clear after the discovery of a new psychological phenomenon, because the methods for effectively analyzing the new phenomenon soon prove to be the traditional methods of psychophysics. Yet, psychophysical methods undergo continual development as new empirical discoveries reveal the need for additional theoretical concepts. The improved psychophysical models provide for new mental measurements—measurements that expose to view otherwise unobservable mental activity. These measurements are the unique contribution of Psychophysics to scientific inquiry.

With respect to the origins of psychophysical theory, the mathematical models of J. F. Herbart (1826) provided concepts necessary for the first measurement of the sensory threshold. However, Herbart's mathematical models of psychological processes, like other theories prominent at that time, lacked rules of correspondence relating observable behavior to the internal mental events that were of primary interest. Because observable behavior did not signal the occurrence of internal mental events, experiments designed to investigate Herbart's theories seemed impossible.

Later, E. H. Weber's (1834, 1846) famous sensory experiments remedied this fault by suggesting that the awareness of a difference in sensation measured the existence of a difference in sensation. Thus was born the concept of a *just noticeable difference (jnd)* in sensation and the idea that self-reports about sensation measured the existence of a mental state. Then, G. T. Fechner (1840) began experiments with visual afterimages and later formulated the first theory of mental measurement. Elaborations of this theory, as well as criticisms, prompted substantial experimentation and theoretical development. Today these early theories form a forgotten past, yet occasionally they reappear as new theoretical contributions.

Two strong cross–currents drove forward the empirical development of psychophysical models. Identified most simply these two streams of development are the experimental procedures that measure response frequency, often referred to as response probability, and those that measure response time, sometimes referred to as reaction time. Measurements of response frequency and response time create a psychophysical past filled with theories that predict one or the other of these two responses. Yet, the persistent reappearance of empirical relationships between these two measures of performance makes clear the need for a theory that both predicts and relates these two response measures. Both response measures are most likely the result of a single process which generates empirical laws relating response time and response probability.

The theory, description, and application of this process are the topics of this book. A new theory must somehow explain the past in terms that shed new light on old results. At the same time, new theories systematize earlier results, stimulate new concepts and introduce new tools for exploring mental phenomena and improving mental measurement. A new psychophysical theory must also suggest new laboratory techniques that test the theory and expose our mental abilities to analysis. A purpose of this book is to provide the theoretical foundation for such advances.

A tenet of mental studies, based to a degree on Weber's idea of the *jnd*, stipulates that in order for the mind to distinguish among the many signals that define its state of existence, it must be capable of discriminating between them. Consequently, discrimination and its measure, sensitivity, were recognized long ago as keys to unlocking the secrets of the mind. Concurrently, the discrimination experiment became the psychophysicist's tool for measuring sensitivity and thereby demonstrating the existence of a mental ability.

Fechner's model for discrimination is the cornerstone of present psychological theories of discrimination and choice. Theories that propose to extend Fechner's analysis of mental events must incorporate into their foundations the accomplishments of Fechner and the many contributions of later theorists who extended his work. To understand these theoretical ideas and their development is to know the fascinating tale of the creation of scientific psychology.

1 The Elements

GUSTAV FECHNER

The brilliant insight that led Gustav Fechner to turn psychology toward scientific study and away from philosophical analysis is the foundation for Experimental Psychology. We pay homage to Ernst Weber (1846) for the discovery of psychology's first empirical constant, and to Wilhelm Wundt (1878) for his vigorous development of psychology as an empirically based discipline. But to Gustav Fechner we owe the honor of building the first bridge between theoretical ideas and their experimental verification.

Fechner's idea was powerful and simple. Suppose one claims to possess the mental ability to feel weight. To ascertain the extent, if any, of this ability one is asked to judge which of two objects weighs more. Fechner assumed that each object, when lifted, produced sensations of heaviness that were mentally added together. Further, the average of two heavinesses became a standard against which an individual heaviness was judged (Fechner, 1860/1966, pp. 85–89). In making a judgment of which of two stimuli weighed more, the heaviness produced by one object was compared against this standard or referent. When an object's heaviness was perceived to be greater than the average, that object was judged to be the heavier of the two. When less, the object was judged to be the lighter.

Fechner's Theory

Fechner, once professor of physics at Leipzig and a creator of *"descriptive statistics"* (Neyman, 1976), supposed that the sensory system evoking the sensation of heaviness obeys the same law of error that affects any physical measuring device. According to Gauss's (1809, 1821/1880) Law of Error there exists a *true* value of the object to be measured and an *obtained* value that is the sum of the *true* value and a randomly determined amount of error. Thus the sensation of heaviness evoked by a constant physical stimulus varies depending upon the amount of error added to the *true* heaviness. Whenever the lighter of the two weights was judged to be the heavier, or vice-versa, an incorrect judgment occurred that was attributed to the role of error.

Measuring the size of this invisible sensory error in units of the visible physical stimuli was one of the most outstanding scientific achievements in the 19th century. Fechner argued that the difference between two stimuli, say weights, is measured in grams. These two stimuli also produce a difference between sensations measured in units of sensory error. Thus, the number of sensory error units separating the two sensations equals the number of grams separating the two weights. Dividing the number of sensory error units into the number of grams determines the unit of sensory error in grams. Thus the difference between the two physical stimuli is measured with respect to sensory error units, that is, with respect to the resolving power of the mind. In this way Fechner, and his Psychophysics, exposed to scientific scrutiny differences between sensations that were previously hidden from public view and observed only by personal awareness.

The theory of this measurement, based upon derivations corroborated by the well known topologist A. F. Möbius (b. 1790–d. 1868), will appear obvious to students of classical statistics. It is illustrated in Fig. 1.1 where the abscissa represents heaviness. On the abscissa are the *true* heaviness values, a and b, for two weights corresponding to stimuli $\mathbf{S}_A$ and $\mathbf{S}_B$. The average of these two values $(a + b)/2$ Fechner referred to as *die schwelle*, or the threshold. Fechner postulated that when the lighter of the two weights $\mathbf{S}_A$ was judged to be the heavier of the two, it was because the *true* heaviness of the lighter weight had added to it enough error to make its apparent heaviness greater than the average heaviness of the two weights. Because the lighter weight was perceived to be heavier than the average it was judged, incorrectly, to be the heavier.

By hypothesis, when stimulus, $\mathbf{S}_A$ was lifted the sensory system added a Gaussian error having mean value zero to the *true* heaviness, a. When an error occurred, the heaviness of $\mathbf{S}_A$ necessarily exceeded the referent located at $(a + b)/2$. Because $(a + b)/2$ also equals $a + (b - a)/2$ Fechner reasoned that the amount of error was larger than $(b - a)/2$. Therefore, the probability of an error of judgment must equal the probability of errors larger than $(b - a)/2$. The shaded area under the Gaussian probability distribution to the right of the midpoint, $(a + b)/2$, represents the probability that $\mathbf{S}_A$ has sufficient error added to its *true* heaviness to exceed the average heaviness of the two weights and produce an error of judgment.

When the larger weight, $\mathbf{S}_B$, was judged to be the lighter of the two, the heaviness was judged to be smaller than $(a + b)/2$ which equals $b - (b - a)/2$. The total heaviness was less than b by an amount at least as big as $(b - a)/2$.

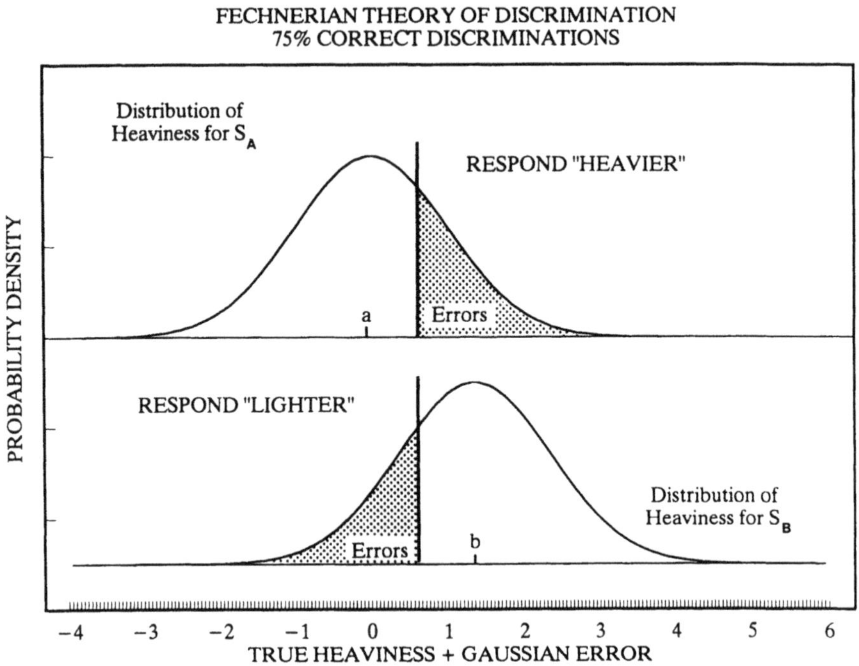

FIG. 1.1. The two Gaussian distributions characterize the distribution of sensory error about two *true* heavinesses, a and b, evoked by stimuli $\mathbf{S}_A$ and $\mathbf{S}_B$. The area to the right of the midpoint, $(a + b)/2$, represents the probability of an error that applies equally to the judgment of either stimulus. The abscissa is scaled in standard deviation units.

Thus the error was of the same magnitude for $\mathbf{S}_B$ as for $\mathbf{S}_A$. Due to the symmetry of the Gaussian distribution and the use of the midpoint as a threshold for judgment, the probability of an error in judging $\mathbf{S}_B$ equals the probability of an error in judging $\mathbf{S}_A$.

To estimate the probability of an error, Fechner determined the relative frequency of an error by computing the ratio of the observed number of such errors divided by the total number of judgments where such errors could occur. Therefore, an error larger than $(b - a)/2$ occurred with a probability that was determined through mental judgments—self-reports about the status of a sensation. Fechner suggested that by repeated stimulus presentations enough judgments could be gathered to provide a statistically reliable estimate of this unknown probability. For this reason the probability of an error in judgment became an important element of the theory of measurement.

Many years of patient observation, and tens of thousands of experimental trials, produced remarkable results. The detailed analysis provided by the following sections reveals the reasons why Fechner's theory remained the foundation of psychological experimentation for so long and why the theory eventually required improvement.

For Fechner estimating the probability of an error of judgment was only a step along the path toward mental measurement. His major concern was to measure the sensitivity of the sensory system in much the same way as determining the sensitivity for any physical measuring device. In the mid-19th century the statistical quantity called the *probable error* served this purpose. As illustrated in Fig 1.2, the distance from the mean of a Gaussian probability distribution to the probable error contains 25% of the distribution. Plus or minus one probable error from the mean defines the interquartile range containing 50% of the Gaussian probability distribution. Beyond one probable error above the mean is 25% of the distribution which corresponds to large positive deviations from the mean. Another 25% of the distribution consists of large negative deviations that exceed one probable error below the mean.

For example, suppose two weights of 300 g and 318 g are correctly discriminated with probability 0.75. The mean heavinesses of these stimuli are shown as points *a* and *b* on the abscissa of Fig. 1.1. In terms of decision making, the average heaviness is a decision criterion that lies midway between the heavinesses corresponding to the two stimuli. The average heaviness corresponds to

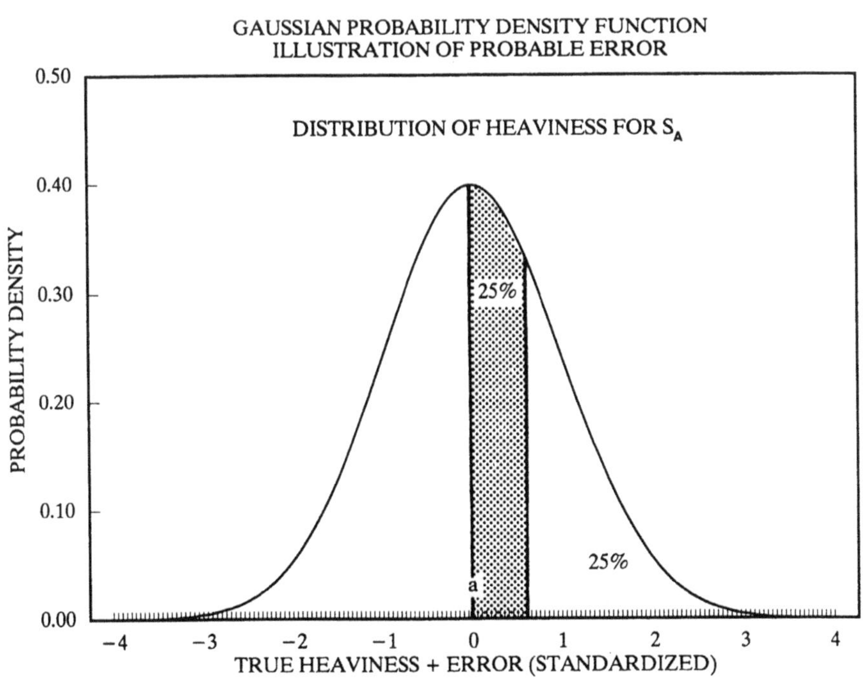

FIG. 1.2. The distance from the mean to the probable error contains 25% of the probability within the Gaussian distribution.

an interpolated hypothetical stimulus of 309 g. The probable error equals 9 g because 25% of the probability distribution of heavinesses exists between 300 g and 309 g. To say that the threshold for judging a difference in weight for a 300 g stimulus is 9 g is to say that a 300 g standard and a 318 g comparison are correctly discriminated with probability 0.75. Thus the threshold stimulus is a stimulus weighing 318 g while the internal threshold, or decision criterion, equals 309 g. Note also that twice the probable error equals the difference between the means of the two distributions of heaviness.

If the probable error is large, then quite extreme deviations from the mean are expected to occur. They occur because with probability 0.50 large positive or negative deviations fall outside the interval defined by plus or minus one probable error. Changes in the distance between the lower and upper limits of the interval defining the interquartile range (plus or minus one probable error), Fechner attributed to changes in the subjects' sensory sensitivity.

To expose the sensitivity of the sensory system to scientific inquiry, Fechner required a numerical measure of the probable error. The reciprocal of this number served as an index of sensory sensitivity. As the interval widened and larger errors became more likely to be added to the *true* heaviness, this reciprocal decreased indicating a reduction in sensitivity. For a small interval few large deviations from the *true* heaviness occurred and the measure of sensitivity increased.

The measurement of the probable error, based on Herbart's and Weber's idea of a threshold, allowed scientists to determine whether one judge could discriminate a smaller difference between two stimuli than another, or whether, for the same judge at different times, the size of the physical difference needed for 75% correct discrimination varied. In this way Fechner showed how to investigate individual differences and also how to examine variations within an individual that might be attributable to such causes as fatigue, time of day, and order of stimulus presentation.

Confusion about the meaning of a threshold crept into some renditions of Fechner's psychophysics. Fechner specifically used the average of two sensations being judged as a threshold for judgment (1860/1966, p. 86). In keeping with the convention for physical measurements derived from the Gaussian Law of Error, and for the purpose of comparing measurements for different individuals, or for the same person at different times, the point beyond which 75% of the heavier judgments occurred became a benchmark for comparing differing sets of observations. By definition a threshold exists at a distance from the standard stimulus that gives rise to 25% erroneous (*lighter*) judgments and thus 75% correct (*heavier*) responses. The comparison stimulus giving rise to this threshold must be twice as far from the standard than the threshold itself because the threshold must lie midway between the mean heavinesses of the two stimuli.

The problem was how to measure the unknown interval between a and the threshold. In theory it should be possible to find two weights that differ just enough to produce 25% errors whenever the lighter weight is judged to be heavier or the heavier weight erroneously judged to be lighter. Then the average of the two weights determines the position of the subject's referent, or threshold. In this case, the distance to $(a + b)/2$ from the *true* heaviness, a, equals the size of the probable error. Therefore twice the probable error equals the distance between the two heavinesses, that is, $b - a$.

Of course, it was as obvious to Fechner as it is to us that the 25% error rate is not a necessary condition to demonstrate the constancy of sensory sensitivity. According to the model of discriminability the average heaviness of two weights defines a midpoint between the sensations evoked by the two stimuli. Suppose two arbitrary weights $\mathbf{S}_A$ and $\mathbf{S}_B$ are selected for comparison and when the lighter weight is compared against the heavier 40%, rather than 25%, errors occur. This implies that the average heaviness, $(a + b)/2$, is much nearer a and b than in the case of 25% errors. By constructing a table of the Gaussian distribution Fechner determined how distant the threshold could be from a to produce various error probabilities. Using these values and the physical difference between the stimuli, he then determined the theoretical size of a comparison stimulus (a *threshold stimulus*) that produced 25% errors.

Fechner's Conjecture

Yet considerably more was extracted from this procedure than just the measure of the threshold for discerning heaviness. Although Fechner (1860/1966) noted that no a priori principle existed to determine how the frequency of correct responses should change for a fixed increment in weight but a variable standard weight, the method could be used to examine one of the most extensively investigated empirical discoveries in physiology, Weber's Law.

In 1834, Weber reported his discovery that following the presentation of a given initial physical stimulus, **S**, the "relative" increment in stimulation needed to produce a *just noticeably different* (*jnd*) comparison stimulus, $\mathbf{S} + \Delta s$, remained constant across a range of the physical values of **S**. With respect to weight Weber (1834/1978) commented:

> If two weights are placed on the hands, and one weighs 32 oz., and the other is gradually reduced until we feel the difference between the weights by touch, we learn the minimum ratio between weights that can be recognized by touch. Now if we repeat the same experiment using weights of 32 drachms instead of the previous weights, i.e. eight times less, we observe that the minimum proportion still recognizable by touch is almost the same as in the previous experiment: for example, if weights of 32 and 26 oz. could be discriminated before, those of 32 and 26 drachms will now generally be discriminable. (pp. 122–123)

> If two weights of 30 and 29 half-ounces are compared by touch, the difference is perceived no more easily than for weights of 30 and 29 drachms. The difference between weights is one half-ounce in the former case and one drachm in the latter. However, since a half-ounce equals four drachms, the weight difference is four times smaller in the latter case than the former. But since the difference is perceived

> no more easily in the former case than the latter, it is shown that what is perceived is not the absolute weight difference but the ratio. The same thing is also evident if we compare weights whose difference is so small that it is certainly not perceived. (p. 131)

> My statements concerning weights compared by touch also hold true for the comparison of lines by vision. For whether you compare longer or shorter lines, you will find that most subjects cannot perceive the difference if one line is smaller by a hundredth part. The difference in length between two lines is greater in absolute terms if you compare long lines, and less if you compare shorter lines. For example, if you compare a line 100 mm. long with one 101 mm. long, the difference will be 1mm. If you compare a line of 50 mm. with one 50 $^{1}/_{2}$ mm., the difference will be equal to $^{1}/_{2}$ mm. The disparity is recognized as easily in the latter case, even though it is twice as small, because in both cases the difference between the two lines is equal to $^{1}/_{100}$ of the longer line. (p. 131)

Weber measured the difference between two stimulus magnitudes, Δs that gave rise to a just noticeable difference in sensation, but Fechner deduced the general principle: The famous constant of sensitivity, $\Delta s/\mathbf{S} = c$ represents a differential equation whose solution yields a logarithmic function relating the magnitude of sensation to the magnitude of the physical stimulus. Fechner named this first psychophysical law Weber's Law.

Many scientists confirmed Weber's surprising finding. As Fullerton and Cattell (1892) observed,

> Since Weber published this result, an amount of experiment and theorizing has been devoted to this subject, which has probably never been surpassed in the history of any science. (p. 21)

Indeed, Fechner argued that the *jnd* was a unit of mental measurement and that this unit was a psychological constant in much the same way that a centimeter was a constant unit of length. Perhaps, a relative difference used to test for discrimination would also yield a probability of correct responding that remained constant as the magnitude of the initial physical stimulus increased. Generalizing Weber's result, Fechner argued that a difference in sensation, $b - a$, sufficient to produce 25% errors for a standard heaviness of amount a produces a Weber ratio equal to $(b - a)/a$. Therefore, any stimuli derived from the same Weber Ratio should also generate 25% errors.

For example, let discriminations between two weights of 300 g and 318 g produce 25% errors (and thus 75% correct responses). On the physical scale the ratio of the increment in weight to the standard equals (318 g − 300 g)/300 g = 0.06. Multiplying these initial weights by a factor of, say 5, yields two weights of 1500 g and 1590 g which share with the previous two weights a common relative difference of (1590 g − 1500 g)/1500 g = 0.06. If the sensory system reproduces this multiplication accurately, then the heaviness of the new lighter weight equals some constant, say c, times the previous heaviness, a. The new heaviness of the lighter weight is *ca*. The new heaviness of the heavier weight is *cb*. The ratio, $(cb - ca)/ca$ still equals $(b - a)/a$. Therefore sensitivity should be unaffected by the multiplication. When the two new stimuli are presented for comparison an error rate of 25% should be observed.

Fechner also recognized that the 25% error rate was not a necessary condition to demonstrate constant sensitivity. Suppose two weights of 300 g and 312 g are selected for comparison. When the lighter weight is compared against the heavier 30%, rather than 25%, errors occur. In 30% of the judgments the sensory measurement system generated an error that produced a heaviness in excess of the average heaviness. In this case, the ratio of the increment in weight, 12 g, to the standard, 300 g, equals 0.04. Fechner postulated that if sensitivity remained constant then any pair of stimulus values generating this ratio should produce 30% errors of judgment. To maintain this error rate such pairs of stimuli are always multiples of the values of the original two stimuli. For example, under the hypothesis of constant sensitivity a 1500 g standard and a 1560 g comparison should produce 30% errors.

This generalization of the idea of constant sensitivity beyond the probable error generating a value of 25% for errors is Fechner's Conjecture. It permitted Fechner to use in his comparison experiments any two weights. Given a pair of comparison weights any multiple of these weights should produce judgments equally in error. Of course, the number of observations necessary to obtain reliable frequency estimates of the probability of an error of judgment was considerable. Still, Fechner persisted and, after many years of experimentation described as having "never been surpassed in extent and thoroughness by any investigation in any science" (p. 21) (Fullerton & Cattell, 1892), Fechner demonstrated both the usefulness of his method and its support for the principle of constant sensitivity.

An example of what Fechner discovered after several years of gathering observations, principally from himself, is shown in Table 1.1 (Fechner, 1860/1966, p. 155).

The experimental method is the Method of Multiplying Stimuli. The entries in Table 1.1 are the numbers of right liftings or right cases, *r*, out of a total of 512 liftings which constitute 256 presentations of a standard weight, *P*, and comparison weight, $P + D$, that correspond to the weights S_A and S_B referred to previously. For the details of this experiment, and a discussion of important methodological points which becomes a focus of later controversy, there is no better source than Fechner (1860/1966).

These data are from experiments begun in 1855, "with the simple intention of making a more exact check of Weber's Law." Fechner (1860/1966) noted that:

> The single handed series was completed in October and November 1856. . . . Each series consisted of 32 experimental days, each having 12 sessions involving lifting the weight 64 times, so that there were in all 32 × 12 × 64 = 24,576 simple

TABLE 1.1
Number of Right Responses for Fechner
Comparison Weights, P + D

Standard	*D = 0.04P*				*D = 0.08P*			
	Left		*Right*		*Left*		*Right*	
P (grams)	*Up*	*Down*	*Up*	*Down*	*Up*	*Down*	*Up*	*Down*
300	352	337	344	318	387	372	386	342
500	339	332	348	335	383	402	413	366
1000	325	343	382	388	383	412	389	422
1500	353	358	371	383	406	416	435	430
2000	378	353	369	382	413	418	414	421
3000	367	343	364	386	426	433	429	438

liftings or cases. With each standard weight P (and periodically interchanged with it) two specific proportional increments, namely 0.04P and 0.08P, were used as additional weights. (p. 153)

There are two ways in which the method [of comparing weights] can be carried out. According to the first procedure one decides which weight is heavier or lighter only after much lifting back and forth of the weight containers. According to the second, one commits himself firmly after each container is lifted once for comparison, or in the case of doubt counts the undecided case as half right and half wrong (p. 78). The total number of cases will be determined by the number of single lifts of the weight and not the number of dual liftings. (p. 79)

Since our method consists of comparison of two magnitudes, a successive presentation should be preferred to a simultaneous one, especially since the latter is hardly possible, in that attention is inevitably turned to one or the other magnitude. The experiment should therefore be arranged, if possible, so that although the observations follow each other closely, each is uninfluenced by the other, and the superposition takes place only in the memory of the observer. The ability to compare magnitudes in this way is, as E. H. Weber has already noted, very curious, and its explanation must await the future progress of inner psychophysics. (p. 73)

As far as procedure is concerned. . . . The heavier weight is switched in completely systematic manner between the four principal cases of the disparate temporal or spatial orders: (1) where it is in the left container and lifted first; (2) where it is in the left container and is lifted second; (3) and (4) the corresponding cases for the right container. We have then, in order to keep the four principal cases clearly apart, the heavier weight:

1. At the left in the container lifted first
2. At the left in the container lifted second
3. At the right in the container lifted first
4. At the right in the container lifted second

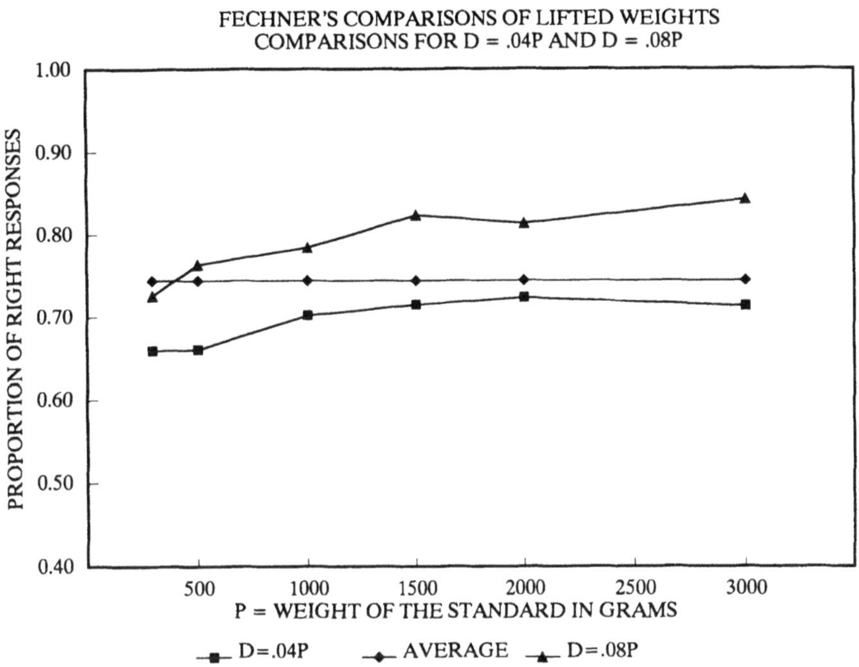

FIG. 1.3. Results from Fechner's 24,576 liftings of weights judged during October and November 1856.

> . . . the temporal order of lifting and the spatial position of the containers influence the determination of the apparently heavier weight. I shall call the effect due to temporal sequence p, and that due to spatial location q. (pp. 93–94)

These results, extracted from a lengthy series of experiments, were taken by Fechner as support for his interpretation of Weber's Law. A chi-squared test for constant response frequencies reveals a marked deviation of the data from strict adherence to Weber's Law. In this regard, a summary of the results reported in Table 1.1 is shown in Fig. 1.3 where, for each value of the standard weight, the proportion of correct discriminations between the standard P and the comparison $P + D$ is shown. As a point of reference the average proportion of correct judgments, 0.744, is shown as a constant value across increases in P. Most likely the nearness of this value to 0.75 is no accident but rather a consequence of selecting stimulus values on the basis of previous research.

Fechner's Conjecture, that the probability of a correct response should remain constant as the magnitudes of the standard and comparison are multiplicitively increased, is obviously false. Whether the value of the increment, D, equals $0.04P$ or $0.08P$ the probabilities of a correct response increase as the weight of the standard increases. In fact, if Fechner's Conjecture holds then the marginal proportion correct for each weight increment estimates the probability of a correct judgment for each of the six comparison stimuli defined by the size of increment. The marginal proportion of correct responses for the $0.04P$ increment equals 0.696 and for the $0.08P$ increment is 0.792. The marginal estimate for the $0.04P$ increment multiplied by 512 should predict all 24 response frequencies in the left section of Table 1. A chi-squared test comparing this prediction with the observed results yields an improbable chi-squared goodness-of-fit value of 44.9 based on 23 degrees of freedom. A similar calculation for the $D = 0.08P$ data on the right-hand side of Table 1.1 yields an even more unlikely value of 80.9, again with 23 *df*.

Had Fechner applied today's statistical criteria, and abandoned his theory because the conjecture based on Weber's Law failed, a terrible loss would have occurred. Fortunately Fechner was not swayed by a modest departure of data from theory. He commented:

> Should these experiments prove the law [Weber's] to hold directly and precisely, all the r's (number of correct judgments) for the different standards should be not only approximately but exactly alike, when the same ratio of additional weight to standard weight holds everywhere. This is not the case. The deviation of the experimental values for those predicted by the law should, nevertheless, be given no more importance as a true deviation than the one we found at the lower limits in the field of photoreception, but should be regarded as an analogous consequence of the law. Just as we had to take into account in the case of light sensations the inner stimulation which exists, even if no external light is admitted, so we have to consider here the weight of the arm and of possible covering pieces of clothing (in my own experiments only with light shirt sleeves) which are present even without the external weight P and are lifted when it is lifted. (p. 153)

Immediately two major theoretical problems appeared that influenced profoundly the future course of psychology. The first of these arose from a difficulty inherent in the procedure Fechner adopted for measuring choice responses. In the Method of Right and Wrong Cases two stimuli are presented for comparison. Often, in concurrence with the method used previously by Weber for establishing the size of a *jnd*, the subject only chose the stimulus of apparently larger physical magnitude. Sometimes, however, the subject simply could not decide which stimulus to choose. In these doubtful cases Fechner treated the response as both right and wrong and added half a response to the right and wrong response tallies.[1] As Fechner (1860/1966) stated, there appears to be "an interval of

[1]Fechner did not simply ignore the occurrence of these responses, nor did he discard and repeat the trials that yielded judgments of equality. However, he did not report the frequency with which these responses did occur, thus prohibiting the reanalysis of his data in the light of present day theories.

uncertainty" (p. 63). Some 40 years later the interval of uncertainty reappeared as a challenge to Fechner's theory.

The second problem derived from a more immediate objection to Fechner's theory of sensory discrimination. In keeping with the use of probable error as a measure of sensitivity, Fechner defined a threshold of difference to be exceeded when discriminations between two stimuli produced 25% errors and 75% correct responses. This definition of a difference threshold suggested to some scientists that two stimuli differing by a smaller amount than the difference threshold should be indistinguishable. After all, they argued, there must be some minimal difference below which a difference between two stimuli cannot be discerned.

Fechner's Transition

Fechner was 60 years old when *Elemente der Psychophysik* appeared in 1860. His distinguished career as a physiologist, physicist, and psychophysicist was to be followed by 11 years as an experimental aestheticist and as, "during his last eleven years [1876–1887], an old man whose attention had been brought back by public acclaim and criticism to psychophysics" (Boring, 1950, p. 283). In the last years of his life Fechner saw the publication of Sully's influential *Sensation and Intuition: Studies in Psychology and Esthetics* (1874) which brought to the English speaking public a clear assessment of the contributions to the study of sensation made by Weber and especially by Fechner. Later Ward (1876) in the first volume of *Mind* would criticize the three experimental methods discussed by Fechner: the Method of Just Noticeable Differences, the Method of Average Error, and the Method of Right and Wrong Cases. As well he repudiated Fechner's explanation of Weber's Law and sought to explain the increases in sensation on the basis of stimulus waves projected in certain brain centers.

Fechner's research and theorizing was not without its positive critics who, like Müller (1878), sought to refine the methods. The more negative critics, like Ward (1876), dwelled upon surprisingly insignificant matters while overlooking the astonishing feat of measurement that Fechner achieved. This first measurement of the amount of internal error in a wholly psychological quantity such as heaviness launched psychology on its course as an empirical science capable of producing measurements of previously unobservable mental events.

2 Challenges and Extensions

CHARLES S. PEIRCE

The first criticism backed by new experimental data came from the father of American philosophical pragmatism (Ayer, 1968) assisted by a graduate student at the University of Pennsylvania. Together they brought a forceful mathematical argument to bear on Fechner's theoretical basis for the Method of Right and Wrong Cases. This remarkable paper, "On small differences of sensation," read before the United States National Academy of Sciences on October 17, 1884, was authored by the eminent Charles S. Peirce and coauthored by Jacob Jastrow.

Peirce drew from Fechner's deeper concerns about thresholds a solid criticism with which to discipline Fechner's errant reasonings. Fechner (1860/1966) argued that:

> There is a certain paradox inherent in the nature of the threshold. A stimulus or stimulus difference can be increased up to a certain point without being felt; above that point it is felt and its growth is noticed. How can something that does not reach consciousness when it is weak begin to affect the mind when it is strengthened? It appears that the sum of nothing adds up to an effective something. (p. 205)
>
> Let us assume that the difference between two weights or, in general, between two experimental stimuli, is so small that it falls below the point where it can be consciously recognized. The question is now: can it influence the number of right and wrong cases? Is not the relationship between them as though no difference existed at all, until the difference has exceeded the boundary where it is perceptible as such; and from there on is not the influence to be evaluated, not by the absolute size of the difference but by the difference from the value where it begins to become truly perceptible? (p. 206)
>
> At first this conclusion seems obvious, for how can a difference which does not affect our consciousness determine our judgment? Nevertheless it is not admissible

> without destroying the principles of this method of measurement and at the same time the principles by which one calculates the probability of error with respect to the error's magnitudes, which is the whole basis for this method of measurement. Also, a more searching examination will lead to conclusions opposite to this apparently obvious assumption. In spite of the fact that a difference is itself unnoticeable, it will nevertheless allow us to find a preponderance of right cases in favor of the heavier weight, or more generally in favor of the greater stimulus, provided a sufficient number of comparisons is made. (p. 206)

> We must take into account that, simultaneously with the difference we are trying to apprehend, there are at work random effects which would on the average make the judgment come out equally often in favor of one or the other. The difference creates an additional tendency to throw the judgment in one direction, partly by causing tendencies which would have been subliminal without it, to become noticeable in favor of one direction, and partly by strengthening the already noticeable tendencies in that direction and prevailing over the contrary tendencies. This effect occurs regardless of the fact that the influence of the added increment, together with the fluctuations which are present at the time, often remains below threshold, in which case judgment remains ambiguous. (p. 206)

> One can see in this way differences that are themselves subliminal can become effective in one direction by summation with other influences. The level of probability . . . depends on the size of the difference so as to allow us to determine a measure of sensitivity from it, as we have already seen. (p. 207)

The counterargument by Peirce and Jastrow (1884) is simplicity itself:

> the errors of our judgments in comparing our sensations seem sufficiently accounted for by the slow and doubtless complicated process by which the impression is conveyed from the periphery to the brain; for this must be liable to more or less accidental derangement at every step of its progress. Accordingly we find that the frequencies of errors of different magnitudes follows the probability curve, which is the law of an effect brought about by the sum of an infinite number of infinitesimal causes. This theory, however, does not admit of an *Unterschiedsschwelle* [threshold]. On the contrary, it leads to the method of least squares, according to which the multiplication of observations will indefinitely reduce the error of their mean, so that if of two excitations one were ever so little the more intense, in the long run it would be judged to be more intense the majority of times. . . . If there is a least perceptible difference, then when two excitations differing by less than this are presented to us, and we are asked to judge which is the greater, we ought to answer wrong as often as right in the long run. Whereas, if the theory of least squares is correct, we not only ought to answer right measurements not that wrong, [*better than chance*] but we ought to do so in a predictable ratio of cases. (pp. 75–76)

To test their hypothesis, Peirce and Jastrow invented a new experimental procedure which later became the most commonly used experimental procedure. Rather than continue to test Weber's Law they sampled sensitivity to a fixed standard stimulus by varying the magnitude of comparison stimuli. Because the experiment is a major turning point in the development of psychophysics and experimental psychology it merits a full discussion.

The experimenters wanted judgments to be based upon an entirely mental process. Therefore, they selected as stimuli weights, which were applied to one end of a lever in order to deliver a known pressure to the pad of a forefinger resting atop the opposite end of the lever. A technique similar to one originated by Weber. Subjects decided whether the experimenter administered an increase followed by a decrease in pressure, or vice versa. A standard, or reference, pressure delivered by a weight of 1 kg was compared to pressures produced by weights of 1015 g, 1030 g, and 1060 g.

To balance the possible effects of differential sensitivity to increases or decreases, the trials were divided into blocks within which only a fixed difference between pressures occurred. Within the block two types of trials could occur. Type R trials consisted of an increase from the reference pressure to the comparison pressure followed by a decrease to the reference pressure. Type B trials began with a comparison pressure followed by a decrease to the reference pressure followed by an increase to the comparison pressure. Thus, at the end of a sequence of the standard followed by an increase and then decrease in pressure the subject is left with the standard pressure plus the error introduced by the intervening increase and decrease in pressure. When the comparison is presented and followed by a decrease and increase in pressure the subject is left with the comparison pressure plus the error introduced by the decrease and increase in pressure. In either case the same error is introduced into the sensation generated by the stimulus.

Peirce and Jastrow (1884) noted that:

> in our experiments there were two opportunities for judging, for the initial weight was either first increased and then diminished, or vice versa, the subject having to say which of these two double changes was made. . . . The advantage of having two changes in one experiment[1] consists in this: If only one change were employed, then some of the experiments would have an increase of excitation only and others a decrease only; and since the former would yield a far greater amount of sensation than the latter, the nature of the results would be greatly complicated; but when each experiment embraces a double change this difference in the amount of sensation caused by an increase and decrease of pressure affects every experiment alike, and the liability to error is constant. (pp. 76–77)

Table 2.1 contains experimental results selected from their report. Quite clearly, as the ratio of pressures decreases the error proportion increases. In their tables the order of the columns, which represent increasing stimulus difference, is from largest to smallest difference rather than from smallest to largest. Perhaps

[1]At the time the word *experiment* also denoted the present meaning of a single experimental trial.

TABLE 2.1
Table of Errors for Mr. Jastrow
Ratios of Pressures

		1.060		1.030		1.015	
Date (1884)		*N*	*Errors*	*N*	*Errors*	*N*	*Errors*
January	24	50	4	50	11	50	18
February	11	50	1	50	7	50	18
	17	50	2	50	10	50	17
	18	50	2	50	11	50	17
	24	50	2	50	8	50	15
Totals		250	11	250	47	250	85
Error proportion			.044		.188		.340
Normal deviates			1.706		.885		.412

the authors did not conceive of the stimuli as increasing along an underlying dimension of difference.

If the Method of Least Squares does apply then the error proportions estimate probabilities from a single Gaussian probability distribution. This hypothesis leads to the prediction that lower error proportions occur as the difference between the reference and comparison pressures increases. Of course, any test of this hypothesis requires some known relation between the stimulus values and the dimension on which the Gaussian distribution is defined. Peirce and Jastrow implicitly assumed a linear relationship.

Figure 2.1 illustrates a Gaussian probability density function centered at the reference pressure generated by a weight of 1000 g. Sensations evoked by this stimulus vary according to the Gaussian Theory of Errors. Consider the case where the reference pressure occurs first and is followed by an increase in pressure and then a decrease back to the reference pressure. Sometimes an error of appreciable size is added to the sensation generated by the 1000 g standard. When this combination of *true* pressure plus error is judged an error occurs because the standard pressure is perceived to be greater than the comparison stimulus pressure. The 0.25 probability of error is shown by the shaded region of Fig. 2.1 where the total amount of error from all sources is assigned to the distribution of the 1000 g-standard's pressure. Because the source of error is the same for both the standard and comparison one would expect the proportion of errors to be the same for presentations of either the standard or comparison. For very large differences between these pressures the probability of an error is slight. As the comparison nears the standard the error probability increases.

Peirce and Jastrow showed that their error proportions derived from a single Gaussian distribution by using the Method of Least Squares to estimate a probable error common to all stimuli. Then, using this estimated value, or a corre-

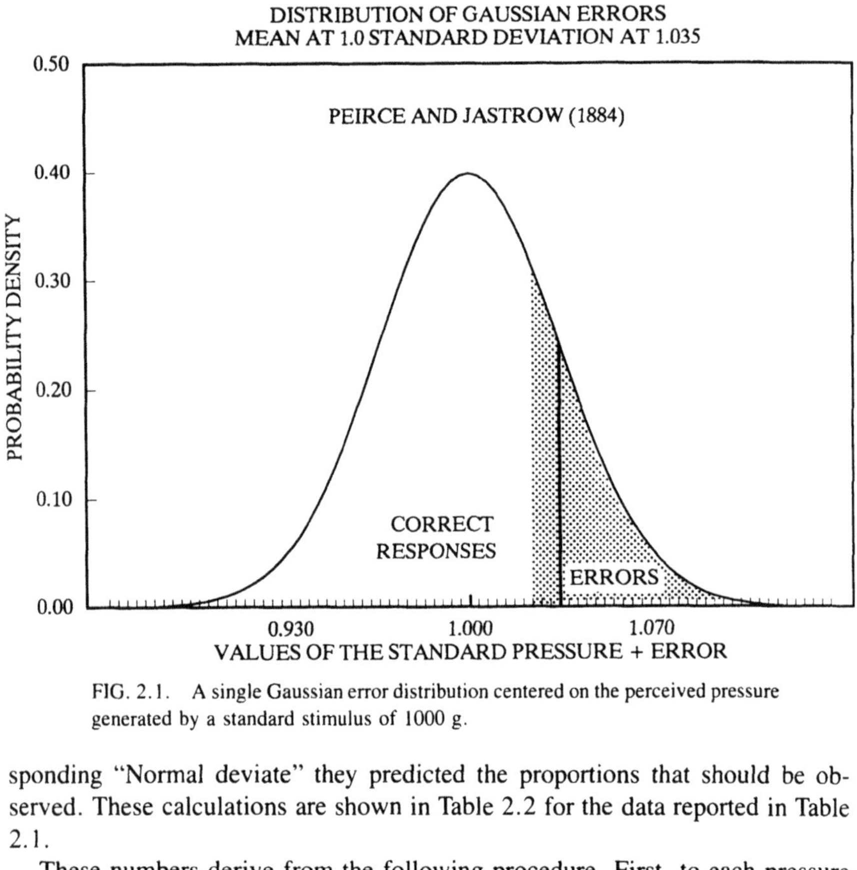

FIG. 2.1. A single Gaussian error distribution centered on the perceived pressure generated by a standard stimulus of 1000 g.

sponding "Normal deviate" they predicted the proportions that should be observed. These calculations are shown in Table 2.2 for the data reported in Table 2.1.

These numbers derive from the following procedure. First, to each pressure ratio there corresponds an error probability and normal deviate. Assuming that these deviates should be a linear function of the stimuli, which are fixed values not subject to error, the Method of Least Squares determines the slope and intercept of the best fitting linear relation. The linear equation relating the ratio of pressures to the predicted normal deviate (PND) is:

$$\text{PND} = 28.557(\text{Ratio of Pressures}) - 28.5557 \qquad (2.1)$$

TABLE 2.2
Least Squares Method for Three Stimuli

Pressure Ratio	*Estimated Normal Deviates*	*Predicted Normal Deviates*	*Observed Error Proportion*	*Predicted Error Proportion*
1.015	.412	.430	.340	.334
1.030	.885	.858	.188	.196
1.060	1.706	1.715	.044	.043

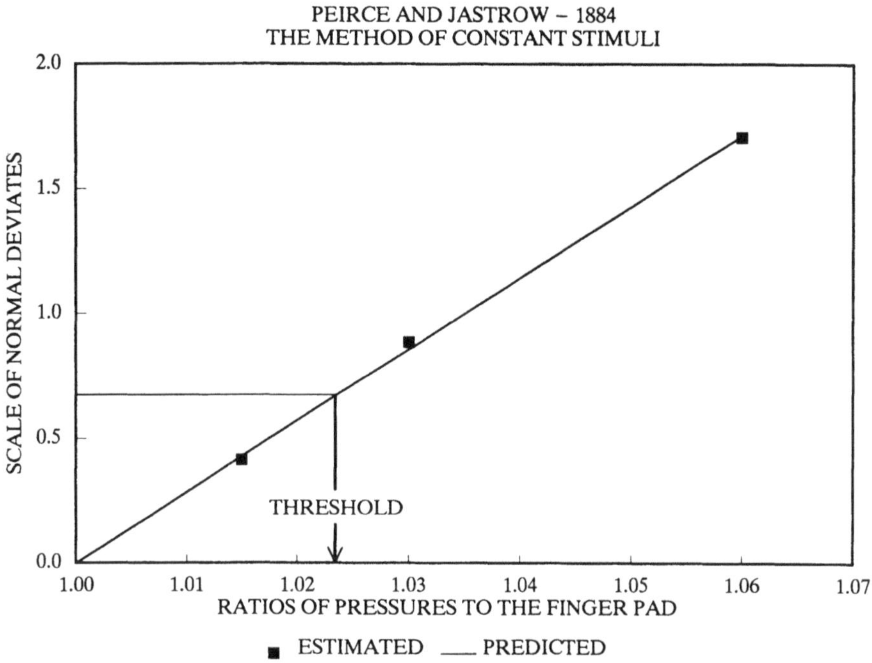

FIG. 2.2. Estimates of Gaussian deviates as a function of the ratio of pressures delivered to the fingerpad (Peirce & Jastrow, 1884).

To solve Equation 2.1 for the ratio of pressures corresponding to the probable error, that is, for the ratio of pressures that gives a probability of an error of 0.25, first convert the probability to a corresponding normal deviate, in this instance 0.674. Then, substituting 0.674 for PND in Equation 2.1 shows the appropriate ratio of pressures to equal 1.0235. The comparison weight necessary to produce responses at the defined threshold weighs 1023.5 g.

Figure 2.2 shows the position of the threshold and experimental stimuli on the abscissa, and the corresponding normal deviates on the ordinate. Substituting for values of the ratio of pressures yields predicted normal deviates that define the associated error probabilities found in a table of the Gaussian (Normal) distribution. As shown in Table 2.2 these probabilities are quite close to the observed response proportions.

In spite of Peirce's attack on the definition of the difference threshold, there is little difference between the positions of Peirce and Fechner. In fact, were it not for Fechner's wish to confirm Weber's Law, he too might have used the experimental procedure that we now attribute to Peirce and Jastrow. Everything considered, Peirce and Jastrow's interpretation of Fechner's work seems less a challenge to Fechnerian theory than a positive increment to its development. The basis for their criticism is at best questionable and the criticism itself may constitute the most colossal misunderstanding in the history of psychology.

Fechner's own experiments attest to the fact that he believed that doubling the size of the difference between the standard and comparison stimuli should produce a decrease in errors consistent with the theory of errors. And, as Fechner (1860/1966) pointed out:

> There exists as yet no a priori principle that determines how the ratio r/n (right judgments/total judgments) should vary with the magnitude of the standard weight P, when the additional weight D is held constant. Rather, this should be looked upon as a matter for a law that should be determined experimentally. *It is possible, on the other hand, to state a priori, according to the principles of probability, how the ratio r/n must vary (assuming a large n) when the standard weight P is held constant and the added weight is varied* [italics added] (with differential sensitivity h remaining the same), or if, in general, anything is changed that determines the apparent gain in weight which will be represented here once and for all by D. The same principles hold good in this case as those by which one can determine the variation of the relative number of errors of observation with a change in their magnitude, assuming constant precision of observation. (pp. 85–86)

There is probably no better statement of the relationship between Fechner's approach to the measurement of psychological phenomena and the methods used in other sciences that also adopted the Gaussian Law of Error as a fundamental principle of measurement. Clearly, Fechner believed that for a fixed standard and various comparison stimuli, one must observe response proportions consistent with probabilities derived from a Gaussian distribution with fixed probable error or standard deviation.

In this regard Table 1.1 provides data for computing the difference between threshold values for the two comparison stimuli at each level of the standard stimulus. For standards of 300 g, 500 g, 1000 g, 1500 g, 2000 g, and 3000 g these differences (in standard deviation units) are 0.192, 0.306, 0.255, 0.363, 0.297, and 0.443. Thus the unwanted increases in response probabilities as a function of the weights of the standards led to increased rather than constant differences between threshold values.

Peirce apparently misconstrued the meaning of the threshold, which today would be recognized as a decision criterion, with a fixed value below which discrimination became impossible. This confusion is the source of a great deal of misunderstanding about Fechner's work and the idea of a threshold. But a further source of confusion, one that continues to haunt Psychology, is the wholesale misunderstanding of probability theory. Some concession to the times is, of course, necessary because a formal basis for probability theory did not appear for another 50 years (Kolmogoroff, 1933). Still, the errors in logic are often alarming and, worse, unintentionally communicated without criticism. Titchener's (1905) otherwise outstanding presentation of Experimental Psychology propogates the mathematical errors introduced in theoretical discussions attributed to other authors. Titchener's failure to correct these errors suggests a disturbing misunderstanding of probability theory.[2]

Psychophysical interests soon focused on rather minor issues such as whether the response proportions measured by the Method of Right and Wrong Cases or the Method of Constant Stimuli came from a Gaussian or other type of distribution. Because the symbol for the cumulative Gaussian distribution probabilities evaluated at the value γ was $\Phi(\gamma)$, this hypothesis came to be called the *Phi-gamma Hypothesis*. Numerous theoretical and empirical arguments pro and con raged between the advocates of the two positions.

Criticism of psychophysics never abated. William James (1890), in a play on Fechner's wonder at how a sum of nothings can add up to something, quipped that the sum of psychophysics added up to nothing. Of the many challengers of Weber's and Fechner's ideas two stand apart from the herd. The first descends from Peirce and argues that if there exists a differential threshold then there must exist a small difference between two stimuli that is imperceptible.

Response Confidence

A second challenge derived from an enlightened extension of the method for measuring psychophysical phenomena. Again Peirce and Jastrow pushed this development forward. Fechner noticed a perceived "interval of uncertainty" when making judgments about a difference between two stimuli. Peirce, requiring empirical measures of the psychological phenomenon, demanded a measure of such uncertainty and found it in measures of response confidence.

Peirce (1877) noted that

> the chance of an event has an intimate connection with our degree of belief in it. . . . Any quantity which varies with the chance, might, therefore, it would seem, serve as a thermometer for the proper intensity of belief. Among all such measures there is one which is particularly appropriate . . . it is the logarithm of the chance. But there is another consideration which must, if admitted, fix us to this choice for our thermometer. It is that the belief ought to be proportional to the weight of evidence, in this sense, that two arguments which are entirely independent, neither weakening nor strengthening each other, ought, when they concur, to produce a belief equal to the sum of the intensities of belief which either would produce separately. Now, we have seen that the chances of independent concurrent arguments are to be multiplied together to get the chance of their combination, and therefore the quantities which best express the intensities of belief should be such that they are to be added when the chances are multiplied in order to produce the quantity which corresponds to the combined chance. Now the logarithm is the only

[2]A typical example arises in Titchener (1905, Volume II, Part 1, p. 47) where the author claims that the height of a probability density function equals the probability of an event. No wonder psychophysics always seemed to be such a mystery! Yet, such basic misunderstandings still occur in some current psychology textbooks.

quantity which fulfills this condition. There is a general law of sensibility called Fechner's psychophysical law. It is that the intensity of any sensation is proportional to the logarithm of the external force which produces it. It is entirely in harmony with this law that the feeling of belief should be as the logarithm of the chance, this latter being the expression of the state of facts which produces the belief. (pp. 708–709)

In view of this early, remarkable, speculation about the nature of our confidence in events it is particularly significant that Peirce and Jastrow (1884) measured response confidence for each judgment on a four point scale. They computed average confidence values for each experimental condition and commented that

The average marks seem to conform to the formula:

$$m = c \log\left(\frac{p}{1-p}\right)$$

where m denotes the degree of confidence on the scale, p denotes the probability of the answer being right, and c is a constant which may be called the index of confidence. (p. 77)

As Jastrow (1888) noted later,

In spite of the obvious arbitrariness and inadequacy of this method it has proved itself surprisingly useful; it ought, however, to be improved in future work. (p. 305)

In another theoretical arena, Peirce's measurement of Jastrow's confidence in discriminative judgments augured the use of new measures of discriminative ability. None received the formal development of the measurement of response probabilities created by Fechner, but new types of data offered the promise of new insights into the human mind.

If a measure of Peirce's genius is that his exceptional theoretical developments occurred in the intellectually remote United States, whereas the focus of psychological fervor was Leipzig, the center of intellectual development in the powerful Austro-Hungarian Empire, then a measure of Fechner's success is that in 1878, 49-year-old Wilhelm Wundt became head of a small, newly founded, Psychological Institute at Leipzig University. Wundt, already a prominent physiologist, published in 1878 a masterful 800-page revision of his *Lehrbuch der Physiologie des Menschen* (1865), published 5 years after Fechner published *Elemente*. Wundt (1878) explained in detail Fechner's theory of discrimination and generalization of Weber's Law to response proportions. So armed, Wundt came well equipped to found the first psychological laboratory.

JAMES McKEEN CATTELL

Response Time

Wundt's laboratory quickly launched a series of experiments designed to investigate responses to stimuli so different from each other that the probabilities of discriminating were at the maximum value of 1.0. To show that there was still a behavioral response to the difference between these perfectly discriminated stimuli, Wundt measured response time. No solid theoretical foundations predicted changes in reaction time. As a result, a door opened for a parade of new reaction experiments, eagerly trotted onto fresh empirical turf.

As early as 1886 James McKeen Cattell (1860–1944), Wundt's student, *und der ganz amerikanische assistant*, published observations on "The influence of the intensity of the stimulus on the length of the reaction time." Cattell, in association with G. O. Berger, began these experiments in Wundt's psychological laboratory in 1884, the same year Peirce and Jastrow published their critical test of Fechner's theory. Through the use of smoked-glass filters they stepped down the intensity of a referent light to create five stimuli of lesser intensities. Two additional stimuli, each brighter than the referent, provided for a total of eight intensities. The observer, either Berger or Cattell, sat in a darkened room and peered through a telescope to where the stimulus would appear. They measured the time between the occurrence of the stimulus and lifting of the hand from an electrical contact. (Cattell mistakenly defined the reaction time to be the time between the *appearance* of the light and the lifting of the hand. He could not measure when the light appeared, a psychological event, only when the light first was physically present.)

The results for 150 trials at each intensity appear in Fig. 2.3. Mean reaction time declines substantially as the intensity of the light increases. Although the results are orderly, Cattell (1886) lamented:

> I cannot, however, formulate a general law from the table. (p. 513)

Later, Piéron (1914, 1920) created the law relating response time to stimulus intensity.

By 1886 the three major response variables, frequency, duration, and the more subjective confidence, were established as measures of psychophysical phenomena. Although each measure varied as a function of experimenter-controlled stimulation, only response frequency derived from a well-formulated mathematical base. Yet each measure had advantages. For example, response time, as shown by Cattell and Berger, varied when response frequency showed no variation because the subject always discriminated the presence of the light. In contrast, the measurement of response time became impractical when discriminating between lifted weights, and thus the measurement of response frequency

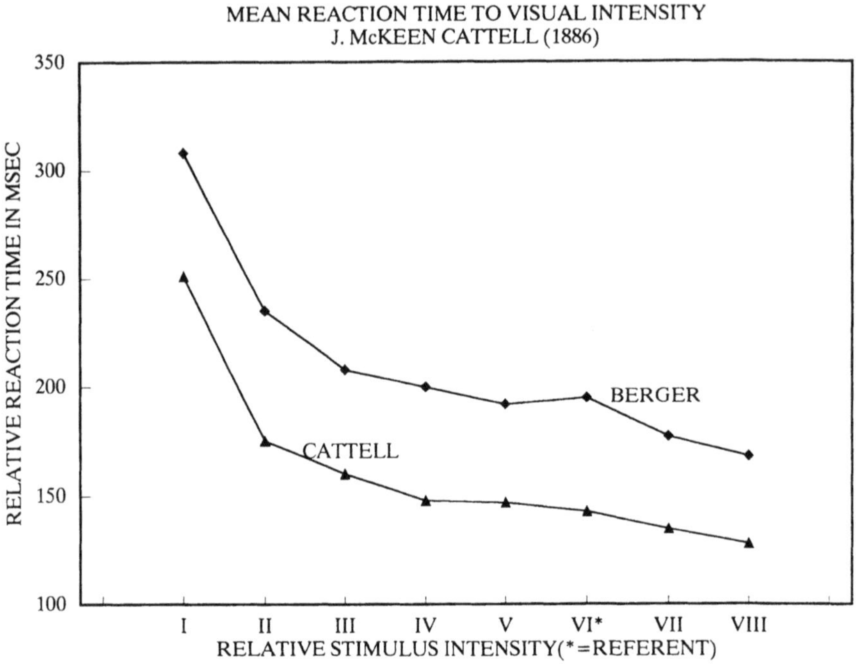

FIG. 2.3. Results from 150 trials per stimulus for eight increasingly intense visual stimuli.

was essential. Due to these differences between experiments the two response measures, frequency and duration, appeared to apply to qualitatively different phenomena. Soon, the belief that different mental mechanisms gave rise to these response measures became firmly entrenched in psychological lore.

Cattell received a PhD degree from Wundt in 1886 and returned to the United States to found the Department of Psychology at the University of Pennsylvania. Soon after, in 1892, G. S. Fullerton and Cattell presented to the English speaking audience an elegant summary of many previous psychophysical results and an extensive application of psychophysical methods. Their monograph entitled "On the perception of small differences: With special reference to the extent, force and time of movement" stands as a milestone signaling the beginning of a less theoretical and more behavioristic psychology that sought to equate the collection of experimental data with science.

The introductory statement of Fullerton and Cattell's (1892) new approach echoes Fechner's ideas for the study of individual differences through the use of psychophysical methods:

> Instruments discover differences which the unaided sense cannot detect. Sensations may appear alike to common observation, while the stimuli differ by amounts which may be measured with the help of rules, clocks, balances, thermometers and

> other instruments. We thus find a psychological problem which admits of scientific study. How great a difference must there be between two magnitudes in order that it may be perceived, and how does this least noticeable difference vary with varying circumstances? The relation between the intensity of a stimulus and the least noticeable difference is the only problem which has hitherto received much attention. But the intensity of a stimulus is only one of a number of factors affecting the error of observation. The time and space relations of the stimulus are not less essential than its intensity; and the variations of the different observers, and of the same observer in different conditions, deserve study. The determination of the relation of one magnitude to another makes psychology an exact science, whereas the study of personal difference may find practical application in medicine and education. (p. 9)

One major research effort measured the duration and accuracy of a bodily movement toward a targeted position, foreshadowing future studies of human performance. Their other experiments measured the relative frequency of responding and the degree of confidence in the response made.

With regard to response confidence and the degree of stimulus difference Fullerton and Cattell (1892) observed:

> Our experiments with lifted weights were made with a view to studying methods for determining the fineness of discrimination and the factors entering into the perception of movement. (p. 118)
>
> We loaded wooden boxes so that they weighed (to $^1/_{100000}$) 100, 104, 108, 112, and 116 gm. . . . The observer sat in an ordinary chair facing a low table. The boxes were placed, two at a time, in a line parallel to the edge of the table, but a little back and opposite the left shoulder. A seconds pendulum swung in front of the observer. (pp. 118–119)
>
> After the observer had lifted the two weights he was required to say whether the second weight was heavier than the first. He was also required to estimate what confidence he felt in his decision. a, b and c were used in the manner already described, a meaning sure of the correctness of the decision, b less certain, and c uncertain. If the observer could discern no difference whatever, he was allowed to say "doubtful (d.), but was asked to use this as seldom as possible, and in our later experiments he was required to guess even when quite doubtful. (pp. 119–120)
>
> In these experiments we used four amounts of difference, our object being to combine the method of just noticeable difference with the method of right and wrong cases, and to study the application of the probability integral. As regards this latter, our experiments agree very well with the mathematical theory. The probable error is nearly the same, whether calculated from a small difference and a relatively small percentage of right cases or from a larger difference and a larger percentage of right cases. (p. 124)
>
> By varying the amount of difference and making it in some cases so large that it was usually perceived correctly, we have combined the method of right and wrong cases

with the method of the just noticeable difference. This shows clearly that the method of just noticeable difference in its scientific form is simply a case of the method of right and wrong cases. (p. 125)

We see from the figures in the table *(XXIV)* that the number of times the observer felt confidence in the correctness of his decision usually increased as the amount of difference was taken greater. (p. 126)

Compared with the percentages of right cases, the fictitious[3] averages are:

100:	104.	108.	112.	116.
% of times right,	69	77	89	94
% of times a and b	61	67	74	84
[# of a's	*23*	*28*	*40*	*47*
[# of b's	*38*	*39*	*34*	*37*[4]

The confidence thus varies nearly as the percentage of right cases, and some reliance may, therefore, be placed on such introspection. We see, however, from the table *[Table XXVI, not reproduced here]*, that different individuals place very different meanings on the degree of confidence. Some observers are nearly always quite or fairly confident, while others are seldom quite confident. . . . those observers who felt the greatest degree of confidence in their judgment had the largest probable error, while those who were most seldom quite confident had the smallest probable error. It is, therefore, absurd to measure the fineness of discrimination of an observer by the difference which he thinks he can notice, as is done when the method of just noticeable difference is used in its usual form. (pp. 126–127)

Experiments on the extent of bodily movement, the force and duration of movement, together with their studies of lifted weights, intensity of light, and the memory for these stimuli, complemented a surprisingly extensive empirical literature reviewed by Fullerton and Cattell. The new emphasis on experimental data generating the "Laws of Psychophysics" rather than theory predicting laws to be confirmed through experiment, characterized the pragmatic approach that fostered new empirical studies. The approach is purely behavioristic in the sense that theoretical arguments do not motivate the experimental designs and methods of data analysis.

Instead, the collection of data proceeds in the hope that some sense will eventually be made of the results. Thus, almost any new experimental approach generated grist for the behaviorist mill. The impossible struggle to develop new psychological insights without the advantage of mathematical theories as sophisticated as Fechner's theory of discrimination, proved the value of creating solid

[3]Analysis of their Table XXI suggests that they meant the conditional probability of right given right or wrong, that is, excluding the doubtful cases.

[4]These results are the number of *a* or *b* confidences averaged across the 10 subjects. They appear as the marginal values from Table XXIV. The total number of trials per subject per weight was 100. Thus the sum of *a* and *b* equals 100 − number of *c* responses − *d*, the number of doubtful cases.

theoretical predictions in order to better understand many apparently discrepant empirical discoveries.

After 15 years of research using various experimental procedures, Cattell (1902) broke new ground by proposing a means of assessing the intensity, and not just the sensitivity, of perception. Because Cattell's article best states, and represents, a new approach to the investigation of mental processes, its opening arguments merit attention:

> Weber's work on the relation between the magnitude of the stimulus and the just noticeable difference and Fechner's deduction of a logarithmic relation between stimulus and sensation have been followed by a large number of experimental researches and an extensive literature. In the experiments and discussion I have shared.[5] *I have argued that in most of our experiments we have measured the error of observation, which determines the accuracy of perception, and not the intensity of sensation* [italics added]. This holds for experiments by the methods of average error and of right and wrong cases. The method of just observable difference appears to me to have been used ambiguously. On the one hand, it may be reduced to the method of right and wrong cases, when the wrong cases are but few and the determination is difficult. On the other hand, the observer may seek to make differences equally noticeable, in which case the results are analogous to those of the method of mean gradation and really estimate differences in sensation. Such estimates are, however, inexact, lacking an objective criterion. Different observers or the same observers at different times vary greatly in their estimates of equally noticeable differences. It seems that these estimates may be based on the known objective relations of the stimuli, and consequently can not be used to determine a relation between the sensation and stimulus. (pp. 63–64)
>
> *It is, however, possible to approach the problem from a new direction* [italics added]. An observer can not decide with any certainty when the difference in one pair of sensations is equal to the difference in another pair, but the time it takes to perceive a difference can be measured. The smaller the difference between two sensations, the greater is the time required to perceive it. *When differences require equal times for discrimination, the discrimination is equally difficult, and the differences are equal for consciousness* [italics added]. (p. 64)

Once the difference between a standard and comparison increased beyond the point at which errors rarely occurred, Fechner's theory of sensitivity had little relevance, and certainly provided no measurements of sensitivity. Yet, as Cattell and Berger showed in 1884, response time dropped dramatically as stimulus intensity continually increased. Now, nearly 18 years later, Cattell (1902) suggests two principles for response time that permit the determination of equivalences between psychological events.

---

[5]On the perception of small differences (with Professor G. S. Fullerton), University of Pennsylvania, 1892. On errors of observation. *Amer. Journ. of Psychol.* 1893.

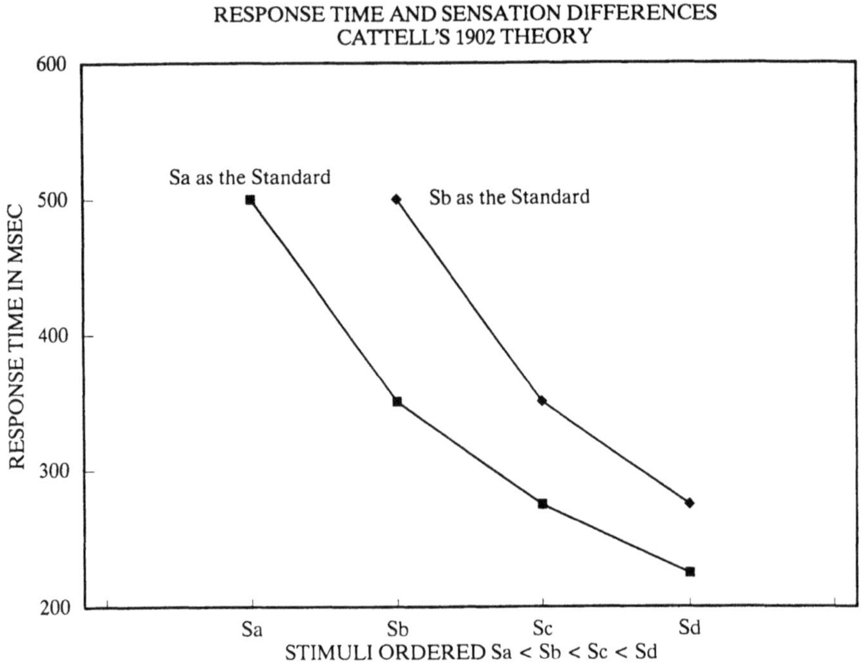

FIG. 2.4. Cattell's use of response time to indicate equal differences in sensation is illustrated by showing the response time differences between two stimuli to be equal.

Cattell's Principles:

Principle I: The smaller the difference between two sensations, the greater is the time required to perceive it.

Principle II: When differences require equal times for discrimination, the discrimination is equally difficult, and the differences are equal for consciousness.

Figure 2.4 illustrates the implications of Cattell's Principles. The abscissa of these hypothetical results shows increasing differences between stimuli whereas the ordinate scales response time. The two comparisons, **Sa** versus **Sa** and **Sb** versus **Sb**, have no difference at all. Therefore, consistent with Principle II, these comparisons are shown to generate equal response times. When **Sb**, a comparison stimulus different from **Sa** is presented, the response time declines, as it must according to Principle I. As the comparison stimulus deviates farther from **Sa** the response time continues to decrease. These results are also consistent with the data of Berger and Cattell if the reaction stimulus is considered to be compared against the ambient luminance of the darkened experimental room.

From Cattell's point of view, the differences between the two stimuli are psychophysically equal when the response times for their comparisons are equal. The psychological difference between **Sa** and **Sa** equals the difference between **Sb** and **Sb**. Therefore the response times should be equal. The generalization, that when comparing any two stimuli equal differences in response time imply equal psychological differences, suggests numerous hypothetical results. For example, the comparison of **Sa** and **Sb** produces the same response time as the comparison of **Sb** and **Sc**. This suggests that the psychological difference between **Sa** and **Sb** equals the psychological difference between **Sb** and **Sc**. Of course, no process is postulated that gives rise to these principles, thus predictions drawn from the principles are somewhat hard to come by.

Cattell's remarks are the beginning of a serious rift in metapsychophysics[6] caused by the unspoken assumption that the mechanism giving rise to Fechner's errors, and that giving rise to response time, were in some way different. Cattell attempted to establish such a difference by arguing that response time depends on sensation whereas errors do not. Then, freed from the tyranny of Fechnerian theory, Cattell proposed that differences in response time measure differences in sensation.

In Europe Helmholtz (1850) measured the speed of nerve conduction and deJaager (1865/1970) and Donders (1868) measured response times in such cognitive tasks as discriminating between two colors or identifying which of several possible signals occurred. Summarizing early research using reaction time Titchener (1905) asserted that:

> We are measuring, primarily, the duration of certain excitatory processes in the living organism, the time taken by certain physiological events. What these processes are, we know only schematically. Nor do we know at all accurately what the variability of the reaction time signifies: whether the addition of extra processes to (or the subtraction of normally constituent processes from) the complex of typical processes, with the substitution of paths that this addition (or subtraction) involves; or a variation of some or other of these typical processes themselves, under the varying influences of connected centres or organs. It is perhaps safe to say, in the rough, that both factors are at work in the ordinary course of the reaction experiment. At any rate, the representative values and measures of variability stand, primarily, for physiological processes. (pp. 362–363)
>
> Some of these processes, now, are psychophysical, i.e., have conscious concomitants. Hence the temptation to say that the 'reduced' reaction time (the reaction time *minus* the time of those physiological events which are not psychophysical) measures the duration of certain mental processes. It is, of course, by no means an easy matter to determine this reduced time, even approximately. But that apart, the question has been raised whether we have any right at all to speak of the 'duration'

---

[6]By metapsychophysics I mean the mental counterpart to metaphysics, the study of the ultimate nature of mind.

> of mental processes in terms of objective time: (see M. F. Washburn, *Psychological Review*, x., 1903, 416 ff. It is a question which, however worthy of consideration, cannot be gone into here. (p. 363)

In spite of reservations regarding the interpretation of response time, the measure afforded a different view of psychophysical processes. Unlike self-reports of a mental state, such as those obtained from the Method of Right and Wrong Cases, response times were a behavioral measure, perhaps uncontaminated by "subjective" factors. Further, response time presumably measured the psychophysical process regardless of its outcome.

VIVIAN A. C. HENMON

By the early 1900s enough accumulated experimental evidence showed response time to change as a function of experimenter controlled stimulus differences to encourage Henmon, Cattell's student, and eventually Jastrow's colleague, to summarize earlier works by Cattell (1887, 1893), Jastrow (1888), Munsterberg (1894), Warren (1897), Martin and Müller (1899), Angell (1901), and Whipple (1901). This evidence also encouraged Henmon (1906) to argue, somewhat boldly, that:

> The extensive literature which has sprung up since the time of Weber and Fechner on the measurement of differences in sensations is an evidence of the interest of the psychologist in this problem and the importance attached to it. In well-nigh all the investigations made to measure such differences the unit of measurement has been the just noticeable difference. The theoretical difficulties involved in the assumption of this unit and in the methods used to determine it are numerous and have been frequently pointed out. It will be the purpose of this introductory discussion to call attention briefly to some of these difficulties. This will perhaps serve to introduce the method of the time of perception as a measure of the differences in sensations, which is used in this study, and to suggest its raison d'etre. This method proposes the substitution of the time of perception as a unit of measurement for the just noticeable difference. (p. 5)

The method required the subject to compare two different stimuli and judge which stimulus, on the left or right side of a displayed card, possessed more of an experimenter defined characteristic. In particular, in the case of comparing horizontal line lengths, two lines appeared side by side on a card 50 mm in front of the subject. The subject released the left- or right-hand response key to choose the appropriate stimulus (what was appropriate remains a mystery).

Figure 2.5 shows the results for Henmon's two subjects. Each subject contributed 400 response times (RTs) per point for a total of 2400 observations. No mention is made of errors. The mean RTs, presumably of all responses to the

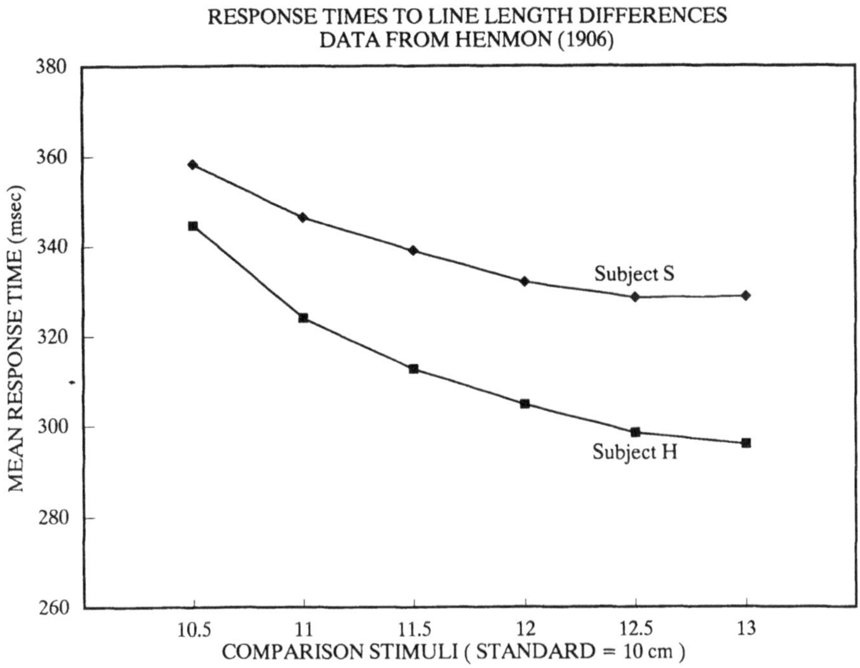

FIG. 2.5. Henmon's response time results for two subjects show how response time declines as a function of the difference between two stimuli.

stimuli, show a marked decline as the magnitude of the difference between stimuli increases, thus confirming one of Cattell's expectations.

Henmon's research represents the first clear demarcation between those experimenters who led the way toward an analysis of sensory function in terms of response time and those who relied on the classical measure of response frequency. Henmon reported only the response time and an occasional number of false reactions, but no choice response frequencies at all! Yet, after reporting extensive response time data for the perception of color and linear magnitudes, Henmon (1906) concluded:

> It seems to the writer that it should be possible to apply to these results Fechner's *Unterschiedsformel* and to correlate the increase in the magnitude of stimuli inversely with the increase in the time of perception, and thus test the psychophysical law. . . . The mathematical difficulties involved in this application are such that the writer prefers to publish the results of the experiments with the general indication of their relations rather than to attempt an application and interpretation of doubtful validity. (p. 51)

Henmon's quest for a new measure of discriminability and hope of discovering a relation between response time and Fechner's decision criterion, is a likely motivator of his later rapprochement (Henmon, 1911). In this elegant experiment Henmon measured response time, response frequency, and response confidence to two simultaneously presented line lengths of either 20.0 mm or 20.3 mm. On 50% of the 1,000 trials per subject, the subject chose which line was shorter and on the remaining trials, which line was longer. The small difference between the line lengths insured that errors played a significant role.

With respect to response confidences the four response confidence categories, (A) pretty confident, (B) fairly confident, (C) with little confidence, and (D) doubtful, led to different numbers of responses in each category depending upon whether the response was correct or in error. Subject BL, for example, generated 820 correct and 180 error responses. Figure 2.6 shows the relative frequency of using each confidence category and how the relative frequency of errors increased as confidence decreased. For the "doubtful" confidence category there were 37 correct responses and 40 errors. These 77 responses amounted to 7.7% of all responses.

Subject BL's average correct and error response times for each confidence category are shown in Fig. 2.7. Henmon (1911) interpreted these results:

> In summary the following conclusions with reference to the relation of the time of judgment to accuracy and confidence may be noted:
>
> 1. The time of judgment increases uniformly as the degree of confidence decreases.
>
> 2. The time of wrong judgments is on the average longer than the time of right judgments, while under each category of judgment the wrong judgments are in general shorter[7].
>
> 3. The time of wrong judgments is more variable than that of right judgments, and there are indications of two types of wrong judgments with reference to time, those that are rendered too quickly and those that are prolonged beyond the optimal time. (p. 195)

The covariation between response time and response probability that Vivian Henmon's experiments demonstrate became a source of curiosity for many succeeding psychophysical theorists.

---

[7]The rather surprising fact that average error time proves to be greater than average correct response time follows from averaging across confidence categories with unequal numbers of correct and error responses. As is easily seen in Fig. 2.6, response confidence categories are not used with equal frequency. Furthermore, as might be expected, the number of errors increases as confidence diminishes. Thus, the average error response time includes few low RTs from Category A but more longer error RTs as confidence decreases. The mean correct RT includes many low RTs from Category A and fewer longer RTs as confidence decreases. Compared to the mean error RT, which contains many long RTs, the mean correct RT contains few long RTs and is therefore lower.

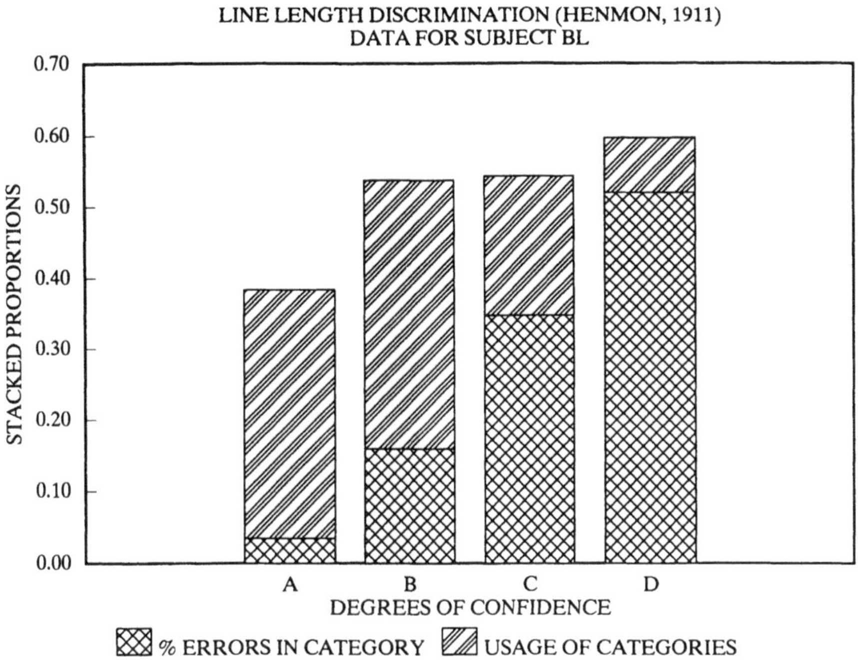

FIG. 2.6. From Category A to D the number of confidence judgments decreases, but the proportion of errors in a confidence category increases from Category A to D.

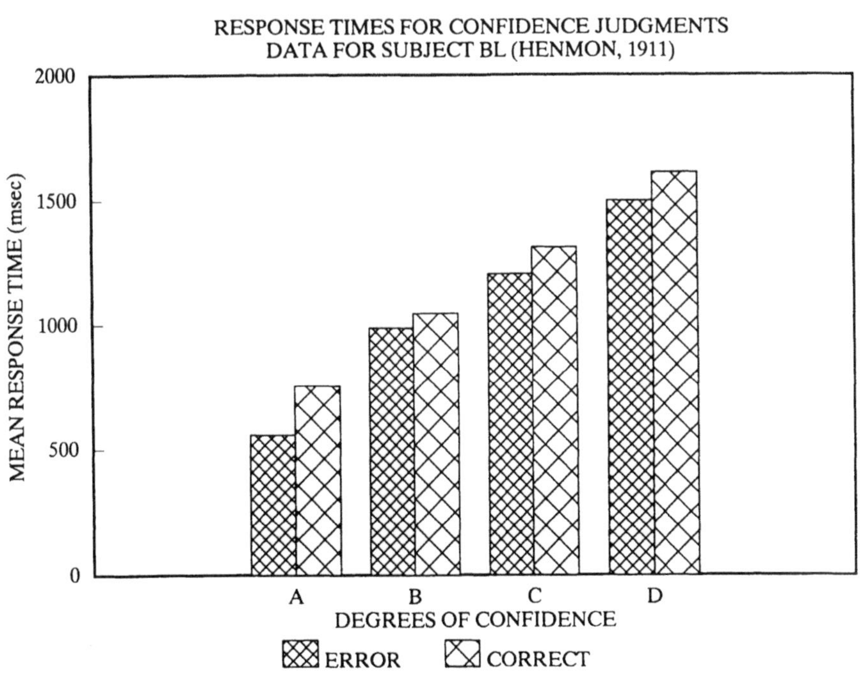

FIG. 2.7. Subject BL's average correct and average error response times show increases as a function of the degree of confidence in the discriminative judgments.

Henmon might have advanced beyond these results by investigating changes in response time and response proportion as a function of increases in the difference between the two line lengths, but this experiment waited for a much later analysis of sensory discrimination. In the meantime, substantial variations in response proportions were known to occur as differences between stimuli increased. These observations, together with the Method of Constant Stimuli, and Fechner's comments regarding an "interval of uncertainty," foreshadowed the creation of the *Psychometric Function*.

3 The Origin of the Psychometric Function

Narrowly focused on sensitivity and the ability to discriminate between stimuli, Fechner overlooked the potential importance of the failure to discriminate. In the Method of Right and Wrong Cases, for example, the subject occasionally volunteered that the standard and comparison stimuli were equal, although they were never equal. In order to resolve such failures of discrimination Fechner added half a response to the total counts for each of the two response alternatives, as if half a correct response and half an error occurred. This artifice obscured the frequency with which failures to discriminate occurred. Fechner (1860/1966) observed that

> From a practical point of view the method of just noticeable differences is the simplest and most direct of the three methods of measurement. . . . One of its major drawbacks is the fact that whatever is termed just noticeable leaves more room for error as a subjective judgment, than is the case with the other methods. This term is in no way absolute, since neither the first point, where a difference in sensation becomes just noticeable, nor the second point, where it disappears, can be determined quite precisely. *One passes through an interval of uncertainty* [italics added] as to whether one is aware of the sensation or not. (p. 63)

F. M. URBAN

The Interval of Uncertainty

Aware of its potential for altering the interpretation of psychophysical data, Urban (1907, 1908, 1910) seized upon the *Interval of Uncertainty* (IU) as a new tool for the measurement of sensitivity. Within the IU subjects FELT that two different stimuli were of equal magnitude. Their inability to discriminate, reasoned Urban, must be the source of the IU. A subject unable to resolve even large stimulus differences would evidence a broad IU, whereas keen discrimination should produce a small interval.

Bringing the Interval of Uncertainty under scientific scrutiny required a measurement of its size. Urban quickly advanced an empirical technique for determining this unknown value. In contrast to Fechner, Urban recorded those cases where the subject judged the comparison stimulus to *equal* a standard. To measure the likelihood of equality, subjects were required to make one of three responses: *smaller*, *equal*, or *greater*.

In order to show how the proportion of *equal* judgments changed as a function of stimulus differences Urban (1907) extended the experimental design invented by Peirce and Jastrow. The new design employed a fixed-standard stimulus and presented comparison stimuli smaller than, equal to, or greater than the standard's physical magnitude. The stimuli, often weights, increased in an arithmetical progression with the standard near the middle of the stimulus range. Moreover, adopting the popular agricultural procedure (cf. Cochran, 1976) of randomizing across unwanted effects, each trial consisted of the standard stimulus followed by a comparison stimulus chosen at random from the various alternatives. This three-choice Method of Constant Stimuli (which Wundt called "the method of three cases") yielded the typical response proportions shown in Table 3.1.

To estimate the size of the Interval of Uncertainty, Urban (1910) first defined a *Psychometric Function* (PF) to be

> A mathematical expression which gives the probability of a judgment as function of the comparison stimulus . . . in imitation of the term biometric function. (p. 230)

TABLE 3.1
Three Probabilities of Responding *Smaller, Equal* or *Greater*[1] to a Range of Comparison Stimulus Weights Given a Standard of 100 Grams

	Comparison Stimulus Values (g)						
Response	84	88	92	96	100	104	108
Smaller	.93	.86	.70	.45	.23	.10	.02
Equal	.05	.12	.19	.26	.24	.08	.08
Greater	.02	.02	.11	.29	.53	.82	.90

Note. From *The Application of Statistical Methods to the Problems of Psychophysics.* (p. 178) by F. M. Urban, 1908, Philadelphia: The Psychological Clinic Press.

[1]These results also appear in Urban (1910, p. 233) where the response categories of *smaller*, *equal*, and *greater* are used. In the 1908 monograph the response categories, for the same data, are given as *lighter*, *equal*, and *heavier*.

As defined by Urban the Interval of Uncertainty ranged from the comparison stimulus value that yielded a probability of responding *smaller* of 0.50 to the comparison stimulus value that evoked a response of *greater* with probability 0.50.

Because the experimenter's set of stimuli seldom if ever contained stimuli that produced exactly 50% *smaller* or *greater* responses, Urban required a technique to estimate the values of such stimuli. The solution came from the Lagrange function, used to interpolate between the actual stimulus values in order to determine theoretical stimulus values defining the upper and lower limits of the Interval of Uncertainty.

The Lagrange polynomials defined a Psychometric Function for each response. If $s_1, s_2, \ldots, s_n$ are the values of n comparison stimuli which give rise to a *smaller* judgment with probabilities $p_1, p_2, \ldots, p_n$, then the Lagrange equation for the *Psychometric Smaller Function* is:

$$
\begin{aligned}
L(s) = {} & \left[\frac{(s - s_2)(s - s_3) \ldots (s - s_n)}{(s_1 - s_2)(s_1 - s_3) \ldots (s_1 - s_n)} \right] p_1 \\
& + \left[\frac{(s - s_1)(s - s_3) \ldots (s - s_n)}{(s_2 - s_1)(s_2 - s_3) \ldots (s_2 - s_n)} \right] p_2 \\
& + \ldots \\
& + \left[\frac{(s - s_1)(s - s_2) \ldots (s - s_{n-1})}{(s_n - s_1)(s_n - s_2) \ldots (s_n - s_{n-1})} \right] p_n
\end{aligned}
$$

where $L(s)$ is the probability of responding *smaller* given a comparison stimulus of size s. A similar function, $G(s)$, defined the *Psychometric Greater Function.*

Urban solved such polynomial equations for the stimulus magnitude that produced a *smaller* response with probability 0.50. Similarly from the Psychometric Function for *greater* responses Urban extracted the stimulus magnitude that produced a *greater* response with probability 0.50. The difference between these two stimulus magnitudes defined the Interval of Uncertainty.

As illustrated in Fig. 3.1 the estimated Psychometric Functions determine the comparison stimulus values that generate a response probability of 0.50. Using Urban's (1910) calculations the Smaller PF gives a stimulus value of 95.20 g and for the Greater PF it is 99.55 g. The difference between these stimulus values, 4.35 g, is the size of the IU. The smooth Psychometric Function shown in Fig. 3.1 results from spline (Bloxom, 1985; Schoenberg, 1946; Schumaker, 1981), rather than Lagrange, fits to the data.

Additional features of discrimination derived from the Psychometric Functions. The stimulus magnitude that corresponds to the intersection of the *greater* and *smaller* Psychometric Functions defines the *Point of Subjective Equality (PSE)*. Any deviation between the value of PSE, for these data equal to 97.36 g, and the actual value of the standard stimulus (100 g) Urban attributed to a mysterious *Constant Error*.

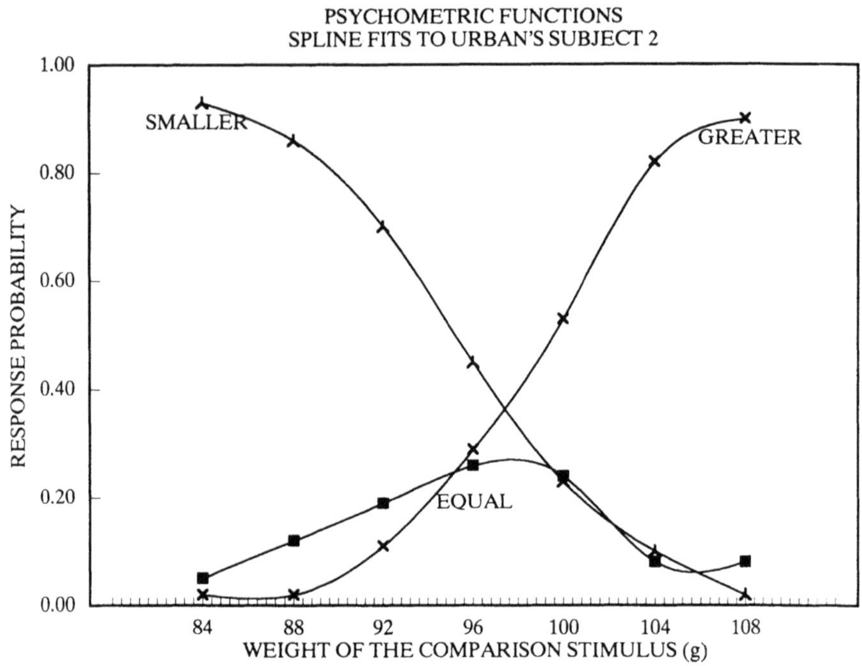

FIG. 3.1. Psychometric functions from Subject 2.

Certainly a large IU implied poor discriminability. Therefore, argued Urban, the reciprocal of IU was an index of discriminability. Insisting that this measure of discriminability needed no theory for its justification, Urban suggested that psychologists, like actuaries who developed biometric functions, should advance their science through the behavioral measurement of Psychometric Functions. Thus, behavior displaced theory.

Although Urban eschewed theoretical motivations for the form of the Psychometric Functions, he ascribed to them properties inconsistent with probability theory. For example, Psychometric Functions for *smaller* and *greater* responses were defined by probabilities. Yet, the Psychometric Function for *equal* responses, also defined by response probabilities, Urban considered to be a Gaussian distribution—implying, of course, that the probabilities for responding *equal* were actually probability densities. In a series of lively exchanges with Urban regarding the form of the Psychometric Function, Culler (1926a, 1927) and Thomson (1920) vigorously criticized the peculiar confusion that Urban shared with Titchener.

Many untested assumptions form the basis for Urban's impressive theoretical achievement. First, fitting a smooth polynomial to the data points assumes that performance changes continually, even for very small stimulus differences. Second, by requiring the subject to respond *smaller*, or *equal*, or *greater* the judgment of *equality* was considered comparable to the judgments of *smaller* or *greater*. If the subject were to respond only *smaller* or *greater* in one experiment and then respond only *equal* or *nothing* in another experiment then any differences due to the type of judgment might be revealed.

Perhaps the most serious theoretical complaint is that the upper and lower limits of the Interval of Uncertainty suggest the existence of two different thresholds for the sensation of equality. Also, the use of a probability of 0.50 to define the limits of the Interval of Uncertainty is subject to obvious criticism. Any particular value must be based upon implicit theoretical ideas that in some way suppose the quality of a sensation to change, to be more like a sense of smaller or greater, when the limits of the Interval of Uncertainty are exceeded. But how can a sudden shift in feeling occur when the probabilities defining the Psychometric Function change smoothly? Once again the theoretical problem addressed by Peirce and Jastrow manifests itself in the determination of the Interval of Uncertainty just as in the determination of any threshold.

WARNER BROWN

Because of its apparent correspondence with the cumulative Gaussian distribution, the form of the Psychometric Function for *greater* responses generated considerable interest. In a series of articles on the Phi-gamma Hypothesis, the hypothesis that the form of this function was indeed that of a cumulative Gaussian, some theoreticians found supporting data. Indeed, Peirce and Jastrow's (1884) article is an important contribution to the debate about the Phi-gamma Hypothesis, which they assumed to be true. Finally, in 1910, Brown, a student of Cattell's, provided definitive data concerning the mathematical form of the Psychometric Function. In this magnificent study on the perception of minimal differences, Brown (1910) noted:

> That Weber's law is so commonly stated in terms of thresholds rather than 40 or 90 per cent. of correct judgments is only one indication of this silent assumption that the threshold is a particularly noteworthy or critical point in the judgment series. The constant employment of the method of just noticeable differences, as well as the retention of "equal" or "no difference" judgments in the method of right and wrong cases, despite the warnings many years ago of Cattell and Jastrow, has fostered the idea that it is desirable to find a point where a difference in sensation first appears, just as it is deemed desirable to find a point in the scale of physical intensities where a sensation quality first appears. The actual distribution of errors does not, however, indicate there is a virtue in any particular number of judgments by which one is warranted in asserting that if there were fewer of them they must be different in kind. So far as the threshold implies that 'sub-liminal' differences provoke a judgment different in quality from those due to larger differences, it is a misleading concept, and unsupported by fact. (p. 19)

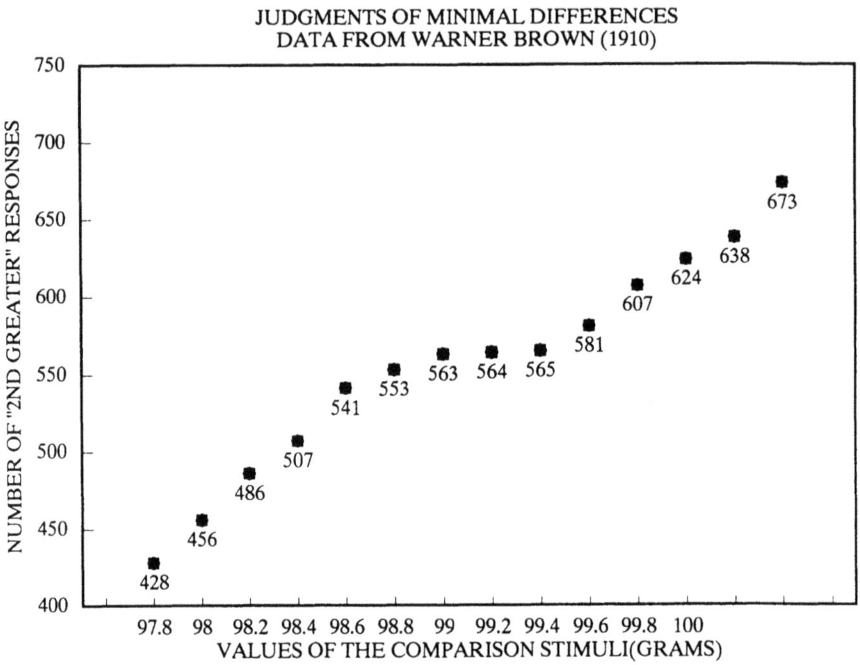

FIG. 3.2. Results from Brown (1910) on the accuracy of discriminative judgments.

Brown showed the accuracy of perception to be amazingly precise when enough observations were obtained to dispel the cloud of statistical error that often shrouded discriminative judgments. Using well-practiced subjects, and in one instance the author's future wife, who dutifully judged some 80,000 pairs of lifted weights, Brown obtained the Psychometric Function shown in Fig. 3.2.

Given a standard fixed at 100 g, comparison weights varied in small steps of 0.2 g to a maximum of 100.4 g from a minimum of 97.8 g. The standard error of proportion based on 1000 judgments per stimulus difference is no larger than 0.01. Brown concluded that although there appears to be a slight flattening of the function between 99 g and 99.4 g, there is no evidence of a discontinuity indicative of any form of threshold. Note in these results that the standard stimulus weighed 100 g but the proportion of *greater* judgments for a comparison weight of 100 g equals 0.624.

Brown's opinions based upon analyses of the many judgments produced during numerous experiments are especially informative. Brown (1910) found fault with the use of the response *same*. For example:

> Evasive answers such as "equal," "don't know," or "no difference" permit the observer to avoid forming a judgment in just those cases where it is most important to discover whether or not a correct judgment can be formed. (p. 28)

> This part of the experiment does, however, furnish abundant evidence to justify one in forcing the observer to state in every case which weight is heavier. The category "same" or "no difference" turns out to be unnatural and unsatisfactory to the trained observer. After about 4000 of these comparisons the observer made the remark "those two are exactly the same, and that is the first pair that has been so." This remark is indicative of the whole situation. . . . The quality of sameness rarely or never occurs pure. It is necessary to decide in each instance *how near alike* the two things are and then decide whether that amounts to equality. In other words one must maintain a mental standard of equality. As one becomes more or less scrupulous this standard is likely to vary; impressions that at one time seemed slightly different now pass for the same. (pp. 31–32)

Thus, Brown concluded that the type of judgment somehow affected performance. If the type of judgment plays such an important role in the determination of a difference, perhaps the judgment of difference depends heavily on which form of judgment is required. Earlier, Müller (1878) argued that results in discrimination experiments depended on the question asked of the observer. Brown investigated this possibility.

In a single experiment using a fixed standard of 100 g different questions were asked of the observer. To the question "Second?" the observer was asked to tell, verbally, whether the second weight was heavier than the first. To the question "Which" the observer was to point to the standard or comparison weight which was heavier. To the question "Same?" the observer was to respond verbally only if the weights felt the same. The results, based upon 200 observations per point appear in Fig. 3.3.

For the first two questions, "Second?" and "Which?", the results are fairly symmetric with respect to the discrimination of differences. For purposes of comparison with other PFs, the PF labelled "Second?" gives the number of judgments of *lighter* as a function of the size of the comparison stimulus. The decline in the number of responses is smooth except in the region from 94 g to 99 g. To the question "Which is heavier?" the subject responded with response frequencies shown by the "Which?" data. The proportion of *heavier* judgments is shown in the figure, but subtracting this number from 200 to obtain a measure of the number of *lighter* judgments shows the results to be remarkably similar to the previous case. However, the number of error responses is somewhat larger for the "Which?" condition. From these two functions the Point of Subjective Equality appears to be about 96 g. Again, there is a region, from 92 g to 99 g, lacking an orderly increase.

When the observer was asked to respond only when the two stimuli felt the same, the maximum number of responses occurred far from the point of subjective equality. The Interval of Uncertainty is very small, but the *Interval of Equality*, if it can be called that, ranging from about 94 g to 99 g is quite wide, and includes the IU. The Interval of Equality corresponds to the regions where the other two functions tend to undulate.

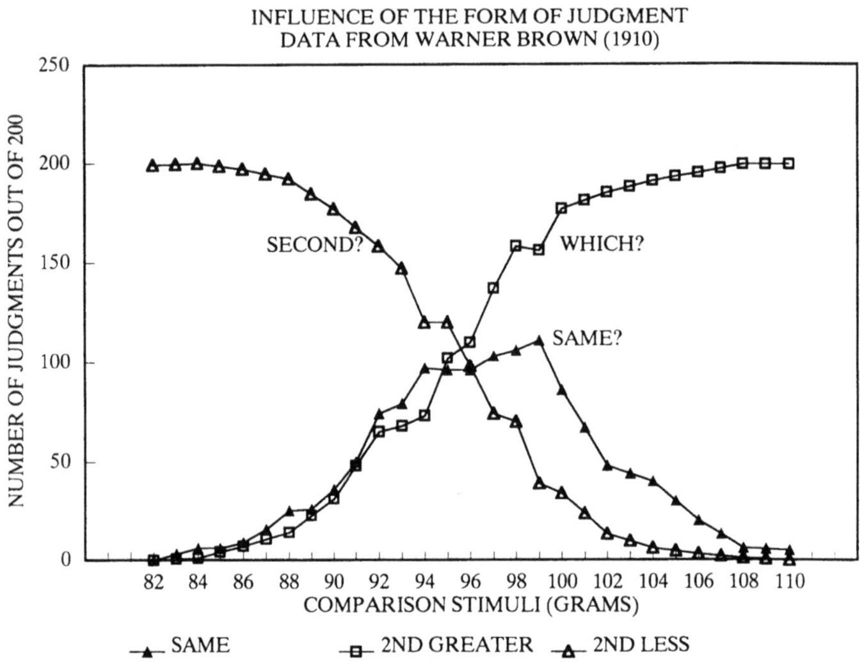

FIG. 3.3. Results from Brown (1910) on the influence of the type of judgment on the form of the Psychometric Function.

In summary of the experiment Brown (1910) concluded that:

> It is evident therefore that there is an advantage for the *second?* (verbal) group over the *which?* (motor) group in regard to the total number of errors. (p. 50)
>
> The results leave no doubt that a difference in the framing of two propositions which are precisely equivalent logically will be a governing factor in making a comparison. (p. 50)
>
> To speak of the "perception of difference" in such a case is to obscure some of the factors in the actual situation. The difference is not merely perceived. The process of comparison involves the active operation of the mind in the expression of a judgment upon the situation in which the difference is only one factor. (p. 51)

Following these very successful experimental studies of discriminating small differences in weight, Brown (1914) published the classic study on the perception of taste. The following paragraph introduces an important theoretical issue and foreshadows the development of a new understanding of the role of the subject in the making of a discriminative judgment.

> The term absolute threshold of sensation occurs very frequently in the writings of psychologists, and it is generally treated as if it were a technical term of precision.

> But it has appeared to the present writer that the concrete experiences upon which the notion of absolute threshold of sensation is founded have received very little consideration, and remain, for the most part, unanalyzed. (p. 201)

> The chief aim in undertaking the present group of experiments was to show that every judgment we make, no matter how simple the subject-matter, is conditioned to a large extent by inward or "subjective" factors. Such factors are more properly designated "central" in contradistinction to the peripheral factors which, in one way or another, compel the individual to make a reaction. The central factors may be thought of as a condition of the organism itself modifying its reaction. Obviously different organisms react differently to the same stimulus, and one reason is to be found in the different capacities which different individuals have. So, too, the same organism does not always react in the same way to stimuli which are apparently identical, and in this case the difference in reaction may be thought of as due to the different condition of the organism. Such a difference in condition constitutes a "central" factor in the reaction. (p. 218)

> In the case of the judgment of difference (Brown, 1914b) it was found that small differences in the task assigned, or small differences in the mode of reaction demanded, had the effect of altering considerably the observer's judgment in situations which were otherwise precisely equivalent. Moreover, a change in the arrangement of the subject-matter so that a particular case appeared in different company so altered the general attitude of the observer that his judgment would be affected. When, for instance, all the differences between the weights which he was judging were small, he could judge very small ones competently; but let the same task be given when coarser differences were being judged, then the central conditions were no longer favorable for such accurate judgment. (pp. 218–219)

> The judgment of difference is a comparatively complicated undertaking. The judgment of sensory stimuli is generally supposed to be much simpler. Is it still true in the simpler case that central conditions play an active part in the formation of judgments? There is no question here of the distinction between sensation and perception or of the association processes necessarily involved in the recognition of a stimulus. The point is whether there are changes in the attitude of the observer which affect his capacity to perform his part in sensation, no matter what that part is. (p. 219)

SAMUEL FERNBERGER

Not only did different instructions cause variations in performance, different experimental procedures appeared to produce different measures of sensitivity! The problem, striking at the heart of psychophysical analysis, came to the attention of Samuel Fernberger via F. M. Urban. To Fernberger (1913) a puzzle existed in the finding "that the comparison of the sensitivity of different subjects by a given method yields perfectly satisfactory results, but that these results do not always agree with those obtained by other methods (p. 1)."

Concluding the article "On the relation of the methods of just perceptible differences (due to Müller, 1878) and constant stimuli" (due to Peirce and Jastrow, 1884) Fernberger (1913) wrote:

> If we compare the observed thresholds of the method of constant stimuli and the calculated thresholds of the method of just perceptible differences, we find that the variations, in the case of both subjects, are always in one direction. The calculated thresholds in the direction of [comparison stimulus] decrease by the method of just perceptible differences are constantly larger than the corresponding values in the method of constant stimuli; and for the threshold in the direction of [comparison stimulus] increase, they are constantly smaller. Thus the limits of the interval of uncertainty are constantly narrowed in the just perceptible difference method when compared with the same interval of the other method. This is a serious fault as it is just this interval that is the measure of sensitivity of our subject. (p. 80)

By 1914, however, the cause of various divergent measurements of thresholds rested with the subjects. Fernberger (1914) credited Woodworth (1914) with the observation that a "reduction in the interval of uncertainty to zero . . . is due to a subjective factor (p. 538)." Furthermore, noted Fernberger,

> One cannot measure sensitivity in absolute terms; one can only say that a given sensitivity has been found to exist *under certain given experimental conditions*. Inasmuch as we are dealing with a total psychological organism and not merely with a sense organ, it is, after all, not surprising that the attitude of the subject should have a profound affect on this sensitivity. (p. 542)

GODFREY H. THOMSON

An Urn Model of Psychophysical Judgment

Urban's attacks on psychophysical theory motivated new approaches to measuring discriminative performance. In fact, long before other psychological theorists such as Watson (1919) and Skinner (1938) hit on the idea of behaviorism, Urban eschewed theory in favor of empirical measures of behavior—an extension of Peirce's philosophical pragmatism to the arena of discriminative performance.

In spite of Urban's avowed disinterest in theory some primitive theory lurks in any jungle of equations. For Urban these equations provided measures of important psychophysical characteristics. For others they became a source of consternation. The others, primarily Thomson and Culler, discovered many errors in Urban's derivations and applications of mathematical methods. Thomson (1913) commented that:

> The purpose of this note is to correct an error in Urban's application of Bernoulli's Theorem for the calculation of the probable error of the Method of Just Perceptible Differences. . . . The two formulae which Urban does use are based on an inverse use of Bernoulli's Theorem, and both are incorrect. (pp. 217–218)

Later Culler (1927) joined in the debate.

> To conclude: we have tried to prove that Urban's formula for the probable error of the limen is fatally defective in no less than three ways.
>
> (1) His own figures, when tested by the actual dispersion of the fractional limens, are fantastically improbable.
>
> (2) The formula yields no single, defined value at all. By the simple expedient of shifting the origin, we can get, over an extensive range, any figure whatsoever.
>
> (3) The quantities h and c, whose mutual dependence is virtually perfect, he treats like wholly unrelated variables, as if they were heads and tails in successive throws of a coin. (p. 311)

Although criticism may expose unwarranted assumptions, or correct erroneous derivations, some criticisms strike at the feelings even of a hardened theorist. As most theorists are quick to point out, criticism does little to bring forth a better theoretical treatment. Thus, in spite of these criticisms, Urban eschewed any theoretical basis for discrimination, but placed at the door to the theoretical pantheon a set of splendid observations awaiting a theoretical treatment.

Godfrey Thomson, a young physics student, met the challenge posed by Urban's data. Thomson set forth a few principles that easily explained the origin of Urban's three-choice data. These principles embodied the important concept of a subject controlled response criterion. By creating a new form of psychophysical theory Thomson produced the first alternative to the theoretical ideas of Fechner, and a psychophysical theory of rare power.

Thomson (1920) sought to explain the familiar form of the three Psychometric Functions found by the many experimenters who used the Method of Constant Stimuli. Taking exception with the view held by Urban and Fernberger, that the Interval of Uncertainty provides a measure of discrimination, Thomson offered an insight that still commands attention.

> This distance (IU) however depends entirely on the subject's readiness to give the answer undecided. It measures therefore rather a moral characteristic than a physical sensitivity. It is a question of quickness of decision on small evidence as much as a question of difference of amount of evidence. If persons A and B differ widely in the number of undecided answers they give, it may just as well be their habits of forming decisions which differ, and not their power of actually discriminating the weights. The moral character of (this measure) is above all seen from the fact that any subject who wishes can reduce it to zero, whatever may be his actual sensitivity, simply by determining that he will never give an answer *undecided*. Difference thresholds therefore are unsuitable for comparing sensitivity of different subjects. (pp. 301–302)

Thomson's theory of discrimination expresses theoretical psychological ideas in a solid mathematical theory. This theory is as powerful as any yet proposed to account for three-choice decisions and occupies a special place of honor in the theoretical pantheon.

A major assumption of the theory is that the process of comparison can be imagined as evoking mental elements corresponding to atomistic sensations, such as lightness or heaviness, a return perhaps to the monadology of Leibnitz, and the sensory elements of Herbart (1824/1850). The proportion, p, of heavy elements or the proportion $(1 - p)$, of light elements, depends, somehow, on the experimenter's choice of the comparison and standard weights.

These stimulus-dependent elements are sampled by the subject; in fact, a fixed sample-size model is proposed. The probability of any combination of light and heavy elements in the sample is derived from the binomial expansion. Thus for a sample of n elements the expansion of $[p + (1 - p)]^n$ generates the various probabilities of obtaining a particular combination of elements. In this way the probability of an event such as $n - 1$ heavy elements and 1 light element is easily determined to be $np^{(n-1)}(1 - p)^1$.

The way a subject used this sample to create a response is a far-reaching, innovative, feature of Thomson's theory. Thomson proposed that a certain number of heavy elements were necessary in order for the subject to respond *heavy*, but the subject chose number of such elements. In this way Thomson introduced the idea of a response criterion as distinct from a sensory threshold. Finally, the goals of the subject enter into the psychometric equation.

In order to respond *heavy* the subject must first establish a *heavy* response criterion, such as $n - 2$ or more sampled heavy elements. A second criterion must be established to control the response *light*. For illustration suppose this *lighter* criterion is less than 3 heavy elements. Given these response criteria a sample containing exactly 3 to $n - 3$ heavy elements falls between these criteria and yields a response of *equal* (or *undecided*).

The predictions of this simple theory are in qualitative agreement with data from several experiments. The quantitative fit of the theory to data such as Urban's in Table 3.1 can take several forms. First it may be possible to determine estimates of the parameters that maximize the probability of the data given the theory. However the equations are quite complex. A useful alternative is to calculate a goodness of fit statistic for every set of parameters. Then that set of parameters that generates the best fit value defines the optimal parameter set. To illustrate the procedure let the probability of sampling a heavy element increase linearly with the magnitude of the comparison. Although arbitrary, this assumption reduces the seven parameters needed to represent seven stimuli to only two. These, together with number of elements sampled, n, and the two response criteria bring the total number of parameters to five.

The chi-squared statistic is a goodness-of-fit statistic that measures the correspondence of Thomson's Theory to the data in Table 3.1. The linear function relating the probability of sampling a heavy element to weight, fit best when the value of n equals 10. The criteria that gave the best fit, not too surprisingly, were

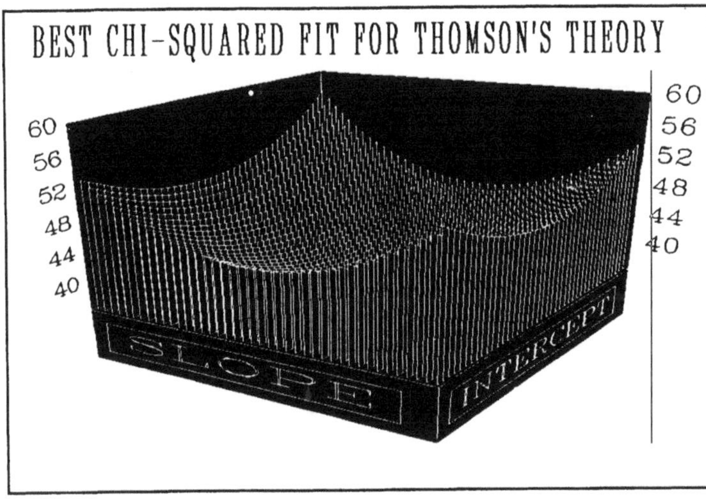

FIG. 3.4. Slope and intercept changes in a linear function relating weight and the probability of sampling a heavy element generate changes in the goodness-of-fit of Thomson's theory to Urban's data.[Data values courtesy of James MacRae and Bruce Weaver.]

at 4 and 6 heavy elements, respectively. Exactly 5 elements produced the response *equal*. Figure 3.4 shows how steeply the goodness-of-fit statistic declines as the optimum value for the slope and intercept parameters are reached. Given that $n = 10$ the minimum value for chi-squared is 38.9. With $14 - 5 = 9$ degrees of freedom, the fit is not impressive.

Yet, as shown in Fig. 3.5, the predicted Psychometric Functions are quite similar to the data from Table 3.1. Perhaps the magnitude of the chi-squared statistic is affected by the small standard error of proportion due to the 400 observations per response proportion. In any case, for the 1920s Urban's proportions appear to be predicted closely by Thomson's theory.

A most significant aspect of this theory is the way in which a skewed Psychometric Function occurs. When the subject uses a criterion of exactly n heavy elements to respond *heavy* an apparent bias toward responding *lighter* must appear. Such results were known to exist, but an explanation defied analysis for many years because the experimental results lacked a theoretical model. Thus, Thomson's ingenious theory provided a handy explanation of such biased Psychometric Functions. More important from a psychologists viewpoint, the subject finally entered into the psychophysical equation as a mediator between purely sensory perception and the eventual response. Thomson's creation of subject-controlled response criteria put in place another cornerstone of modern psychophysical theory.

Yet, important questions remained unanswered. For example, what mecha-

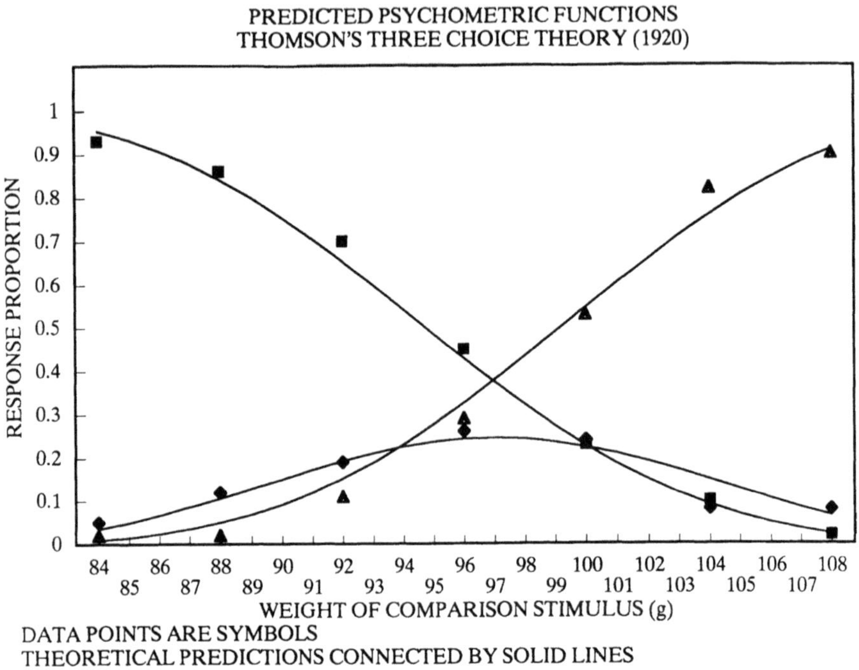

FIG. 3.5. Thomson's theoretical Psychometric Functions derived from a fixed-size Stimulus Sampling Theory appear to agree well with Urban's observed Psychometric Functions.

nism gave rise to the probability that a sensory element was heavy? What determined the probability that a sensory element was sampled? Was the sampling scheme determined by the subject? Furthermore, with so many parameters of the psychophysical process controlled by the subject, did the stimulus serve only to influence the unknown sampling probability? Could this unknown value be determined, somehow, from a subject's experimental results?

4 The Law of Comparative Judgment

LIONEL L. THURSTONE

These questions lost significance in the light of L. L. Thurstone's beautiful and justly famous extension of Fechner's theory of discrimination. Unlike his predecessors, who were largely concerned with measuring sensitivity, Thurstone (1927a, 1955) proposed a method for measuring psychological differences between stimuli. In this regard he was the first to postulate the existence of a *Psychological Continuum*[1] onto which a stimulus projected a distribution of possible values. A single stimulus presentation produced a sample value drawn from the distribution of such psychological values.

The theory of measurement depended upon a theory of stimulus comparison. Thurstone supposed that when two stimuli are compared, each stimulus generates a psychological value. These values are subtracted to determine the size of their psychological difference. The statistical properties of such differences depend upon the parameters of the corresponding probability distributions for each stimulus and any covariation in values that may be inherent in the stimuli or their method of presentation.

Figure 4.1 illustrates the joint probability distribution for two stimuli $\mathbf{S}_A$ and $\mathbf{S}_B$ that generate psychological variables S_A and S_B defined on the Psychological Continuum Ψ_s. These variables are properly considered as random variables with probability density functions that Thurstone conveniently assumed to be Gaussian with means μ_A and μ_B and variances σ_A^2 and σ_B^2, respectively. Their joint representation in a two-dimensional space shows the case where both means and variances of the psychological values are equal. In this case $\mu_A = \mu_B = 0$ and $\sigma_A = \sigma_B = 1$. The joint representation has a second advantage in that any

[1]Author's interview of Dr. Paul Horst.

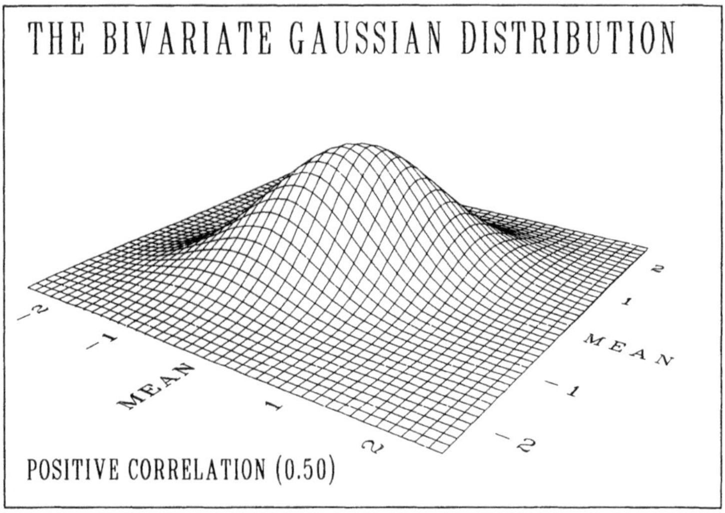

FIG. 4.1. The bivariate Gaussian density function with a correlation of 0.50 shows increases in values of S_A that tend to occur with increases in S_B.

covariation between S_A and S_B is readily apparent. In Fig. 4.1 the correlation is apparent in that increases in values of S_A correspond to increases in S_B.

Given these jointly varying random variables Thurstone argued that a comparison between stimulus values occurred by subtraction of one value from another. For example, a subject comparing $\mathbf{S}_A$ and $\mathbf{S}_B$ first obtained a sampled value, S_A, then a sampled value, S_B, and then compared these values by computing $D_{AB} = S_A - S_B$. A difference greater than zero indicated the sampled value of $\mathbf{S}_A$ was larger than $\mathbf{S}_B$. The subject should conclude, and respond, that $\mathbf{S}_A$ was greater than $\mathbf{S}_B$.

The decision rule for the subject is conveniently represented in the following summary. Let the random variable D_{AB} be the difference between S_A and S_B. Then,

$$D_{AB} = S_A - S_B \tag{4.1}$$

and,

$$\text{if } D_{AB} > 0 \text{ Respond } \mathbf{S}_A > \mathbf{S}_B$$

$$\text{if } D_{AB} < 0 \text{ Respond } \mathbf{S}_A < \mathbf{S}_B.$$

Figure 4.2 shows the distribution of the difference, D_{AB}. Fortunately for Thurstone the difference between any two Gaussian distributed random variables is also Gaussian distributed (but this is the only family of random variables with finite variance for which the difference is a member of the same family).

The value zero represents a decision criterion that partitions the Gaussian distribution of differences into two parts. The area to the right of zero equals the

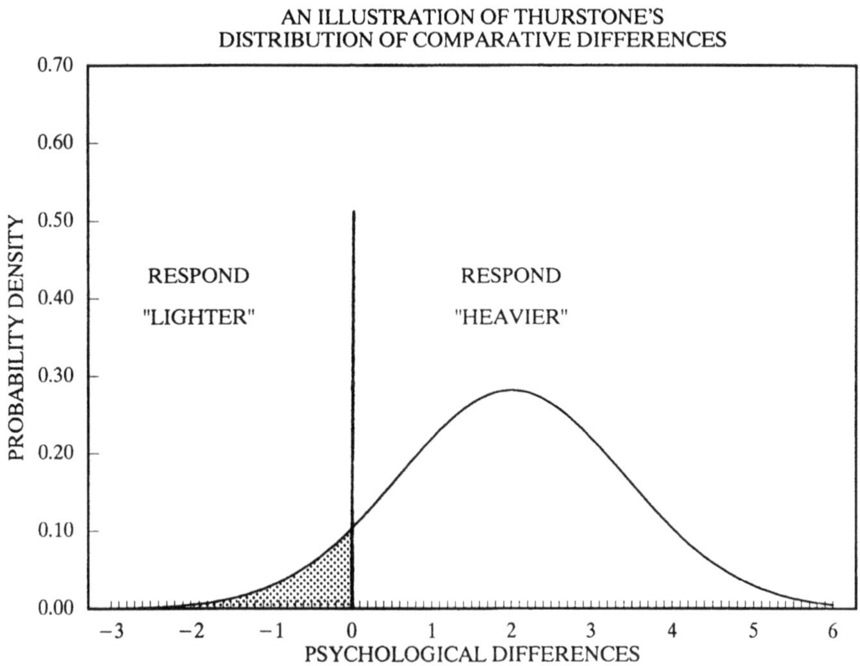

FIG. 4.2. The difference between two Gaussian random variables is itself Gaussian. The area to the right of the decision criterion equals the volume under the joint distribution above the part of the S_A, S_B plane where $S_A - S_B > 0$.

probability p_{AB} that D_{AB} is greater than zero while the area to the left yields the probability, $1 - p_{AB}$, that D_{AB} is smaller than zero.

The random variable D_{AB} is Gaussian but its parameters depend upon the expected values and variances of S_A and S_B and any covariation between them. In the most general case where the variances of S_A and S_B are unequal, and there is covariation between the psychological values of S_A and S_B, and the variance of the discriminable error D_{AB}, is expressed as:

$$Var\ \{D_{AB}\} = Var\ \{S_A\} + Var\ \{S_B\} - 2\ Cov\ \{S_A, S_B\}.$$

Thurstone used the standardized difference, $E\ (D_{AB})/[Var\ (D_{AB})]^{(1/2)}$, to create the

Law of Comparative Judgment:

$$E\{Z_{AB}\} = \frac{E\{S_A\} - E\{S_B\}}{[Var\{S_A\} + Var\{S_B\} - 2\ Cov\{S_A,S_B\}]^{(1/2)}}. \tag{4.2}$$

The mean difference between the two stimuli is the mean value of the Gaussian distribution of differences, which equals $\mu_A - \mu_B$. The Law of Comparative

Judgment expresses this difference with respect to the standard deviation, $[Var\{\boldsymbol{D}_{AB}\}]^{(1/2)}$. The expected value of $\boldsymbol{D}_{AB}$, scaled with respect to the standard deviation of differences, provides a measure of this psychological difference between the mean psychological values of the two stimuli.

The difference between zero and the mean of the distribution of differences is easily measured in units of the standard deviation of a Gaussian distribution. To compute the number of such standard deviation units, note that the area between zero and the unknown mean defines a probability equal to $p_{AB} - 0.50$. This is the probability that a Gaussian random variable is $\boldsymbol{Z}_{AB}$ units away from the mean—a value easily determined by reference to a table of the cumulative Gaussian distribution. For example, when the probability of choosing $\mathbf{S}_A$ over $\mathbf{S}_B$ equals 0.75 the value of $p_{AB} - 0.50$ equals 0.25. Consequently the mean difference between the psychological values for the two stimuli, $E\{\boldsymbol{Z}_{AB}\}$, equals 0.674 standard deviation units. This standardized unit measured the psychological differences between $\mathbf{S}_A$ and $\mathbf{S}_B$.

Thurstone's remarkable addition to psychophysical theory came too late for Fechner to appreciate the insight that exposed the inner psychological events that remained so elusive for so long. The insight was, of course, that no physical values of the stimuli were needed to measure the difference between the mean psychological values. The stimuli may be weights or line lengths, but the great advantage of Thurstonian theory is that the stimuli may just as easily be flavors, songs, or political candidates. Thus, psychophysical measurement was no longer limited to stimuli whose physical magnitudes must be objectively measured.

Reflecting on the origin of the Law of Comparative Judgment Thurstone (1952) observed that:

> The determination of a limen was the basic problem in old-fashioned psychophysics. In order to be scholarly in this field, one was supposed to know about the old debates on how to compute the limen for lifted weights to two decimal places with a standard stimulus of one hundred grams. One could hardly worry about anything more trivial. Who cares for the exact determination of anybody's limen for lifted weights? In teaching this subject I felt that we must do something about this absurdity by introducing more interesting stimuli. Instead of lifted weights we used a list of offenses presented in pairs with the instruction that they should be judged as to their relative seriousness. The subject checked one of each pair of offenses to indicate which he thought was more serious. Instead of selecting one of the offenses as a standard, we asked the subjects to compare every stimulus with every other stimulus. It was now apparent that the classical method of constant stimuli is a special case of the more complete psychophysical method of paired comparison.[2] We followed this same procedure with a list of nationalities that were also presented in pairs. For each pair of nationalities the subject was asked to check the one which he would prefer to associate with. I did not realize at the time that it

[2]Titchener (1901b) and Bock and Jones (1968) credit J. Cohn (1894) with the first published study.

> was going to be necessary to rewrite the fundamental theory of psychophysics in order to cope with data of this type. (pp. 306–307)

In an earlier discourse on Psychology as a quantitative rational science Thurstone (1937) commented:

> In building a unifying rational structure for psychophysics, I began by considering a simple table showing the frequency with which each of several stimuli is judged stronger than each other stimulus. The first question was, then, to ascertain whether such a table reveals some underlying order or law. I found a few simple postulates that would lock the whole table of experimentally determined proportions. These included a new concept in psychophysics, namely the discriminal error. The result was the law of comparative judgment. Armed with this law, we can investigate the relations of the several psychophysical methods. (p. 228)

In Thurstone's view, the variance found in the judgment process depends upon the stimulus and its magnitude, not on the physiological measuring device. Thus, error does not simply add to an unknown *true* psychological stimulus value. Rather each stimulus becomes a source of variability. Further, rather than the variance remaining constant across stimulus magnitude, as was commonly assumed during the 19th century, increases in stimulus magnitude should produce increases in variability. When using the Method of Constant Stimuli, for example, the greater the magnitude of the comparison stimulus, the greater the variance. Thurstone (1937) remarked:

> An interesting result is that we can ask new questions which had not been thought of before. If the discriminal error is an attribute of each stimulus, what should be the consequences when this value is larger than in the conventional limen determinations? If the analysis is sound, then the so-called psychometric curve should be positively skewed for coarse discrimination. Experimental results on brightness discrimination of rats verified the theory. The new psychophysical concepts are applicable to the measurement of affective values, such as in aesthetics, and to many other forms of cognitive and affective discrimination. These concepts and laws are quantitatively formulated, and they can be tested experimentally. Psychophysical experimentation is no longer limited to those stimuli whose physical magnitudes can be objectively measured. (p. 228)

Although this resolution of a theoretical enigma signalled a major victory for Thurstone's attribution of variance to the physical stimulus, a more difficult challenge lay ahead. The major difficulty lurked in judgments about the seriousness of crime. The crimes have no physical measure. Thus, Fechner's method of comparing a difference between the means of two Gaussian distributions to a physical difference in order to scale the difference in sensation simply fails for lack of a physical measure. Thurstone solved this vexing problem and opened a window on those inner feelings that commanded the attention of psychophysicists and psychologists alike.

TABLE 4.1
Response Proportions and Mean Differences in Seriousness

	Abortion	Adultery	Assault	Bootlegging
Proportion of Judgments Where the Column Crime Is More Serious Than the Row Crime				
Abortion	.500	.323	.211	.128
Adultery	.677	.500	.242	.172
Assault	.789	.757	.500	.379
Bootlegging	.872	.828	.621	.500
Estimated Differences in Seriousness				
Abortion	0.00	−0.46	−0.80	−1.14
Adultery	0.46	0.00	−0.70	−0.95
Assault	0.80	0.70	0.00	−0.31
Bootlegging	1.14	0.95	0.31	0.00
Average	0.60	0.30	−0.30	−0.60

The paired-comparison data in Thurstone's (1927a) now famous article on the measurement of social value prompted the first major advance in psychological measurement since Fechner created the first model for measuring sensitivity. Data for four crimes appear in Table 4.1 where a cell entry estimates the probability that the crime heading a column is judged to be more serious than the crime identifying a row. For example, subjects considered Abortion a more serious crime than Adultery with probability 0.677.

Thurstone's (1927a) Law of Comparative Judgment created a method for determining the mean difference in seriousness between any two crimes. This difference equals the number of standard deviation units separating the mean seriousnesses of the crimes defined by a cell's row and column. In perhaps the most significant integration of theory and data since the experiments of Peirce and Jastrow (1884) , Thurstone computed the difference in seriousness within each cell of the paired comparison matrix. These differences, obtained by assuming equal standard deviations for the distributions of seriousness, also appear in Table 4.1.

In the simplest form of the Law of Comparative Judgment there is no covariation between stimuli and all stimulus differences are assumed to have equal variance. Further, the sum of mean seriousness values can be set to equal zero. Using this convenient localization, the average of difference measures within a column is easily shown to equal the average seriousness for the crime heading the column. For example, in Column 1 Abortion is compared against all crimes. By assumption the comparison of a stimulus with itself produces a mean difference equal to zero. Averaging together the differences in Column 1 gives, theoretically,

$$E\left[\frac{1}{4}\sum_{j=1}^{4} D_{1j}\right] = E\left[\frac{1}{4}\sum_{j=1}^{4}(S_1 - S_j)\right] \quad (4.3)$$

$$= \left[\frac{1}{4}\sum_{j=1}^{4}(\mu_1 - \mu_j)\right]$$

$$= \left[\frac{1}{4}4(\mu_1) - \frac{1}{4}\sum_{j=1}^{4}\mu_j\right]$$

Because the sum of all means can be set to zero the right hand term vanishes,

$$= \mu_1 - 0$$

$$= \mu_1.$$

The average values at the bottom of Table 4.1 are obtained by simply adding together the estimates within a single column, including 0, and dividing by the total number of estimates. These means define scale values of the four crimes with respect to a psychological dimension called the "seriousness of crime." The seriousness values are shown in Fig. 4.3. These values are plotted against the scale values for the standard value within the pair, namely: Bootlegging (−0.60), Assault (−0.30), Adultery (0.30), and for Abortion (0.60). Obtaining these values signals a major theoretical advance because by measuring the amount of an unobservable mental quantity such as seriousness, psychophysics lays claim to membership in the ranks of scientific disciplines.

Far more is offered by this advance in mental measurement than just measuring the relative amount of seriousness associated with various crimes. First, within each cell the mean difference in seriousness depends on the mean seriousness of the standard against which a comparison is judged. Thus for a particular row, when a variable comparison is judged against a fixed standard the scale values within the row should be a linear function of the mean seriousnesses for standards. For Row 1 :

$$E(z_{k1}) = (\mu_k - \mu_1)/\sigma \quad (4.4)$$

letting $\mu' = \mu/\sigma$ gives,

$$E(z_{k1}) = \mu_k' - \mu_1'$$

$$= \mu_k' - .60$$

Therefore, the within-cell measurements plotted against the marginal estimates of mean seriousness, μ_k, produce a function with Slope 1 and Intercept equal to minus the mean of the comparison stimulus.

Figure 4.3 illustrates this prediction for the four crimes found in Table 4.1.

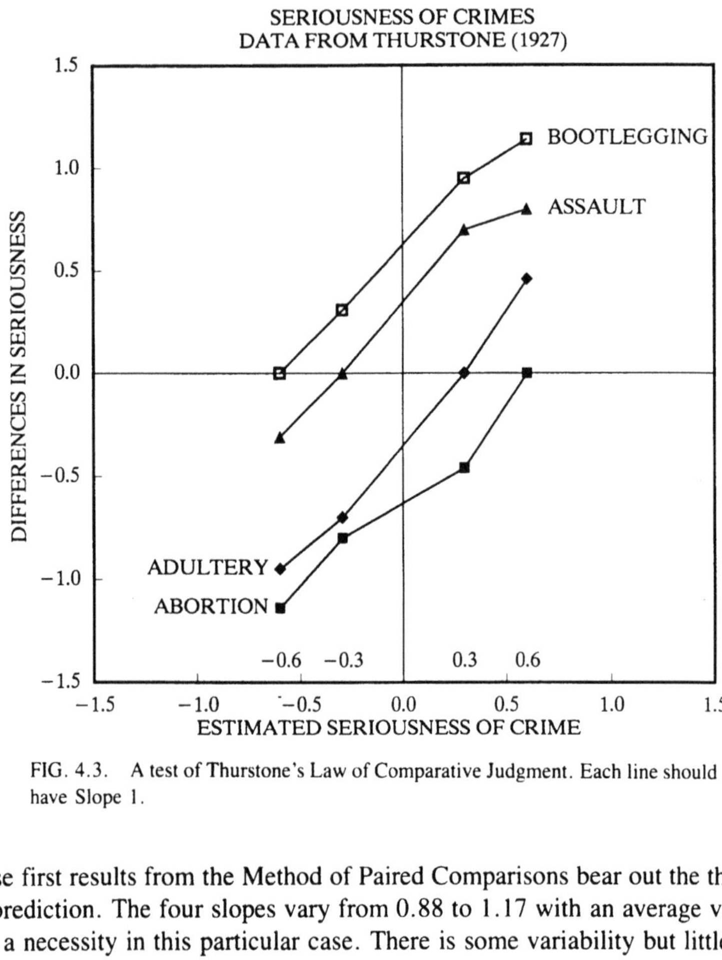

FIG. 4.3. A test of Thurstone's Law of Comparative Judgment. Each line should have Slope 1.

These first results from the Method of Paired Comparisons bear out the theoretical prediction. The four slopes vary from 0.88 to 1.17 with an average value of 1.0, a necessity in this particular case. There is some variability but little question that the theory and data are in good agreement.

In contrast to Fechner's conception of a stimulus, the stimuli in Thurstone's experiments do not generate errors that cause the subject to judge Assault to be more serious that Bootlegging on one trial but Bootlegging to be more serious than Assault on another trial. Indeed, subjects are quite consistent in their judgments. How, then, can the probability of judgment be a reflection of the individual's view of the seriousness of crime?

This theoretical enigma is a major impediment to applications of Thurstonian methods to the individual. Individual choices are often quite stable, either equal to zero or to one. With no intermediate proportions, the techniques for estimating differences between stimuli are useless. If the individual choice proportions are either zero or one, then the proportions obtained from experiments must arise from mixing together a proportion of p subjects who always choose Assault to be more serious than Bootlegging, and a proportion $1 - p$ who always view Bootlegging to be more serious than assault. How are the paired-comparison proportions, obtained by mixing performances across subjects, to be interpreted?

An answer to this fundamental question resides in the theoretical view of the individual as reflecting an opinion about the seriousness of a crime. The strength of the opinion is spread along a continuum of seriousness because of the individual's personal experiences, including religious belief, social training, and whether the individual has been a perpetrator or victim of crime. All these factors affect a person's evaluation of seriousness. The many random events that define personal experience insure that there will be societal variability in the seriousness of a crime. Each individual expresses the effects, or in Murray's (1938) terms, the "*press*," of society. For Thurstone the individual reports a societal, or external, event. For Fechner the individual reports the status of an internal event. In each case, however, the subject provides a self-report concerning a mental state.

Derived Psychometric Functions

Imagine that the Method of Constant Stimuli is used and that each crime is selected at random for comparison against an unknown median crime. The seriousness value of the median crime must be zero. The probability that a crime is judged to be more serious than the median crime determines its distance from the median crime. Now let the seriousness values obtained from the Method of Paired Comparison act in the same capacity as physical stimulus measures do for establishing a Psychometric Function. The seriousness value for the j-th crime is then a measure of the standardized Gaussian distance of the crime from the society's median crime. For example, the crime Perjury has a seriousness value of −0.06. The corresponding probability that Perjury is more serious than the median crime equals 0.48. Plotting the probabilities for each crime as a function of the crime's seriousness value defines the *Derived Psychometric Function*. This function yields the probability that a crime is judged more serious than a hypothetical median crime that splits the seriousness dimension into those crimes of lesser or greater seriousness.

Figure 4.4 shows the Derived Psychometric Function for the crimes investigated by Thurstone. By assumption of equal variance for the seriousnesses of each crime the various response probabilities are determined by reference to a single Gaussian distribution. Thus the form of the Derived Psychometric Function must appear as a cumulative Gaussian distribution. A measure of the variability of crimes within the society comes from the number of crimes outside the interquartile range, that is, outside the limits defined by the probable error. In this regard, only four crimes Vagrancy, Receiving Stolen Goods, Homicide, and Rape exceed these limits.

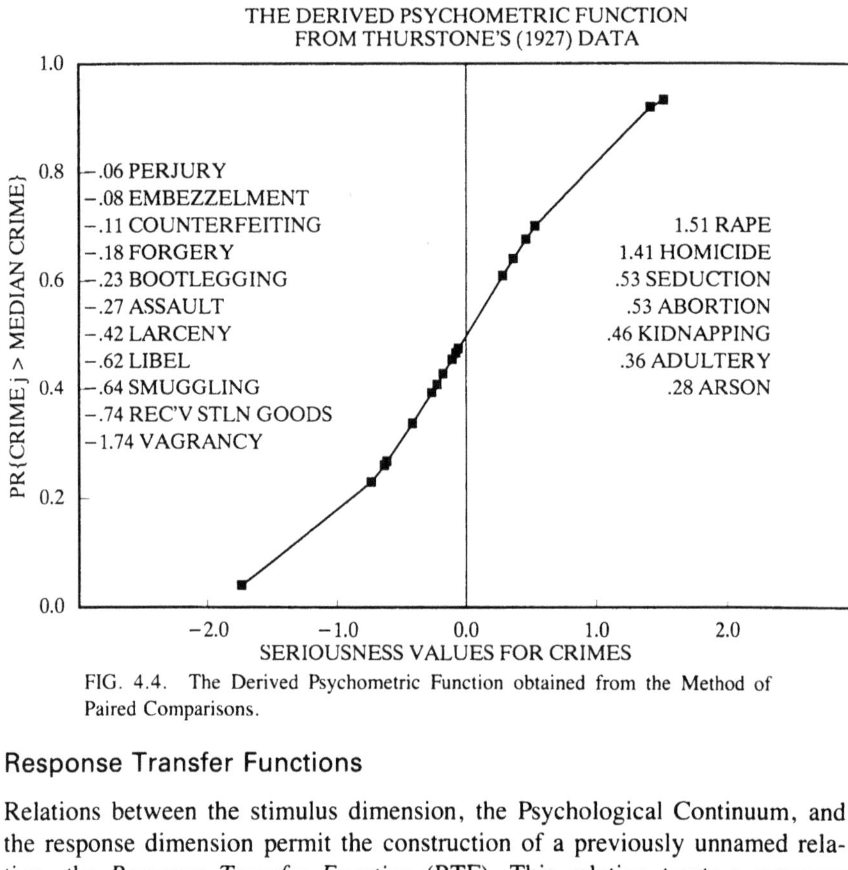

FIG. 4.4. The Derived Psychometric Function obtained from the Method of Paired Comparisons.

Response Transfer Functions

Relations between the stimulus dimension, the Psychological Continuum, and the response dimension permit the construction of a previously unnamed relation, the *Response Transfer Function* (RTF). This relation treats a response variable such as response frequency or response time as a function of the psychological values that characterize the stimuli. The Derived Psychometric Function is one example of a Response Transfer Function. The graph of response probability as a function of mean seriousness shown in Fig. 4.4 illustrates this idea. Each choice probability is plotted as a function of the seriousness of each crime. Each function is similar to a Psychometric Function. Each shows how the response proportions change across the increasing magnitude of seriousness.

This important function provides a transition from the scaled values of stimuli, which were thought to depend upon some form of *Psychophysical Function* such as the logarithmic relation derived by Fechner from Weber's Constant, to the actual measurement of performance. These various relations are illustrated in Fig. 4.5 where the ordinate represents a performance variable such as response proportion. One axis of the base measures the physical magnitude of the stimulus, if any, whereas the other measures the psychological scale values of the physical stimuli. The function relating the scale values of the stimuli to the

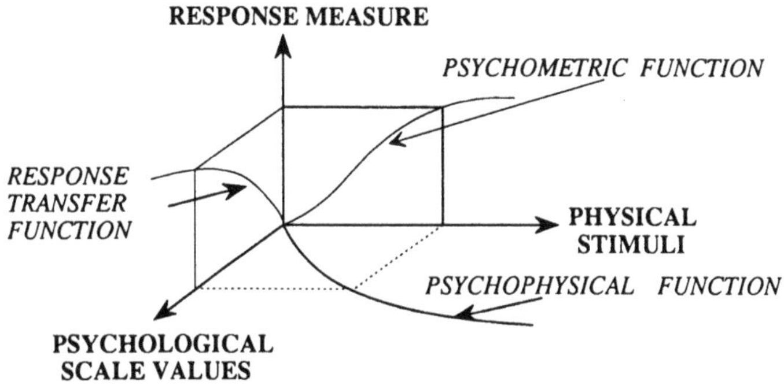

FIG. 4.5. The Response Transfer Function produces response probability as a function of psychological values.

physical values, if any, is the Psychophysical Function. The function relating the response measure to psychological scale values is the Response Transfer Function.

The impressive advances achieved by Thurstone's theory of judgment motivated further insights into psychophysical processes. Yet, in the arena of sensory judgments, where Fechnerian theory remained a cornerstone of psychophysical judgment, what did Thurstone's theory contribute that is different from Fechner's? Fechner argued that when two stimuli are compared, one sensory value is judged against the average of the two. A decision is based on whether the result is greater or less than the average. Although this process is different from that proposed by Thurstone, is there any substantial difference between the two theories?

To investigate this question let X_A and X_B be the Gaussian distributed random variables generated by presentations of stimuli S_A and S_B, respectively. Let x_A and x_B be the particular values of X_A and X_B on a single trial. For Fechner the average, on a particular trial, $(x_A + x_B)/2$ is the decision criterion. Any value of X_A greater than this leads to the judgment that S_A is greater than S_B. For Thurstone if $x_A > x_B$ then, again, S_A is judged greater than S_B. Notice that Fechner's probability of judging S_A greater than S_B is

$$\begin{aligned} \Pr\{X_A > (X_A + X_B)/2\} &= Pr\{2X_A > X_A + X_B\} \\ &= Pr\{X_A > X_B\} \end{aligned}$$

which is the same choice probability as for Thurstone's theory. Thus, in this simplest of cases, when the criterion for judgment is the average of the sensory values sampled on that trial, the two theories are formally equivalent.

If instead, as supposed in an earlier presentation of Fechner's theory, the average of the two random variables X_A and X_B is the criterion then the two theories differ because generally,

$$Pr\{X_A > (X_A + X_B)/2\} \neq Pr\{X_A > [E\{X_A\} + E\{X_B\}]/2\}.$$

Seemingly trivial differences sometimes generate surprising consequences.

If the simplest form of Thurstone's Law of Comparative Judgment and Fechner's theory are equivalent does the Law of Comparative Judgment somehow relate to Weber's Law? Because Fechner's application of Weber's Law to the Method of Right and Wrong Cases provided the first measurements of mental events the Law of Comparative Judgment should be compatible with these earlier results. A simple computation proves that if the probability of choosing $\mathbf{S}_A$ over $\mathbf{S}_B$ equals p, then the probability that $k\mathbf{S}_A$ exceeds $k\mathbf{S}_B$ also equals p. In this regard the Law of Comparative Judgment agrees precisely with Fechner's predictions.

Thurstone also showed that if Weber's Law is true then the Psychometric Function for *greater* responses should not form a cumulative Gaussian distribution. The reason is simple and probably disturbed Fechner, as it did later theorists. As the magnitude of a comparison stimulus increases the psychological quantity, say heaviness, also increases. This implies that the mean heaviness increases. If the size of the sensory error depends on the magnitude of the stimuli, as assumed in the application of Weber's Law, then the variance must also increase. Thus for a fixed standard the variance of the comparison stimuli will increase as comparisons become larger than the standard, and decrease for values smaller than the standard. As a result the Psychometric Function should appear skewed and not match a cumulative Gaussian distribution. Thus the Psychometric Function could only approximate a cumulative Gaussian.

The very success of Thurstone's Law of Comparative Judgment and the concurrent application of the Method of Paired Comparisons drove further analysis of the Psychometric Function from the theoretical arena. Except for an occasional study by students of theoretical psychophysics, the Psychometric Functions simply disappeared. Disguised as a *Generalization Gradient*, it came to be a hotly contended item in learning theory when studies of discrimination yielded generalization functions for animals and humans similar to those obtained in previous psychophysical experiments. But, by and large, the Psychometric Function held the attention of the psychological community only until a new phenomenon emerged to take its place. And, except for the remarkable article by Thomson (1920), no satisfactory explanation for its form ever emerged.

Coincident with these developments in psychology, statisticians began to determine the "best" form for the biometric function. Berkson (1944) argued that the Logistic distribution, similar in its form to the Gaussian distribution, was easier to apply to data and gave consistently better fits to *Biometric Functions* than did the Gaussian or several other distributions. Later Bock and Jones (1968) argued that the Gaussian, Logistic, or Arc Sine distributions all gave comparable fits to Psychometric Functions when deviation statistics defined the criterion for goodness of fit. With no theoretical reason to distinguish between them the choice of which function to use remained arbitrary but dominated by the Gaussian distribution's historical precedent.

JOHN P. GUILFORD

Psychophysical Functions

Theoretical results were poor substitutes for experimental proof of the validity of the Law of Comparative Judgment. As a test of Thurstone's theory in the context of those boring, but classic, lifted weights, Guilford (1928, 1931, 1954)[3], applied the Law of Comparative Judgment to these traditional psychophysical stimuli. Using the Method of Paired Comparisons Guilford presented to a single subject weights varying from 185 g to 215 g. Subjects compared each weight against every other weight 100 times. The data are reported in Table 4.2 where each entry is the proportion of cases where the column weight was judged heavier that the row weight. The symmetry of the table is apparent. The proportion for judging a comparison weight of 190 g to be heavier than a standard of 185 g equals 0.65. The corresponding judgment when the 190 g weight is the standard and the 185 g weight is the comparison yields a proportion of 0.35. The fact that these two proportions sum to 1.0 indicates perfect symmetry. Close inspection will reveal no deviations from this symmetry.

The symmetry is also evident in the graphic display of Guilford's data. Figure 4.6 shows the response proportion surface based on Table 4.2. A fixed-standard weight, such as 185 g yields the response proportions shown at the back of the image, with values beginning at 0.48 and rising to 0.99. At the front, the standard weighs 215 g and the comparisons of 185 g to 215 g are judged to be heavier with proportions ranging from 0.01 to 0.56.

To simplify the application of the theory, the response proportions for each cell are converted to Thurstonian standardized difference values, based on the Case V (equal variance) assumptions. Then averaging the results for each column yields the psychophysical values of heaviness shown in Fig. 4.7. These

---

[3]Guilford (1928) generated many thought provoking ideas in this short article; however, the article is marred by a weak mathematical and statistical development. For example, the article proposes that it is best to determine first the average proportion for each column of the paired-comparison matrix and then compute scale values. The article attempts to determine whether Guilford's or Thurstone's method yields results compatible with a logarithmic Psychophysical Function, but relies for this analysis on a misunderstanding of the concept of correlation. Guilford claims to compute a correlation between the physical weights of the stimuli, which have no variability whatever, and the variable heaviness values derived from two different procedures. There is no covariation at all between a random variable and a "fixed effects" variable, hence, no correlation. Regrettably this misapplication of correlation still occurs.

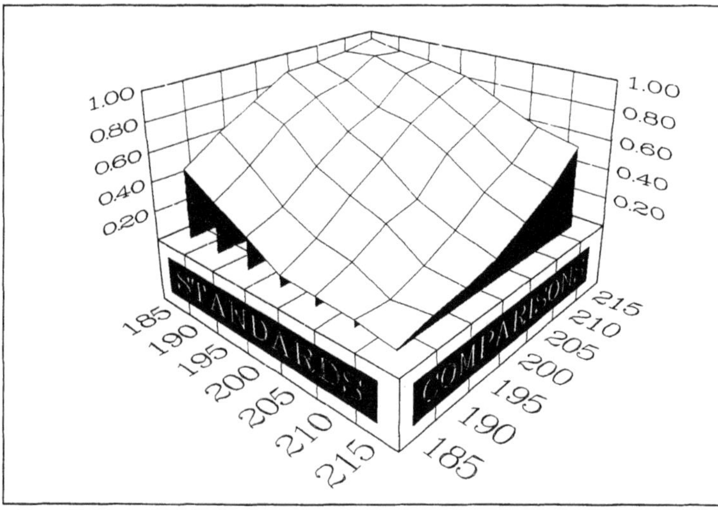

FIG. 4.6. The response surface for responding *heavier* illustrates the proportion of *heavier* responses for each combination of a standard and comparison.

values form a nearly perfect linear relation with the physical weights, showing that the sensory system reproduces an accurate representation of the relative differences between the stimuli. Plotting these calculated heaviness values against the physical magnitude of the stimuli yielded the first view of the Psychophysical Function (Guilford, 1931). Later, in a moment of inspiration, Guilford (1954) plotted the proportion of *heavier* responses as a function of these heaviness values, thereby generating a Response Transfer Function.

The ordinate of Fig. 4.8 is the derived probability that a weight will be judged heavier when compared to the hypothetical median weight. Shown along the abscissa are both the Thurstone heaviness values for the various weights and their corresponding physical values. The derived response probabilities increase

TABLE 4.2
Response Proportions from Guilford's Paired Comparisons of Lifted Weights

	Comparisons(g)						
Standards(g)	*185*	*190*	*195*	*200*	*205*	*210*	*215*
185	.48	.65	.78	.92	.93	.95	.99
190	.35	.50	.69	.80	.85	.97	.94
195	.22	.31	.48	.63	.72	.89	.91
200	.08	.20	.37	.52	.67	.78	.86
205	.07	.15	.28	.33	.54	.64	.74
210	.05	.03	.11	.22	.36	.46	.62
215	.01	.06	.09	.14	.26	.38	.56

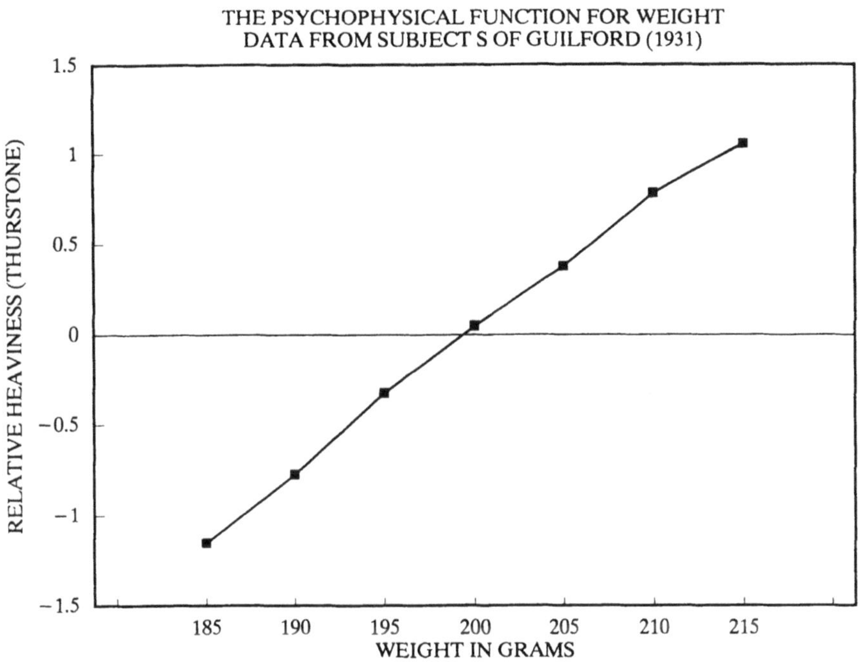

FIG. 4.7. Thurstone's Law of Comparative Judgment provides a method for determining the Psychophysical Function relating heaviness, a psychological variable, to weight, a physical variable.

steadily with increases in the magnitude of the comparison stimulus, hinting at a smoothly increasing cumulative Gaussian distribution. As an estimate of the probable error, note that the interquartile range extends from about 192 g to 208 g, suggesting that for these stimuli a comparison 8 g greater than a 200 g standard will be correctly discriminated 75% of the time.

The Psychophysical Function, the Psychometric Function, and the Response Transfer Function for Guilford's data are all shown in Fig. 4.9. This figure is a concrete rendition of the theoretical idea expressed in Figs. 4.3, 4.4, and 4.5. In this instance, the physical values of the stimuli permit the construction of the Psychophysical Function. Were the stimuli not represented on a physical scale, only the Psychophysical Continuum would provide the ordering of the stimuli.

A goal of Thurstonian theory is to establish measurements of psychological values for stimuli that have no physical measurements. The beauty of flowers, appeal of paintings, aroma of culinary delights, or bouquet of vintage wines are each measurable by the Thurstonian method. The psychological values order the otherwise unmeasureable stimuli. According to the theory, the differences between these psychological values are just as meaningful as the psychological values measured for stimuli with known physical characteristics. If repeated comparisons of physical stimuli provide the same psychophysical order as their

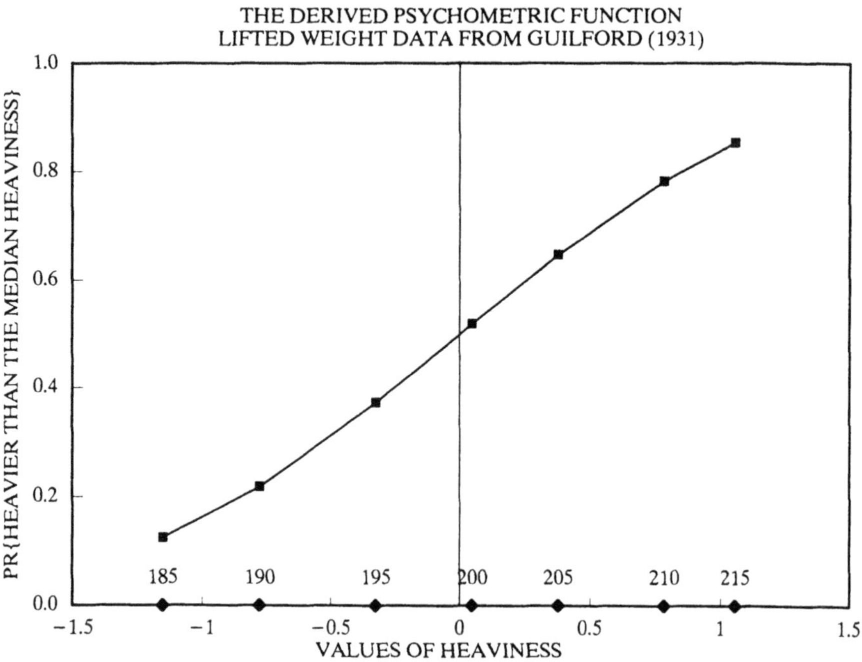

FIG. 4.8. The Derived Psychometric Function for lifted weights. Data from Subject S reported by Guilford (1931).

physically measured order, then there is good reason to believe that the measures of the physical world reflect accurately the sensation that was the impetus to measure the physical stimuli in the first place. If repeated measurements of more "psychological stimuli" produce similar results then there is reason to believe that the psychological values accurately reflect the internal state of the subject.

Figure 4.9 also serves to show that there are many underlying processes that influence the subject's eventual judgment. Certain physiological processes are necessary to transfer the stimulus onto the Psychological Continuum. These processes determine the Psychophysical Function. Thereafter, the decision process and the subject's control of response criteria, as in Thomson's theory, determine the response. The relation between the psychological values and the dependent response measure is the Response Transfer Function. Thus the Psychophysical Function, $\boldsymbol{P}$, defines the psychological value, $\mathbf{S}$, of a physical stimulus, $\mathbf{S}$. The Response Transfer Function, $\boldsymbol{R}$, defines values of the dependent response measure, r. Symbolically,

$$r = \boldsymbol{R}[\boldsymbol{P}(\mathbf{S})]. \qquad (4.5)$$

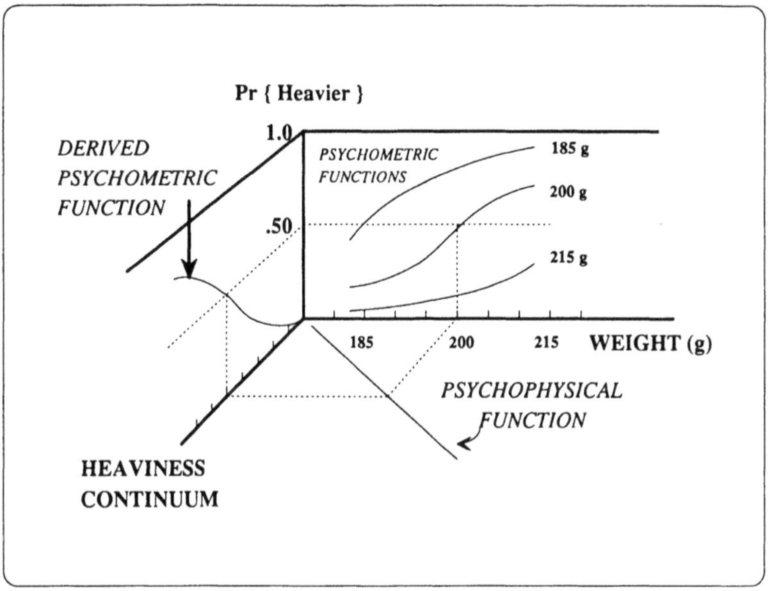

FIG. 4.9. The three-dimensional space defined by the dimension of the physical stimuli, the Psychological Continuum and the response measure illustrate the relations among the Psychometric Function, Psychophysical Function, and the Response Transfer Function.

Urban described the relation between stimulus magnitude and a dependent response measure as the Psychometric Function. By ignoring any theoretical mechanisms that gave rise to the observed response measures, Urban hoped to produce an atheoretical account of psychophysical performance. Compared against the remarkable achievements of Thurstonian theory, Urban's approach offers little insight into the forces that form judgments. Instead the evidence clearly favors the theoretical approach of Fechner, Thomson, and Thurstone that launched psychology into scientific investigations of mental phenomena. These are the theories that created new tools for the measurement of psychological events.

5 Equality and Affective Value

SAMUEL W. FERNBERGER

The many successes of Thurstone's two-choice theory encouraged a flurry of new experiments at the expense of investigations of other psychological judgments. Yet other forms of judgment remained an appealing arena for scientific investigation. And, in spite of the successes of two-choice theory, a few determined scientists persisted in the belief that equality judgments provide valuable insights into psychological processes. Among these was Samuel Fernberger whose early interest in the process that evoked a subject's responses prompted outstanding studies of the role of attitude in the formation of a subject's response (Fernberger, 1914). Later, in his historical and introspective account of the process of comparing, Fernberger (1918) pointed out that

> very few investigators have raised the questions, as to what it is that we are really measuring with the psychophysical methods and what are the processes involved in such measurements. . . . Our present investigation is an attempt to answer the question regarding conscious processes which are present in determinations by the method of constant stimuli. (p. 1)

A decade later, insisting that equality judgments require a program of further psychophysical research and that the use of equal judgments is influenced by the subject's attitude, Fernberger (1930) concluded:

> Difficult as they are to handle experimentally, can we afford to eliminate the equality judgments? As early as 1906 Titchener (1905) said: "If we are dealing with mind as mind presents itself to us for examination, we cannot ignore these judg-

ments." George's[1] and my own introspective studies show that judgments of true subjective equality occur frequently for all observers. (p. 111)

What do we wish to obtain by the psychophysical methods after all? The old idea that we are measuring the sensitivity of a particular sense organ has been abandoned. We now recognize that we are measuring the sensitivity of the entire psychological organism—his sense organs, his concentration, his attitude, his acceptance and understanding of instructions, his degree of practice, and what not besides—to a given stimulus situation. It is recognized that all of these things may change and it is for this reason that we obtain a large number of observations which are then treated statistically, rather than a single observation. (p. 111)

GEORG VON BÉKÉSY

A Quantum Theory of Sensation

Although some psychophysicists shied from the use of equality judgments because of the many questions surrounding their use, judgments of equal loudness became the *modus operandi* in hearing. In fact, at just the time Fernberger issued the call for more extensive studies of equality judgments, von Békésy published two major psychophysical contributions to the study of hearing, each based on judgments of equal loudness.

In the first of these von Békésy (1929) sought quantitative information on the nature of nerve excitation and used judgments of equal loudness to determine just noticeable differences of amplitude. In the second, and more significant theoretical article, von Békésy (1930) introduced the Quantum Theory of Sensation, a theoretical basis for the judgment of equal loudness. The theory descends directly from Leibnitz's Monadology, benefits from Herbart's analysis of mental elements, and owes much to Weber's concept of the just noticeable difference.

The acoustic stimuli that von Békésy heard differ somewhat from those presented to other senses in that the "threshold" for the discrimination of loudness is on the order of 5% of the stimulus magnitude, far larger, for example, than the 1% reported by Weber for discriminating line lengths. Von Békésy (1930/1960) argued that,

the error function for loudness judgments can be determined from the quantum-like variations of sensation. Consider in this relation a comparison of loudness of two successive tone pulses. If the excitation produced by these two stimuli is the same, there will be the same number of stimulated cells. If the excitation changes in magnitude, as indicated in Figure 7-34 [see Fig. 5.1], the sensory magnitude will be altered, provided that the change is sufficient to involve one cell more or one cell less. If a given stimulus always produces the same excitation, then, as the figure

[1]Presumably this refers to S. S. George (1917).

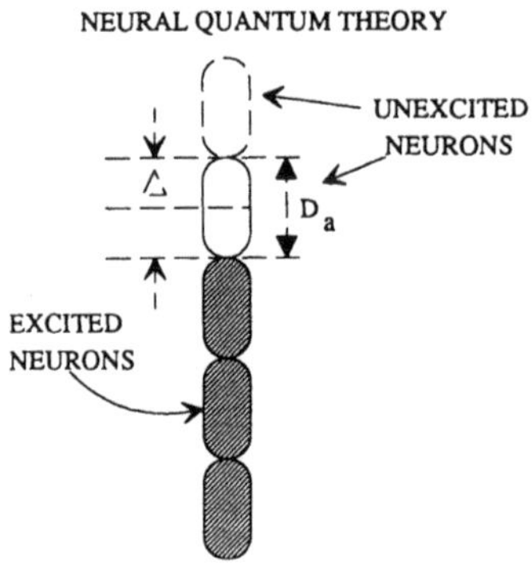

FIG. 5.1. The quantum character of excitation. A change of Δ or $D_a - \Delta$ does not affect loudness, because the number of excited cells is unaltered. [Adapted from von Békésy 1960, Fig. 7-34].

> shows, the excitation can be increased by Δ or decreased by $D_a - \Delta$ without any alteration in the number of stimulated cells. Here are two values of the stimulus for which the loudness remains the same. (p. 240)
>
> If we assume that the excitation level $D_a - \Delta$ can take all possible values with equal probability, then an increase of excitation Δ will give a number of judgments of no change that is proportional to $D_a - \Delta$, and a number of possible[2] judgments proportional to D_a. Hence the probability that an excitation change of Δ will not alter the apparent equality of loudness is $w = 1 - \Delta/D_a$. Here the absolute value of Δ lies between the limits $0<|\Delta|<D_a$. (pp. 242–243)
>
> If $\Delta>D_a$, there will always be an increase in loudness, so that for this condition the probability will always be zero. The same considerations hold for negative values of excitation, for then the sign of D_a is simply reversed. Thus instead of the rectangle of Fig. 7-35 [see Fig. 5.2] we obtain for the error curve a triangle whose apex is at 0 and that attains a value of 100 per cent at $w = 1$, while the base of the triangle has a length on the abscissa of two difference limens. (p. 243)
>
> The experiments showed that with unpracticed subjects the error curve at the beginning included large variations from time to time. After a fair amount of practice, however, the results became fully reproducible and adhered closely to the form shown in Fig. 7-37 [see Fig. 5.3]. (p. 243)
>
> After further practice of about a month some of the subjects gave a curve like that of Fig. 7-38 [see Fig. 5.4], which agrees with the calculations. (p. 243)

Here, then, is a new proposal for the kind of Psychometric Equal Function first produced by Brown (1910). Certainly this function deviates markedly from the empirical functions found in visual, length, or weight discrimination experiments. For the data in Fig. 5.3 the interval of uncertainty must be exceptionally wide, certainly no smaller than from −12% to +10% of the standard, although the interval narrows considerably with practice. For certain exceptional subjects

[2]Perhaps positive is meant here.

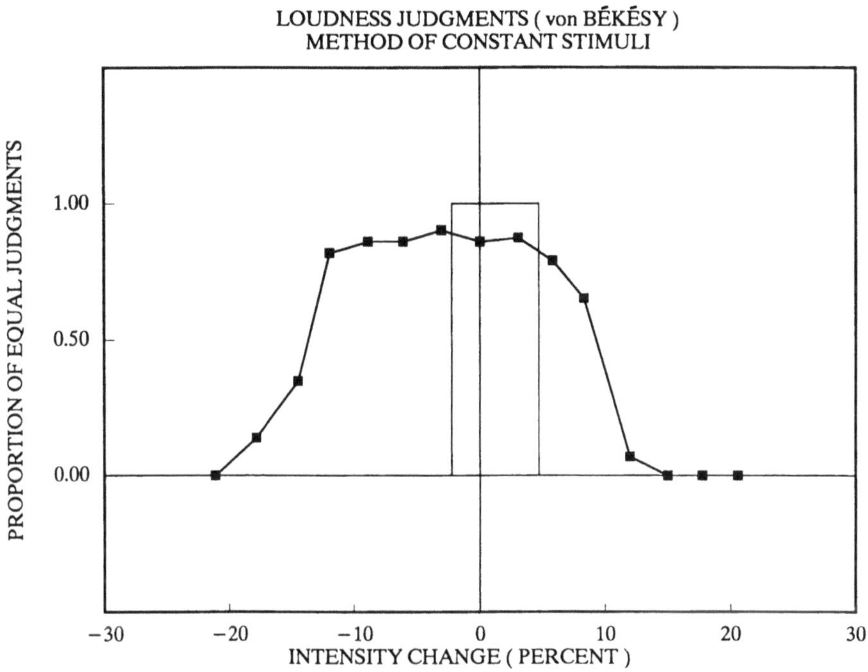

FIG. 5.2. Theoretical (rectangle) and actual distribution of equal judgments as a function of the difference of intensity between standard and comparison stimuli. The standard was 300 cps at 40 db above threshold. [Adapted from von Békésy, 1960, Fig. 7–35].

selected by von Békésy, practice reduces the smallest estimate of the interval to −4% to 2.5% of the value of the standard.

The striking linear increases in response proportion appear at odds with previous studies. Yet the greatest puzzle is the omission of the response probabilities for *louder* and *softer* judgments. If response probabilities are tied directly to the number of elements stimulated, then how can the subject ever produce judgments of *softer* when the comparison stimulus is larger than the standard, or "Louder" when the comparison is smaller than the standard? Yet in previous studies using the Method of Constant Stimuli these errors occur with high probability.

In spite of these reservations, the Quantum Theory of Sensation and the discrimination process proposed by von Békésy filled a theoretical vacuum created by the absence of any other theory capable of predicting how the proportion of equal judgments increased nearly linearly as a function of differences between the standard and comparison stimuli.

On the basis of Cattell's Principle, all equal judgments derived from a common cause should produce equal response times. For some reason, perhaps the belief that the temporal nature of auditory signals might distort time measurements, von Békésy did not collect response time data. Another puzzling feature

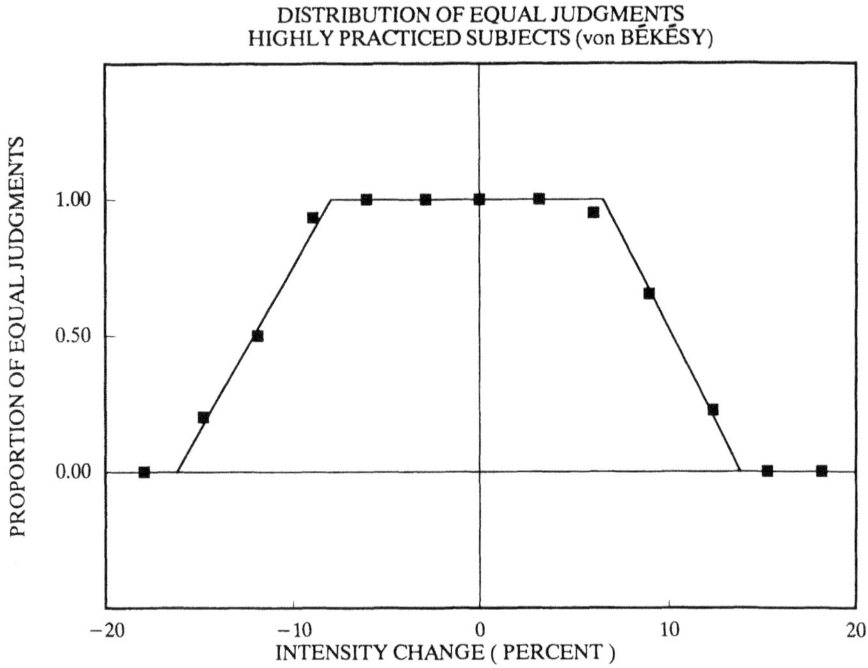

FIG. 5.3. Distribution of judgments for practiced subjects [Adapted from von Békésy, 1960, Fig. 7–37].

of von Békésy's results is the consistency of subjects' performances, whereas decades of previous studies show *equal* judgments to depend on the attitude of the subject. Perhaps von Békésy's highly practiced subjects (see Fig. 5.4) simply produced the required results through a change in attitude? Did practice actually improve sensitivity or simply allow for a change in attitude that somehow resulted in a smaller value of the just noticeable difference.

WILLIAM N. KELLOGG

The apparent malleability of *equal* judgments demanded an analysis of these responses that looked beyond the results of Brown, Fernberger, George, and von Békésy. As if on cue, W. N. Kellogg (1931) trotted out a vigorous experimental analysis of the role of three, versus two, responses by extending the work of Cattell, Urban, Henmon, and Brown.

Kellogg measured the choices, response times, and confidences of five graduate students who judged differences between luminances displayed on opposite halves of a split disk. Four visual stimuli, $\mathbf{S}_1$, $\mathbf{S}_2$, $\mathbf{S}_3$, and $\mathbf{S}_4$, increasing in luminance were adjusted for each subject to appear to increase linearly in bright-

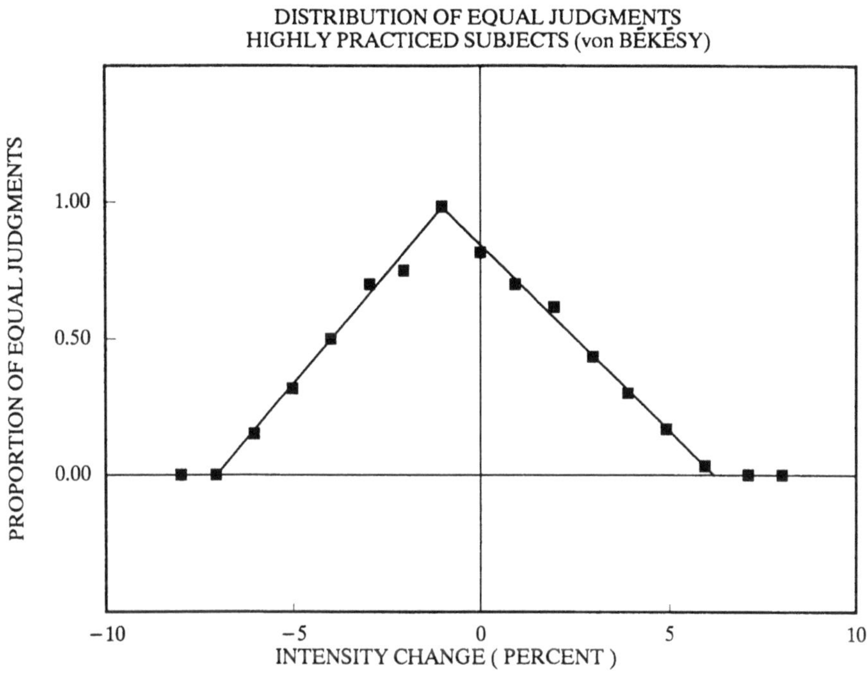

FIG. 5.4. Distribution of judgments for certain highly practiced subjects [Adapted from von Békésy, 1960, Fig. 7–38].

ness. Then they were paired to create seven stimuli: (S_1, S_4), (S_2, S_4), (S_3, S_4), (S_4, S_4), (S_4, S_3), (S_4, S_2), (S_4, S_1). The members of a pair indicate the left- and right-hand intensities, respectively. For the middle pair the two sides were equal at 21.682 metre-candles.

After some 4000 judgments of brightness differences, but without informing the subjects, Kellogg began to record for each trial the subject's response time together with previously recorded response choices. Figures 5.5 and 5.6 contain results for response proportions and mean response times. The labelling of the abscissa, consistent with Kellogg's report, identifies the less intense side and the number of luminance-intervals between the left- and right-hand sides of the disk. For example, L3 identifies the presentation of (S_1, S_4) and R2 signifies the occurrence of (S_4, S_2). The presentation of the equally intense stimuli, (S_4, S_4), is labelled E.

The values defining the 2-left and 2-right curves must add to 1.0 for each luminance difference. For these two-choice responses where the subjects were to indicate which side was darker, the Psychometric Functions cross somewhat below the pair of equally intense stimuli. The corresponding *Chronometric Functions*, in Fig. 5.6, both increase to a maximum on either side of the equally

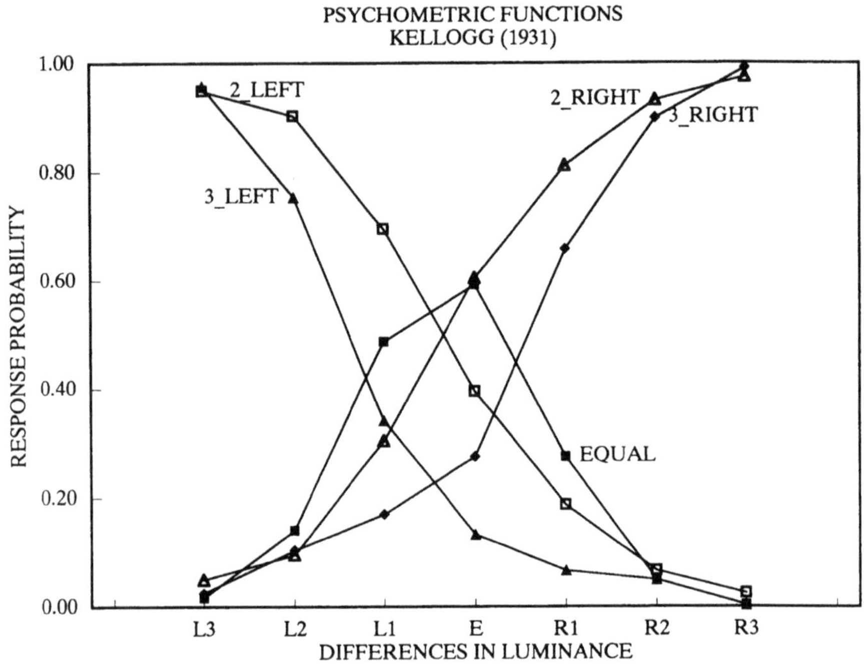

FIG. 5.5. Psychometric Functions for two- and three-choice studies of brightness discrimination.

intense stimulus pair, then a reduction in response time occurs. This confirms and extends Henmon's and Cattell and Berger's earlier result.[3]

Kellogg also extended Urban's results by measuring response times when the subjects responded either *smaller*, *equal*, or *larger*. The Psychometric and Chronometric Functions for this three-choice experiment appear in Figs. 5.5 and 5.6, where the results are labelled 3-left and 3-right and *equal*. In Fig. 5.6, the Chronometric Function for *equal* responses very clearly decreases dramatically as the differences in luminance decline. On the other hand, the Psychometric Equal Function shown in Fig. 5.5 increases as the luminance differences decrease. When the two sides of the split disk are of equal luminance the proportion of correct *equal* responses reaches its peak at 0.592 and the mean *equal* response time is smallest at 1320 msec. In general, as Kellogg (1931) noted, "the psychometric time-curves for all categories tend to be inversions of the psychometric relative frequency curves" (p. 81).

Finally, because response confidences were collected only after every 42

[3]The Psychometric Functions are based on all observations contributed by the subjects for a stimulus-response combination. Chronometric Functions are based on only those averages contributed to by all subjects.

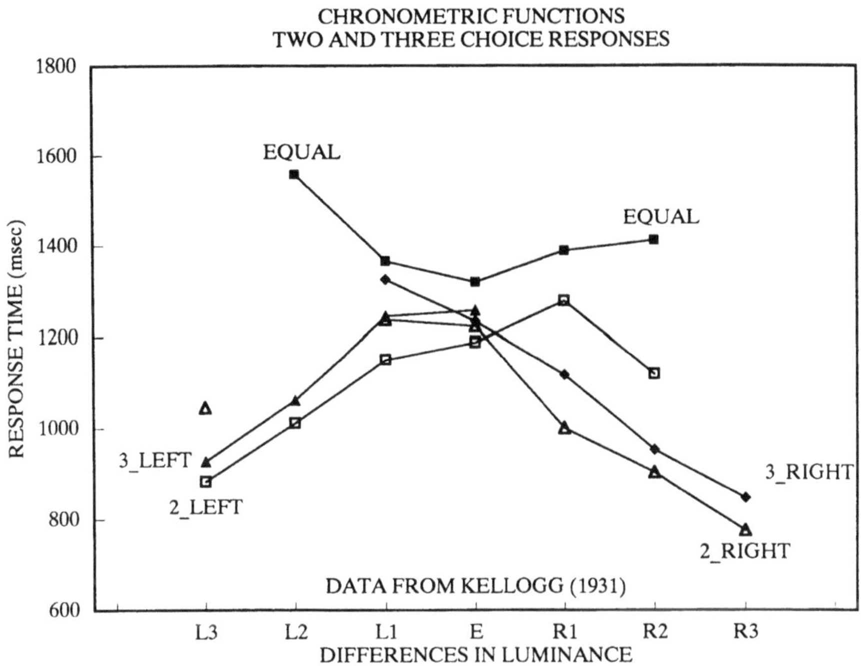

FIG. 5.6. Chronometric Functions for both two-choice and three-choice methods of investigating brightness discrimination.

trials, they are difficult to interpret. Nevertheless, the averages show a crude inverse relationship between response time and response confidence, with longer times coincident with less confidence.

Response times are generally faster, and correct responses more frequent in the two-choice experiment. Kellogg (1931) also noticed, however, "that when the difference between the stimuli reduces to zero and then assumes a negative (reversed) value, the time of judgment tends to continue its increase" (p. 81). The summary of these findings produced

Kellogg's Principle: Within any given category of judgment, the more the stimuli deviate from the ideally perfect condition typified by the category name, the longer the judgment-times tend to become. (p. 81)

Reflecting on the very orderly results for equal judgments Kellogg (1931) pointed to the contradiction between these results and Cattell's Principle I that as stimuli become increasingly similar response time must increase.

> The association between stimulus-differences and judgment times proves beyond a doubt that the greater average time consumed in making equality judgments . . .

> cannot be ascribed to the fact that these judgments are more likely to occur when the difference between stimuli is small. For regardless of the frequency of occurrence of the equal judgments their time is greatest when the difference between stimuli is large. *What then causes the equal judgments as a group to require a longer time than those of the other categories* [italics added] ? (pp. 81–82)

Kellogg's closing question about the increased time required for *equal* judgments became the focus of Fernberger's attention. Already convinced that the attitude of the subject determined the propensity to make *equal* judgments, Fernberger, collaborating with Glass, Hoffman, and Willig (1934) in "Judgment Times of Different Psychophysical Categories" returned to the problem of *equal* judgments and the role of attitude of the subject. Fernberger and Irwin (1932) and Carlson, Driver, and Preston (1934) found that the Method of Constant Stimuli produced equal mean response times for the three-response categories of *smaller*, *equal*, and *larger*: results quite obviously different from Kellogg's data and far removed from the expectations of Cattell's Principles.

Carlson, Driver, and Preston suggested that in those experiments where the *equal* judgments took longer, generally, than the other judgments there was an attitude set to discourage the use of the *equal* category. Fernberger et al. (1934) discovered that

> the judgment times for equal judgments become significantly longer, on the whole, than judgments of either less or greater if the attitude is created in the observers of not giving equal judgments unless absolutely necessary. This attitude can be created by the use of appropriate instructions. (p. 293)

With respect to von Békésy's equal loudness judgments, which are also a form of equality judgment, nothing in Neural Quantum Theory suggests how changes in response time occur as the comparison stimulus slides through the single quantum that defines the interval of equality. Yet Kellogg's results suggest that the response time for *equal* judgments should be fastest in the middle of an interval of equality that is defined by the size of the neural quanta, and Fernberger's experiments confirm that the attitude of the subject influences the results obtained from *equal* judgments. Therefore, some non-physiological aspect of the subject's performance enters into these judgments.

THE PSYCHOPHYSICS OF AFFECTIVE VALUE

In spite of the attitudes of subjects, the Psychophysicist's physical stimuli mapped nicely onto hypothetical psychological dimensions such as *heaviness*, *brightness*, and *extent*. The major determinant of performance remained the difference between stimuli. The larger the difference the fewer the number of errors, or the fewer the reversals in judgment. Even the vexing *equal* judgments showed orderly changes as the difference between physical stimuli increased.

Nearly 100 years after Weber's monumental studies, the orderly changes in responses to well-known psychophysical stimuli suggested that stimuli less well defined physically might also produce an orderly pattern of results. In this regard, the attractive but somewhat quiescent study of aesthetic judgments, creatively explored in *Zür Experimentalen Äesthetik* by Fechner (1871) provided qualitative stimulus differences that teased psychophysicists eager to apply new methods to old problems.

The renewed interest in the study of the sense of pleasantness, or unpleasantness, uncovered an old psychophysical problem that Fechner researched vigorously (see especially Fechner, 1876). In these studies stimuli varied in an unknown physical fashion but gave rise to a sense of pleasure, called by Titchener (1902) "affective tone," that varied from stimulus to stimulus and, being personal, from person to person. Beebe-Center's (1932) scholarly *Pleasantness and Unpleasantness* reviewed, formulated, experimented with, and advanced new ideas concerning the way in which judgments concerning pleasantness occurred.

Shortly after the publication of Beebe-Center's outstanding contribution, Hunt and Volkmann (1937) claimed that

> the similarity between certain phenomena in psychophysics and certain phenomena in affectivity is due to the operation of general principles of judgment in both fields. . . . many phenomena in the field of affectivity can be explained by invoking the principles of judgment rather than the principles of a unique affective process. (p. 92)

JOHN F. DASHIELL

Concurrently, and in response to the question of whether psychophysical judgments might be the same as Beebe-Center's judgments of "hedonic tone," John Dashiell (1937) posed an empirical question:

> Does the time taken to make a choice vary consistently with the amount of difference in the affective values of the alternative stimulus-objects? (p. 57)

The experimenter investigated this hypothesis by choosing classic stimuli to produce affective values; Milton-Bradley coated papers of seven saturated hues: red, green, blue, orange, yellow, purple, and orange-yellow (shade no. 2). The stimuli, pasted in pairs on gray cardboard backgrounds appeared in a tachistoscope. Nine subjects compared all different colors according to Kowalewski's (1904) presentation sequence for a full paired comparison matrix.[4] Each subject

[4]Also see Titchener (1901 Vol. I, Part 1, p. 92). Essentially, the smallest stimulus differences presented first (in order of increasing stimulus magnitudes), followed by the next largest stimulus difference in order of increasing stimulus magnitudes, etc.

chose the member of the pair that was most appealing. Dashiell replicated the design three times to generate a total of $(7 \times 6) \times 3 = 126$ response choices for each of nine subjects and, for each judgment measured both choice and response time.

Perhaps due to the small number of observations (3 per subject per cell) and perhaps because of inconsistency of the subjects' choices, the data were not analyzed using the usual Thurstonian methods for paired comparisons. In fact, the choice proportions are not reported at all. Instead, for each subject, the stimuli were ranked according to the number of times a particular stimulus was most appealing during the entire 126 trials. The rank, an ordinal value, became the measure of affective value.

Figure 5.7 summarizes the judgment time results. The abscissa refers either to the difference between ranks (DR) or to the rank, from 1 to 7, of the most appealing member of a stimulus pair. As the difference between ranks increases, judgment time declines steadily, in much the same way as increasing differences in line length or brightness produce decreases in response time. In this regard the data are similar to results from psychophysical studies of physically defined stimuli.

Also shown in Fig. 5.7 is response time treated as a function of the most to least appealing colors. Mean response time is fastest for those comparisons involving the most appealing color. The mean response time then increases to a maximum at the color ranked fifth, and then declines for the sixth- and seventh-ranked stimuli. This inverted U-shaped function may be due to the changes in average affective difference.

For example, the most appealing stimulus when compared against all other stimuli generates an average absolute rank difference of $(1 + 2 + 3 + 4 + 5 + 6)/6 = 3.5$. This is the same mean difference in rank as occurs for comparisons involving the least appealing stimulus. Such extreme stimuli produce the largest mean differences in rank. For the second or sixth most appealing stimuli the mean absolute rank is $(1 + 1 + 2 + 3 + 4 + 5)/6 = 2(2/3)$. The third- and fifth- ranked stimuli yield an average absolute difference of only $(2 + 1 + 1 + 2 + 3 + 4)/6 = 2(1/6)$. The average rank difference for the color ranked fourth equals $(3 + 2 + 1 + 1 + 2 + 3)/6 = 2$. If the RT function reflects only these considerations of average rank, the function should be perfectly symmetric. However, the mean response times are lower for the most appealing than for the least appealing stimulus. Perhaps affective difference is not the only factor affecting judgment times.

Mean judgment time also appears to increase as the rank of the most appealing member of a stimulus pair decreases. This increase may reflect the fact that comparisons between $\mathbf{S}_1$ and other stimuli produce the largest average difference in rank. Comparisons between $\mathbf{S}_2$ and colors ranked less than $\mathbf{S}_2$ produce the next smallest average difference in rank. Continuing this argument leads to the extreme where $\mathbf{S}_6$ is compared only against $\mathbf{S}_7$ which involves a single difference in rank equal to one. Again, changes in judgment time appear to depend upon the average difference in ranks.

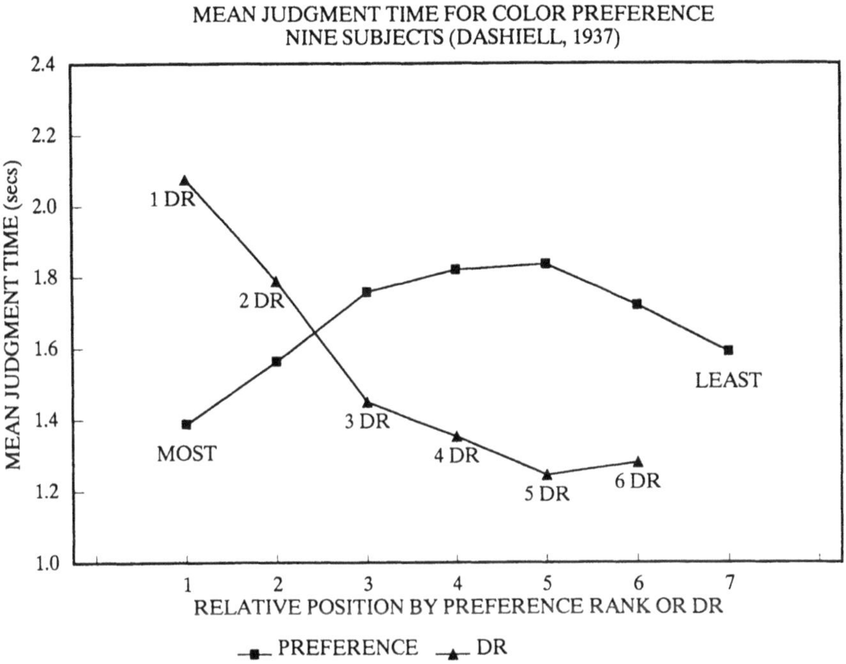

FIG. 5.7. Mean judgment times as a function of affective rank of color stimuli and as a function of the difference (DR) between affective ranks.

Figure 5.8 reveals an important additional fact. The judgment times are conditioned on a fixed less-appealing stimulus. For example, all comparisons against S_7 appear together. For this Chronometric Function the comparisons range from S_1 to S_6. The fastest judgment time occurs when S_1 is compared against S_7. As the comparison stimulus changes the judgment time increases until the comparison of S_6 against S_7 produces the largest judgment time within this Chronometric Function. When S_7, the least-appealing stimulus, is presented the Chronometric Function is generally faster than for any other comparison. Careful examination of the image in Fig. 5.8 will reveal the general finding that Chronometric Functions for the less-preferred stimuli produce the fastest judgment times.

Dashiell (1937), emphasizing these considerations, concluded:

> It would, therefore, seem reasonable to consider that esthetic judgment-time is a function of affectivity both directly and indirectly. It is a function of it indirectly, in that *the affective reaction determines the relative values of the stimuli, the distances between which determine the judgment-time*. It is a function of it directly, too, in that *the latent time of an affective reaction to a given stimulus taken alone varies with its place in the subjective scale of values*. For example, a color, tone or other stimulus that evokes a hedonic response that places it, let us say, extremely high in

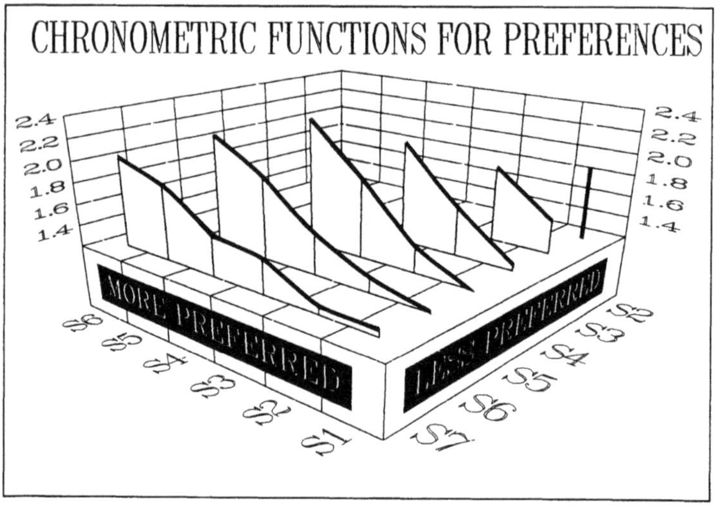

FIG. 5.8. Mean judgment times conditioned on the rank of the least appealing stimulus member. Data from Dashiell (1937).

an affective continuum, will tend to permit very rapid choosing responses between itself and other stimuli (which lie for the most part at a distance on the continuum); but it will also tend to arouse a direct affective response that is also rapid. (pp. 66–67)

VOLKMANN'S EQUATION

Response Time and Confidence

While psychophysical analysis wrestled with new issues raised by the analysis of affective values, questions about the relations between response time and response confidence were answered in more traditional two-choice experiments. Volkmann (1934), seizing the opportunity to elaborate the relations between judgment time and confidence hinted at by the results of Kellogg (1931) and Henmon (1911), presented at the 1934 meeting of the American Psychological Association, a paper entitled "The relation of the time of judgment to the certainty of judgment." To quote Volkmann, *in extenso*:

V. A. C. Henmon has reported a negative correlation between the time of judgment and the confidence with which a judgment is made. The present experiment carried this relation further. Each of three observers gave 100 series of comparative judgments upon the inclination of lines visually presented, using the categories *greater* and *less*. The experiment employed the method of constant stimulus differences. A chronoscope measured the time of judgment. The entire procedure was repeated,

> with eleven categories of certainty of difference replacing the categories *greater* and *less*. The times from the first half of the experiment and the certainties from the second display an interesting relation. The following equation presents one suggestive analysis:
>
> $$t = \frac{a}{2v - 1} + b$$
>
> In this equation, t is the time of judgment, a and b are constants, and v is the amount of certainty with which the judgment is given. The greater the difference in certainty between the two categories, the shorter the time of judgment. (pp. 672–673)
>
> Given certain postulates concerning the certainty of judgment, the relation of time to certainty illuminates the recent findings of Kellogg and Fernberger, and appears as a fundamental relation. (p. 673)

This function, Volkmann's Equation, appears in Fig. 5.9. The abscissa represents the amount of certainty and the ordinate the corresponding judgment time. The value b is both the minimum judgment time and the function's asymptotic value. The role of the parameter a is to stretch the ordinate—large values of a

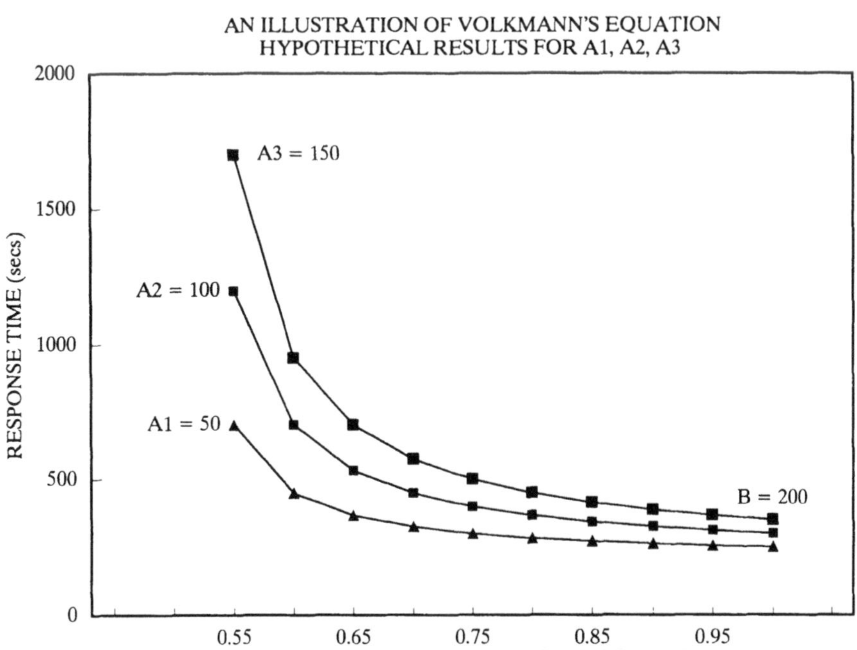

FIG. 5.9. Volkmann's Equation is an empirically derived relation between judgment time and response certainty.

produce larger judgment times, as can be seen from the three curves produced by values of a varying from 50 to 150 ($A1 = 50$, $A2 = 100$, $A3 = 150$). The most critical feature of this relation is the influence of the reciprocal value of the certainty v. Clearly v is greater that 0.50. As v increases the value of the denominator increases, thereby driving the value of the judgment time down.

Furthermore, this equation is easily rewritten to produce a relation between certainty and response time. Solving for v yields,

$$v = 0.5 + 0.5\left[\frac{a}{t - b}\right] \tag{5.1}$$

If t is large then the minimum certainty equals 0.5. This particular value must depend upon the undisclosed nature of the numerical values assigned to the observers' certainty judgments. As t declines the certainty increases.

Volkmann's empirically derived relation relates response time and certainty in much the same way that Peirce and Jastrow related response probability and certainty. The combination of the two laws must certainly produce a derived relation between response time and response probability.

DONALD M. JOHNSON

Speed Versus Accuracy

Although intriguingly simple, Volkmann's Equation required support from an experiment in which the certainty of judgment was measured on the same trial as the judgment time. Henmon (1911), of course, reported a negative relation, but this somewhat incidental account required further elaboration. Yet, no formal investigations of any relationship between speed of responding and the confidence of the associated response occurred until D. M. Johnson (1939) published the insightful and theoretically accomplished article, "Confidence and Speed in the Two-Category Judgment."

Johnson searched for a relation between the time of judgment and confidence by using the Method of Constant Stimuli to provide a precise view of any relation that might exist. Specifically, the standard stimulus was the left 50 mm of a line of variable length. To span a broad range of stimulus differences the experiment used 15 comparison stimuli, always the right side of the line, covering a 20 mm range. The comparison stimuli were lengths of size 40, 42, 44, 46, 47, 48, 49, 50, 51, 52, 53, 54, 56, 58, and 60 mm. Although the three subjects always responded either *larger* or *smaller*, the subjects also indicated the confidence in each judgment by marking a continuous scale that Johnson treated later as a linear scale of confidence ranging from 0 to 100. In this way Johnson obtained for each judgment a measure of response choice, time, and confidence.

Hints of a relationship between speed and accuracy of performance found in

Garrett's (1922) "A Study of the Relationship of Accuracy to Speed," and in Hollingworth's (1939) experiments on the speed and accuracy of typing while chewing, led to the suggestion that faster responses were in some way unconsidered and the subjects less certain, perhaps, about their responses. To influence the speed of a response D. M. Johnson (1939) required the subjects to respond under three different instructions.

> In this experiment the left side of the line will remain constant while the right side will vary. As before you are to make your judgments by pressing either the right key or the left key. Press the right key if you think the right side of the line is longer, the left key if you think it is shorter. (p. 30)
>
> Confidence is to be reported by making a pencil mark on a graphic scale.
>
> Let the right end of the line represent complete certainty that the right of the line is longer than the left. Let the left end of the scale represent complete certainty that the right side of the line is shorter, i.e., that the left side is longer. The center of the scale represents entire lack of certainty in the judgment. Intermediate points on the scale represent intermediate degrees of certainty. Instead of reporting your confidence in words you can represent the precise degree of your confidence by marking an appropriate point on the scale. (p. 30)

These instructions were labelled by Johnson as *usual* to contrast them with the following instructions regarding response speed.

> In this part of the experiment you are to make your judgments as *rapidly* as you can. Do not sacrifice accuracy but *emphasize speed.* (p. 30)
>
> In this part of the experiment you are to make your judgments as *accurately* as you can. Speed is unimportant. Take your time and *make no errors.* (p. 30)

Shown in Fig. 5.10 are mean response times, proportions of correct responses, and average response confidences for Subject V responding under the *accuracy* conditions. Both confidence and response proportion depend directly on the size of the deviation of the comparison stimulus from the 50 mm standard. Response time, as was true of Kellogg's and Henmon's data, increases as stimulus difference approaches zero. The results suggest

Johnson's Principle: Judgment time becomes greater as the category threshold is approached.

A surprising relation between response time and confidence lurked beneath the surface of these results. Johnson discovered that by dividing the confidence spectrum into intervals of 0.20 and plotting the mean response time for responses

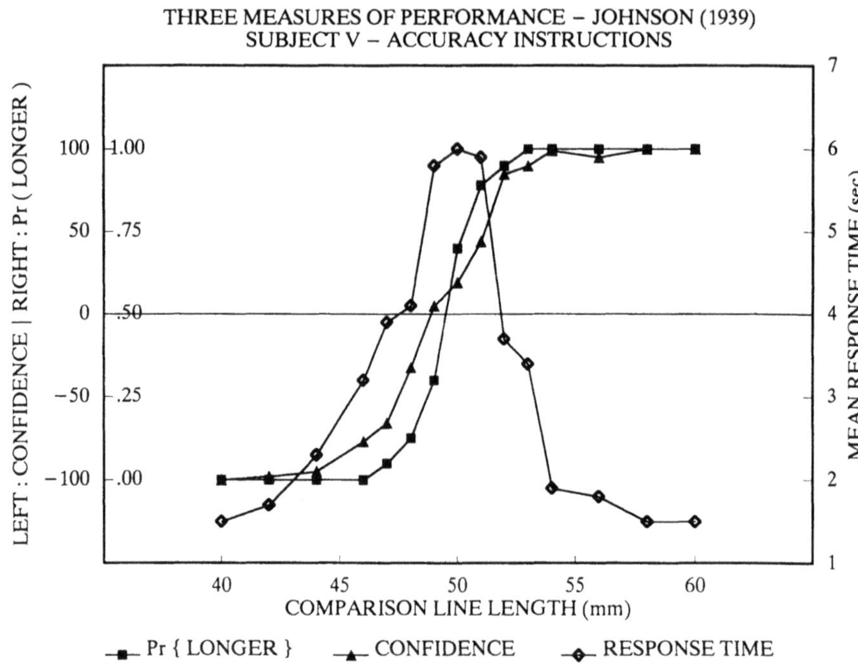

FIG. 5.10. Average results for Johnson's Subject V under the Accuracy instructions.

contained within each interval, a good account of the relation between response time and confidence followed from Johnson's Equation:

$$T = a(10)^{b(1-C)} \tag{5.2}$$

where a = the intercept constant,
b = the slope constant,
and C = Confidence with which the judgment is made.

Converting to the logarithm of response time produces a linear relation

$$\log(T) = \log(a) + b(1 - C) \tag{5.3}$$

that reveals large differences due to the three different instructions. Figure 5.11 illustrates logarithmic response time as a function of grouped confidence values. For the three conditions, *accuracy*, *usual*, and *speed*, and the average data the linear relation is obviously quite good as evidenced by the fit of the linear regression lines to each condition.

Although the logarithmic relation between Confidence and RT is quite acceptable, the striking feature of these data is the parallel form of the three-speed versus confidence functions. The various instructional conditions caused, primarily, a change in the intercept value. The change from condition to condition

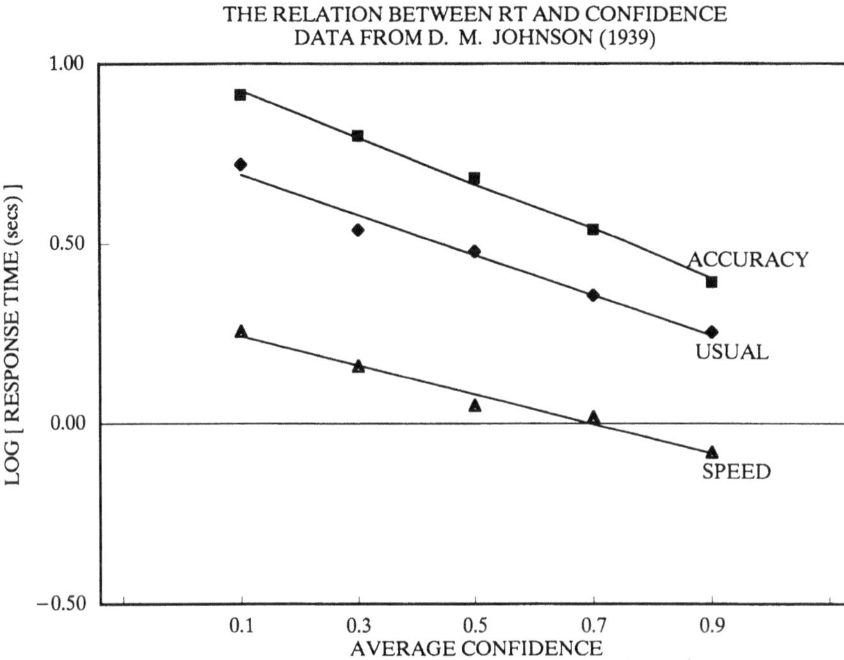

FIG. 5.11. Johnson's logarithmic relation between response time and response confidence is apparent in this logarithmic plot of data for Subject K.

suggests that increases in response time are due to a general feature of the subject's interpretation of the instructions that affects all response times within the condition regardless of the degree of confidence. Thus RT appears to be under the willful control of the subject! Somehow, the subject *controls* response time to meet performance demands.

Of course the fact that RT also varies as a function of the difference between the standard and comparison stimulus suggests that RT depends upon two factors. The psychophysical difference between the stimuli is a prominent part of changes in response time. Yet when the subject is pressed to respond with speed the response time decreases as if the stimuli were separated by greater distances! Responses to stimuli separated by greater differences should exhibit greater accuracy. Yet the response choices show greater error, rather than less. This discovery is the origin of the *speed-accuracy trade-off*: Subjects respond faster but with greater inaccuracy, or slower with greater accuracy.

Choosing to emphasize the importance of changes in RT due to a psychophysical difference D. M. Johnson (1939) concluded that

> the time of judgment increases rapidly as the difference between the standard stimulus and the comparison stimulus decreases slowly. This is in accord with

> Henmon's [1906] results, although his times were much shorter. Another way of putting it is that judgment time becomes greater as the category threshold is approached. (p. 140)

Thus Johnson reinterpreted Kellogg's conclusion that response time decreases as the stimuli more perfectly match the response category. In both Kellogg's and Johnson's views, stimulus difference per se is not the primary determinant of response time. Both conclusions place a new emphasis on the response category as a major determinant of changes in response time.

Yet, as Peirce and Jastrow (1884) argued many years before, how can boundaries, whether for the interval of uncertainty or response categories, act as thresholds for responses whose response probabilities change smoothly across the threshold. In fact, no theory existed that could predict how response time changed as a function of stimulus difference. Moreover, any theory based solely on stimulus difference fails to account for Johnson's results. As if to fill this theoretical void Dorwin Cartwright brought forth an ingenious new idea.

DORWIN CARTWRIGHT

Cartwright (1941b) investigated two theoretical concepts in a brilliant series of experiments. First, Cartwright accepted the idea of stimulus difference as an important determinant of psychological differences. Second, he suggested that increases in decision time also result from a conflict between different, perhaps implicit, responses. The conflict is greatest when a stimulus falls upon the border separating the two responses. The border is defined by that stimulus which leads equally often to the two different responses, such as a stimulus placed at the Point of Subjective Equality.

This speculation may be tested by comparing peaks for a Chronometric Function with the 50% point of the associated Psychometric Function. Cartwright confirmed this prediction using results reported in Figs. 5.12 and 5.13. In Fig. 5.12 are average vocal response times and response proportions from an experiment in which the subjects learned to recognize a single geometric figure shaped like a "less than" symbol (<) having a 90° angle at the vertex. Then, the subject judged whether test stimuli with smaller, equal, or larger vertices were the *same* as, or *different* from, the angle presented during the experiment's learning phase.

The results are clearly similar to those reported by Henmon, Kellogg, and Johnson. The frequency of responding *same* and the mean vocal response time both increase as the angle of the test stimulus approaches the 90° standard. The surprising result is shown in Fig. 5.13. These data are from subjects who first learned to recognize perfectly stimuli with angles ranging from 60° to 100°. Afterward the subjects identified those comparison stimuli, ranging from 10° to 160°, that corresponded to the learned set of stimuli. The decision times increase

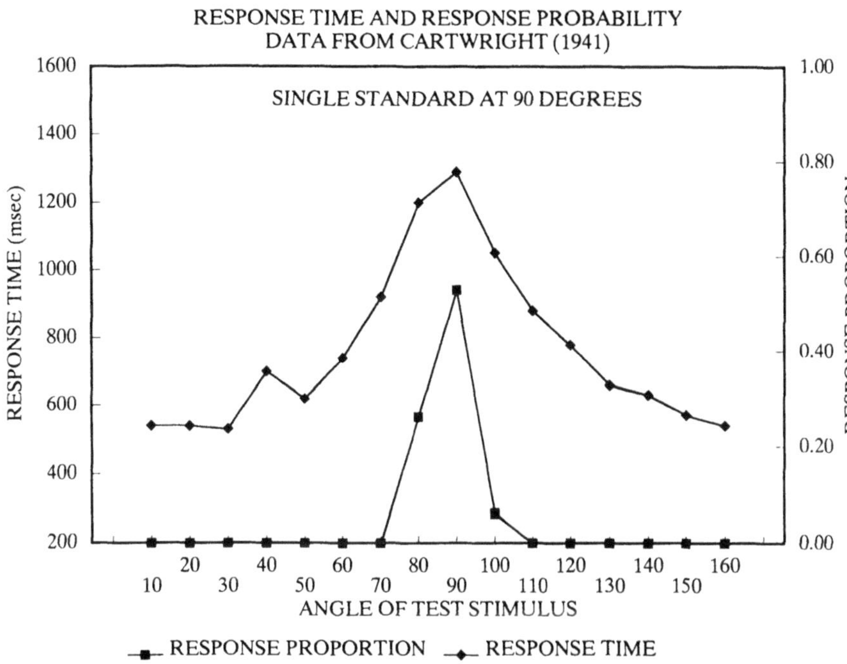

FIG. 5.12. Response times and proportions for test stimuli ranging from 10° to 160° given a standard at 90°.

dramatically near the edges of the range of standards, and the proportion of *same* responses reaches a maximum of 1.0.

Cartwright, supporting Johnson, concluded that the critical determinant of judgment time is the distance of the test stimulus from the response category boundary. This interpretation is strengthened by the reduction in response time as the test stimulus nears the middle of the range of standards. The reduction in response time appears to be about the same as that for *different* responses elicited by stimuli at a comparable distance outside the upper and lower bounds around the standards.

DORWIN CARTWRIGHT AND LEON FESTINGER

An Opponent Process Theory of Discrimination

The results prompted Cartwright to create the first discrimination process to predict both the observed response times and proportions. Today this topological theory of judgment remains a monument to the influence of the Gestalt psychologists Wolfgang Köhler and Kurt Lewin. Examining the relation between decision

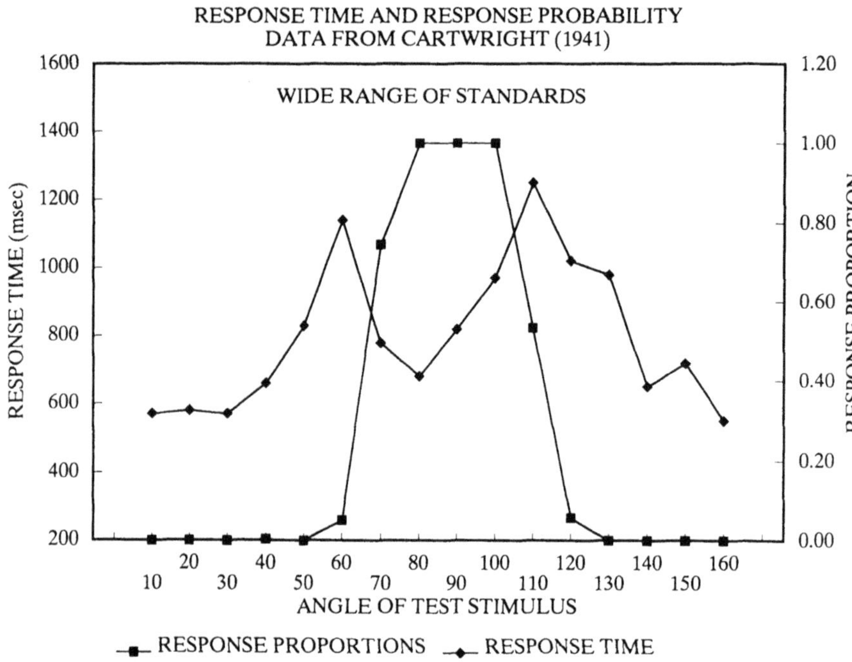

FIG. 5.13. Response times and proportions for test stimuli given a range of learned standards.

time and response category Cartwright, and Festinger who was responsible for the mathematical development, proposed that increases in the length of decision time result from a conflict between different response forces, and that such conflict increases as a stimulus nears a response threshold separating two ranges of stimuli.

In the usual psychophysical experiment stimuli that lie at the edge of the categories are associated with maximal decision times. Because nearly all relevant experiments employed only two categories, a test stimulus equal to the standard generally fell on the border between the ranges of stimuli yielding judgments of *greater* and *lesser*. According to the theory, therefore, a comparison stimulus equal to the standard should yield the longest decision time, a prediction often confirmed under these conditions.

A different prediction may be expected when three categories are employed. Then there are three ranges of stimuli for the three responses and two borders separating the stimulus dimension into the three response regions. The test stimulus equal to a standard falls in the middle of the central range which yields a judgment of *equality* with high probability. Under these conditions decision time should be maximal on either side of the two borders and should decline as a test stimulus recedes from a response boundary.

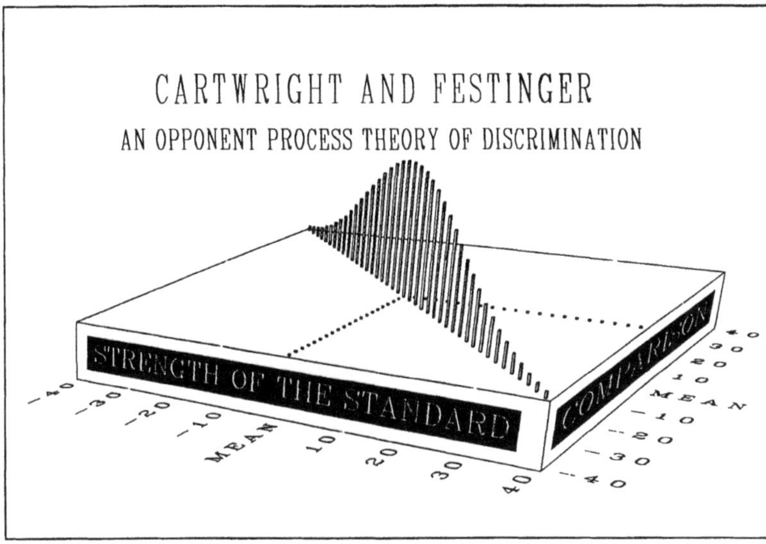

FIG. 5.14. The two Gaussian distributed forces toward responding ***longer*** or *shorter* are perfectly negatively correlated. Thus a strength s for responding *shorter* coexists with a strength $-s$ for responding *longer*.

The theory is solidly Herbartian, postulating a process in which response forces compete against each other and must overcome a restraining force before one or the other response occurs. Because the theoretical principles embodied in this theory are of primary importance to understanding response time mechanisms, the theory merits scrutiny.

In a critical departure from previous psychophysical theories, Festinger (1943) and Cartwright and Festinger (1943) assumed that forces toward one response or another, such as responding *longer* or *shorter*, will fluctuate in time about a mean value. Further, these two Gaussian distributed variables have equal variance and, as a simplification, are perfectly negatively correlated. An illustration of these relations and the distribution appears in Fig. 5.14.

If any size of difference between forces resulted in a response, all responses would have equal response times. To lengthen response times, Cartwright and Festinger proposed that the difference between these competing response forces must overcome a restraining force that is also Gaussian distributed and fluctuates over time. The subject must wait to respond until the absolute difference between stimulus values exceeds the restraining force.

Figure 5.15 clarifies the operation of the response and restraining forces. The figure presents a plane with axes for $\boldsymbol{Z}$, the difference between response forces, and $\boldsymbol{R}$, the restraining force. One can imagine on top of this plane the joint probability distribution for $\boldsymbol{Z}$ and $\boldsymbol{R}$. The authors conveniently assumed $\boldsymbol{Z}$ and $\boldsymbol{R}$

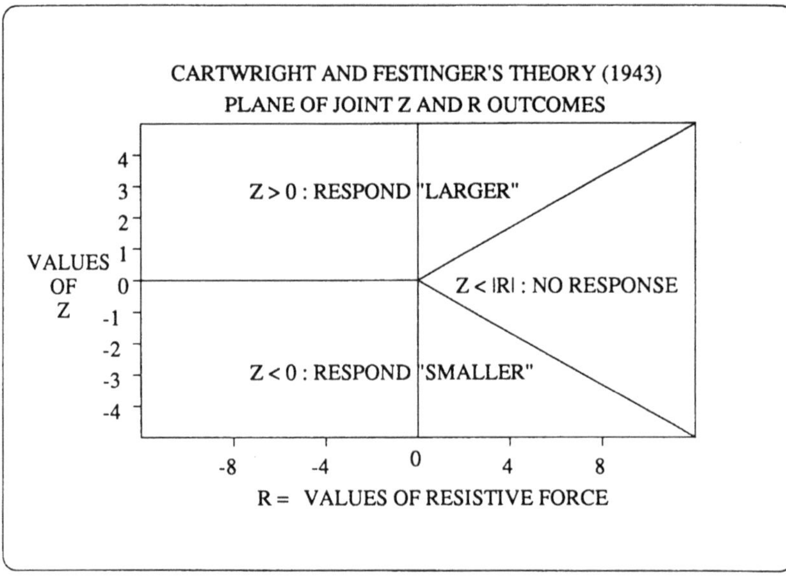

FIG. 5.15. Points in the plane represent all joint occurrences of differences between response tendencies and restraining forces. The joint density function above the shaded area contains the probability that no decision will occur for a sample of $\boldsymbol{Z}$ and $\boldsymbol{R}$.

to be independent random variables. Therefore the joint distribution is a bivariate Gaussian distribution equal to,

$$f_{Z,\boldsymbol{R}}(z, r) = f_Z(z) f_{\boldsymbol{R}}(r). \tag{5.4}$$

The plane in Fig. 5.15 contains all the points of the sample space for a single sampling of response and restraining forces. Regions of the plane correspond to the three possible events of responding *longer*, *shorter*, or making no response and sampling again. Note that only when the absolute value of the response force exceeds the restraining force does a response occur.

To calculate the probabilities of these various events, integrate the joint density function in the following ways: The probability that a sampled difference, $\boldsymbol{Z}$, will be both positive and exceed the restraining force is,

$$\begin{aligned} P_L &= Pr\{\boldsymbol{Z} > 0 \cap \boldsymbol{Z} > |\boldsymbol{R}|\} \\ &= \int_{-\infty}^{0} \int_{0}^{\infty} f_{Z,\boldsymbol{R}}(z, r) dz dr + \int_{0}^{\infty} \int_{r}^{\infty} f_{Z,\boldsymbol{R}}(z, r) dz dr. \end{aligned} \tag{5.5}$$

The probability that a sampled difference is smaller than $\boldsymbol{Z}$ but exceeds the restraining force is

$$P_S = Pr\{\mathbf{Z} < 0 \cap \mathbf{Z} > |\mathbf{R}|\}$$

$$= \int_{-\infty}^{0} \int_{-\infty}^{0} f_{\mathbf{Z},\mathbf{R}}(z, r)dzdr + \int_{0}^{\infty} \int_{-\infty}^{-r} f_{\mathbf{Z},\mathbf{R}}(z, r)dzdr. \quad (5.6)$$

The probability of re-sampling equals one minus these probabilities, which is the volume of probability density above the wedge shaped area corresponding to restraining force values that exceed the sampled stimulus difference.

$$1 - P_L - P_S = Pr\{Resampling\}$$

$$= \int_{0}^{\infty} \int_{-r}^{r} f_{\mathbf{Z},\mathbf{R}}(z, r)dzdr. \quad (5.7)$$

These probabilities change depending upon the positioning of the bivariate Gaussian distribution.

Although the originality of the theory tickles the imagination, the mathematical analysis leaves room for considerable explication. Today some immediate consequences are derivable without specific knowledge about the parameters of the various fluctuating force distributions. For example, let p be the probability that the response force, $\mathbf{Z}$, will exceed a randomly sampled value of the restraining force. Then $p = P_L + P_S$. With probability $1 - p$ a sampled value of $\mathbf{Z}$ does not exceed the restraining force and another sample of response forces is taken. In order for a response to occur sampling must continue until a value of $\mathbf{Z}$ exceeds the restraining force.

This process generates a Geometric distribution of waiting times for a decision to occur. The probability distribution is similar to a declining exponential distribution. To calculate properties of the response waiting time distribution, that is, the latency of the response once the stimulus is presented, assume that a sampling of forces occurs during a period of time Δt. In particular let $\mathbf{N}(\mathbf{n})$ be the latency of a decision requiring $\mathbf{n}$ samples. Because each sample requires time Δt the $\mathbf{n}$ samples require a total time of $\mathbf{n}\Delta t$. The probability of waiting for $\mathbf{n} - 1\ \Delta t$ time periods and then deciding at the end of the $\mathbf{n}$th time period is,

$$Pr\ \{\mathbf{N} = \mathbf{n}\} = p(1 - p)^{(\mathbf{n}-1)} \quad \mathbf{n}>0.$$

The average time to decide, the expected decision latency, equals,

$$E\{\mathbf{N}\} = \sum_{\mathbf{n}=1}^{\infty} \mathbf{n}(\Delta t)p(1 - p)^{(\mathbf{n}-1)} \quad (5.8)$$

$$= \Delta t/p.$$

As the value of p increases the probability of exceeding the restraining force increases, and the expected latency decreases.

The factors affecting the mean values of this distribution follow from traditional psychophysical ideas. Judgments occur only when the restraining force is overcome. Responses occur more quickly for restraining forces with a small mean than for those with a larger average value. First, as the difference between stimuli increases, the mean of the difference between response forces increases. Thus for a fixed restraining-force distribution, increases in stimulus difference result in a higher probability of exceeding the restraining force, thereby driving down response time. Because the time of a judgment depends upon a factor other than stimulus difference, the response times that form a Chronometric Function depend upon the subject's willingness to respond, that is, upon the subject's control of the restraining force. Response times, then, depend upon the attitude of the subject—upon the willingness of the subject to respond.

What about the response probabilities? Once again the Geometric distribution plays an important role. At any time $\boldsymbol{N} = \boldsymbol{n}$, a decision has not yet occurred with probability $(1 - p)^{(n-1)}$. At time $\boldsymbol{N} = \boldsymbol{n}$, the decision *longer* occurs with probability P_L. Let N_L be the time when the process stops with a *longer* decision. Then

$$Pr\,\{N_L = n \text{ and decide } longer\} = P_L(1 - p)^{(n-1)}.$$

Summing these probabilities across all possible time intervals yields the probability of deciding in favor of the alternative *longer*. That is,

$$\begin{aligned} Pr\{longer\} &= \sum_{n=1}^{\infty} P_L(1 - p)^{(n-1)} \qquad (5.9) \\ &= P_L/p \\ &= P_L/(P_L + P_S) \end{aligned}$$

where $P_S = Pr\{shorter\}$. Thus the probability of the decision *longer* depends on the probability $P_L + P_S$ that the restraining force is exceeded by a sampled difference larger in absolute value than the restraining force. In a similar manner the unknown probability of responding *shorter* is,

$$\begin{aligned} Pr\{shorter\} &= \sum_{n=1}^{\infty} P_S(1 - p)^{(n-1)} \qquad (5.10) \\ &= P_S/(P_L + P_S). \end{aligned}$$

When there is no restraining force the sum of P_L and P_S equals 1.0 and a decision occurs after the first unit of time Δt. The value of P_L equals the probability that $\boldsymbol{Z} > 0$ and the value of P_S equals the probability that $\boldsymbol{Z} < 0$.

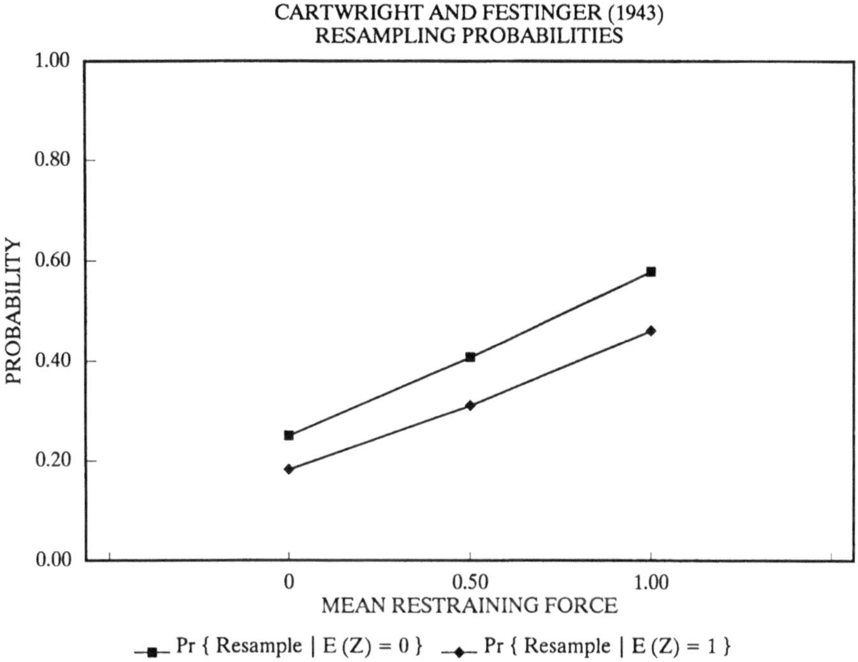

FIG. 5.16. The probability of resampling the stimulus display increases as the mean resistive force, $E\{\boldsymbol{R}\}$, increases but decreases as the expected stimulus difference, $E\{\boldsymbol{Z}\}$, increases.

Because the distribution of the restraining force causes the values of P_L and P_S to change, the subject's control of the restraining force changes response probability and response time.

Computing the values of P_L and P_S by direct integration of the joint distribution of $\boldsymbol{Z}$ and $\boldsymbol{R}$ proves to be more complicated than Cartwright and Festinger indicated. However, some rather surprizing predictions offer ways to examine the influence of these slippery parameters. To illustrate the influence of increases in restraining force and stimulus difference, numerical integration (using MATHCAD™) generated response probabilities for P_L, P_S, and the probability of resampling, $1 - (P_L + P_S)$, for each set of parameter values listed in Table 5.1. The Gaussian distributions have unit variance. Thus a change in mean value from zero to one is a change of a full standard deviation unit.

What is easily seen in Fig. 5.16 is the increase in the probability of resampling as the mean restraining force increases. When the mean stimulus difference equals zero, so that responding *longer* or *shorter* is equally likely, the probability of resampling increases to 0.578 from 0.250 as the mean restraining force increases to 1.0 from 0. When the mean stimulus difference increases to 1.0 the probability of resampling, naturally, declines, but the increases in resampling

TABLE 5.1
Table of Probabilities for Cartwright and Festinger's Opponent Process Theory of Discrimination. Both ***Z*** and ***R*** Are Gaussian with unit variance

$E\{Z\}$	$E\{R\}$	P_L	P_S	$1 - (P_L + P_S)$	$Pr\{longer\}$	$Pr\{shorter\}$	$E\{N\}$
0.0	0.0	.375	.375	.250	.500	.500	1.33
	0.5	.296	.296	.407	.500	.500	1.69
	1.0	.211	.211	.578	.500	.500	2.37
1.0	0.0	.710	.108	.182	.132	.868	1.22
	0.5	.611	.080	.310	.116	.884	1.45
	1.0	.487	.052	.461	.097	.903	1.86

probabilities are still substantial, from 0.182 to 0.461 as the mean restraining force increases from 0 to 1.0.

The probabilities of resampling affect mean decision time. Figure 5.17 shows that as the mean restraining force declines, the probability of sampling a sensory value in excess of the restraining force increases, leading to a faster decision. The declines in decision time should be reflected in corresponding declines in mean response times. Thus, in this example long response times are those associ-

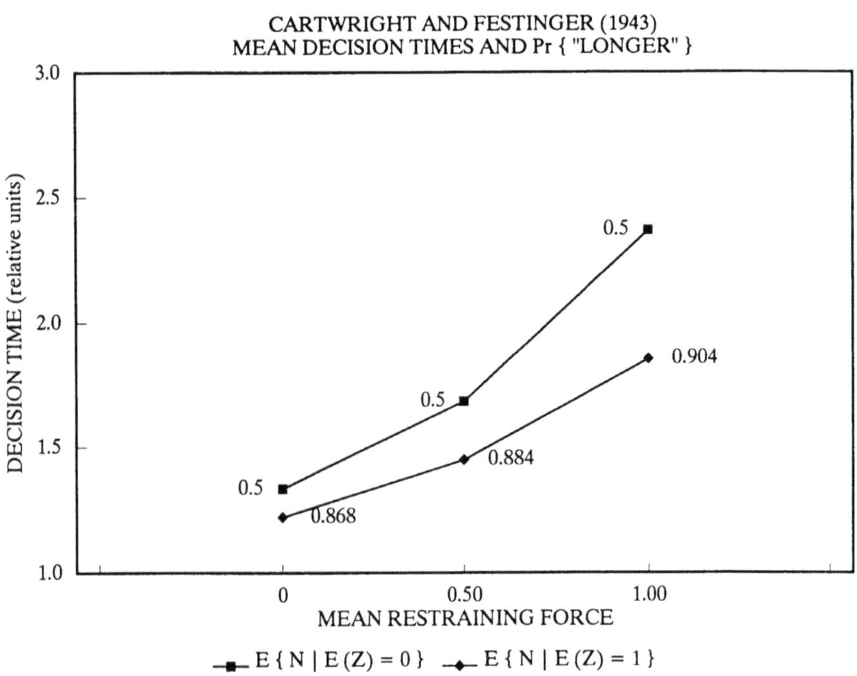

FIG. 5.17. Mean decision time, $\Delta t/p$, as a function of mean restraining force for two different mean stimulus-difference values of 0.0 and 1.0.

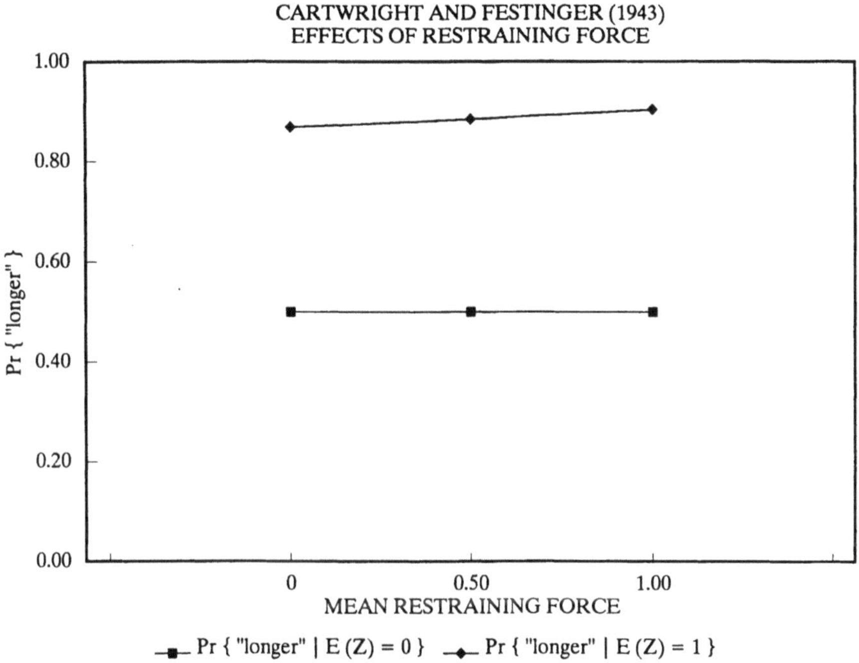

FIG. 5.18. In spite of the dramatic increases in the probability of resampling as a function of the mean resistive force, the probability of responding *longer* depends primarily on the average stimulus difference $E\{\mathbf{Z}\}$.

ated with the most difficult discriminations, where the mean stimulus difference is zero. As the mean stimulus difference increases the response time declines. Overall, reducing the restraining force leads to faster decision times.

The question that remains unanswered is whether the process of resampling leads to a higher or lower probability of a correct response? Whenever there exists a restraining force, more than a single sample is needed for a decision to occur. If additional samples are required then $(P_L + P_S)$ must be less than 1.0. When the additional samples increase the value of a correct *larger* response, $P_L/(P_L + P_S)$, there is an increase in decision time which improves performance.

The degree of this trade of speed for accuracy is shown in Fig. 5.18. The probability of responding *longer* hardly changes at all as a function of changes in mean restraining force. When the average stimulus difference is zero, the probability of deciding *longer* or *shorter* equals 0.50 for all restraining forces, as is to be expected. But, when the mean stimulus difference is 1.0, large by Thurstonian standards, the probability of responding *longer* increases from 0.868 to 0.903 as the mean restraining force increases from 0 to 0.50 to 1.0—a small increase in accuracy for decision times that increase from 1.22 Δt to 1.45 Δt to 1.86 Δt, as shown in Fig. 5.17. These are very fast decision times but they illustrate the relation between decision probability and decision time.

Certainly Johnson's results suggest that a speed for accuracy trade-off must be a consequence of any theory of judgment that accounts for both response time and response probability. Cartwright and Festinger (1943) believed so and noted:

> Since the precision of the relative frequency ogive [the psychometric function] is used as a measure of sensitivity in some psychophysical experiments, sensitivity as thus measured is affected by certain attitudes on the part of the subject. (p. 608)

Does it follow, now, that the response theory must be the same as Thurstone's Law of Comparative Judgment? Values of $\mathbf{Z}$ greater than zero lead to the decision *longer* and values less than zero lead to the response *shorter*. These decision probabilities, $P_L/(P_L + P_S)$ and $P_S/(P_L + P_S)$, may be determined experimentally but the values of P_L and P_S remain unknown. Thus, an investigation of whether the Law of Comparative Judgment applies to these judgments must wait until there is a way to estimate the values of $P_L + P_S$. Then the individual values of P_L and P_S can be determined and Thurstonian stimulus differences derived.

Greenberg (1962) noticed that the mean latency is the same for both responses, regardless of their response probabilities! To prove this unlikely assertion let the probability that the sampling process stops at time $\mathbf{N} = \mathbf{n}$, given the response is *longer*, be

$$Pr\ \{\mathbf{N} = \mathbf{n} \mid \text{respond } longer\}.$$

Let $\mathbf{N}_L$ be this stopping time. Then the conditional probability that the response *longer* occurs at time $\mathbf{n}$ is,

$$Pr\ \{\mathbf{N}_L = \mathbf{n}\} = \frac{Pr\{\mathbf{N} = \mathbf{n} \text{ and respond } longer\}}{Pr\{\text{respond } longer\}}. \qquad (5.11)$$

Applying the argument yielding Equation 5.9 gives,

$$\begin{aligned} Pr\ \{\mathbf{N}_L = \mathbf{n}\} &= \frac{P_L(1 - p)^{(\mathbf{n}-1)}}{P_L/(P_L + P_S)} \\ &= p(1 - p)^{(\mathbf{n}-1)}. \end{aligned}$$

Similarly, for the decision *shorter* the stopping time $\mathbf{N}_S$ has probability,

$$Pr\ (\mathbf{N}_S = \mathbf{n}) = p(1 - p)^{(\mathbf{n}-1)}.$$

These conditional distributions are identical to each other and to the marginal distribution of $\mathbf{N}$ itself! Thus the mean decision times must also be equal,

$$\begin{aligned} E\ \{\mathbf{N}\} &= E\ \{\mathbf{N}_L\} \\ &= E\ \{\mathbf{N}_S\} \\ &= \Delta t/p. \end{aligned} \qquad (5.12)$$

Therefore, mean decision times for deciding in favor of *longer* or *shorter* are the same as the overall mean decision times shown in Fig. 5.17. The mean decision time does not depend upon the probability or responding *longer* or *shorter*, only on the probability of exceeding the restraining force.

The observed response times consist of the amount of time taken for a decision to occur plus other components not directly related to the parameters of the decision process. As Cartwright and Festinger pointed out, the mean response time consists of two additive components, one for the decision time, a second for the nondecision components, and equals

$$\begin{aligned} E\,\{\boldsymbol{RT}\} &= E\,\{\boldsymbol{DT}\} + E\,\{\boldsymbol{K}\} \\ &= (\Delta t/p) + E\,\{\boldsymbol{K}\}. \end{aligned}$$

As the probability of exceeding the restraining force decreases the value of $p = (P_L + P_S)$ declines and the response time increases.

The theory proposed by Cartwright and Festinger contains many elements that work together to produce simultaneous changes in both response time and choice probability. The unwieldy Gaussian distributions hamper analysis of the theory. Yet, the theory stands as the first to treat response time and probability as a consequence of a single judgment mechanism.

The onset of World War II brought additional theoretical developments to an abrupt halt. The demands placed on psychology by World War II fostered the creation of new techniques for the analysis of attitudes and opinions to the forefront of applied psychology. Cartwright served in Washington, D. C. with the Department of Agriculture's Division of Program Surveys developing attitude measuring devices under the leadership of Rensis Likert (Dorwin Cartwright, personal communication, April, 1989). Many research programs during the war, as well as important theoretical advances in psychometrics and the scaling of attitudes, remained hidden until a suitable time for publication occurred after the War's end (Festinger, 1947; Guttman, 1947).

6 The Psychological Ideal

WALTER C. SHIPLEY

The Crossover Effect

Research surged forward following World War II. Shipley, Coffin, and Hadsell (1945) reported a paired-comparison study of color preference in which subjects chose the most preferred of two colors. As in Dashiell's (1937) experiment the choice times proved to be linearly related to the affective distances between colors. Shipley noticed that response-time deviations from the regression line relating mean response time to an affective distance value (calculated by a method attributed to Guilford, 1936, p. 237), tended to be positive when the two stimuli were among those less preferred. Alternatively, negative deviations, indicating faster than expected responses, occurred for choices among the more preferred colors.

Thus, concluded Shipley et al. (1945)

> mean speed of judgment was found to be determined by two factors:
>
> (1) An affective-distance factor accounting for about 74 percent of the variability.
>
> (2) A positive value of the stimulus factor, contributed to about equally by both comparison stimuli, accounting for an additional 16 percent of the variability. (p. 214)

The two-factor theory suggested that if subjects were to judge the least, rather than the most, preferred stimulus, then the deviations of response times from the regression line relating RT to affective value would reverse themselves. Those stimuli least preferred would be responded to more quickly than those most preferred. Shipley, Norris, and Roberts (1946) confirmed the hypothesis.

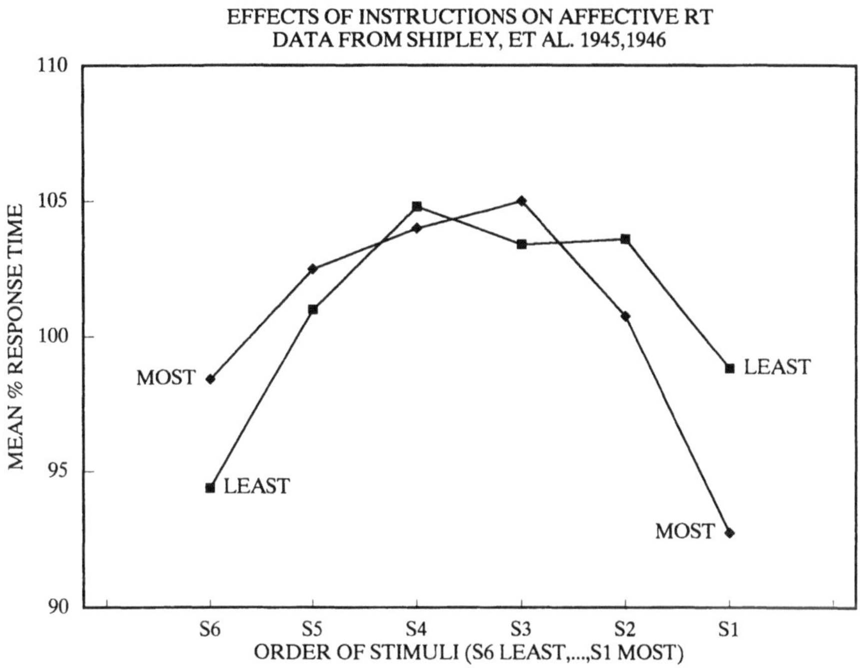

FIG. 6.1. Mean response time as a function of stimuli ordered according to preference for two different instructions about what to choose.

A reanalysis of both studies, presented in Fig. 6.1, illustrates that mean response time, expressed as a percentage of subjects' mean RT, increases toward the middle range of stimuli and then decreases again. The obvious point, however, is that when subjects are instructed to choose the most preferred stimulus of a pair, the RT is smaller when the stimulus preferred most is one of the pair. When the stimulus preferred least is one of the comparison pair, the mean response time is longer.

Data from the second study (Shipley et al., 1946) shows this relation to reverse itself when the subjects choose the least, rather than most, preferred stimulus. The crossover of the two response-time curves (whence the denomer, *crossover effect*) indicates that simple affective difference is not sufficient to account for the changes in response times. Thus, as in the Dashiell experiment, something more than the difference in affective value, some other aspect of judgment, influences response time.

The same conclusion cannot be drawn from the affective values estimated by the method suggested by Guilford (1936). The correlation between estimated affective values from both studies is 0.99. This poses the question of whether only the response time is affected by this second factor or whether Guilford's method of determining the values of affective distance hides a similar finding.

HAROLD SCHLOSBERG AND RICHARD L. SOLOMON

Animals are easily taught to exhibit a learned preference while still discerning differences between stimuli. For example, Schlosberg and Solomon (1943) overtrained rats to perfectly discriminate the color White, the reinforced stimulus, from Black. Using a Lashley jumping stand, they measured the latency to jump toward the lighter of two stimuli. One comparison stimulus varied from white to black whereas the other alternative always remained black. Figure 6.2 shows the latencies to increase substantially as the brightness difference between the comparison and black standard decreases.

The animal, however, always jumps to the lighter, preferred, member of the pair. Perhaps the increase in response time is also due to the same phenomenon as shown in Dashiell's data and discovered by Shipley. White is the preferred choice that is always reinforced and must be considered a preferred stimulus. As the two comparison stimuli deviate more from the preferred stimulus, the response time increases.

Support for the idea that the direction of the response with respect to the ordering of the stimuli is important comes from comparisons of the off-white,

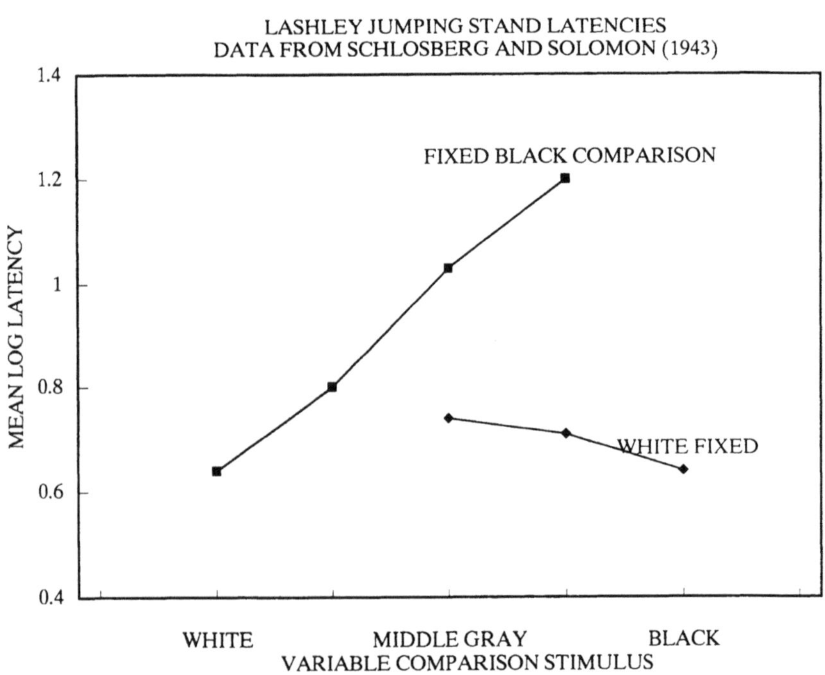

FIG. 6.2. Latency data from Schlosberg and Solomon (1943) show that responses to a reinforced member of a stimulus pair are faster than responses to a pair containing a nonreinforced stimulus.

off-black stimuli. When the fixed comparison is white the response times are substantially faster to the medium-gray and dark-gray stimuli than when the fixed comparison is black.

The entire set of paired comparisons was not run because the experimenters trained the rats through an errorless procedure whose purpose would be destroyed by occasional leaps toward a darker stimulus. Instead, test trials differing from the white-black training stimuli were inserted into the usual sequence of white-black errorless discrimination trials. This procedure excludes potential data on the critical question of whether these rats would produce choice results similar to the results of Shipley et al. (1946).

CLYDE H. COOMBS

The Theory of Ideals

Among the difficulties facing Thurstone's Law of Comparative Judgment was an obvious, but occasional, error in predicting when three stimuli should obey the property of transitivity. Suppose stimulus $\mathbf{S}_A$ is preferred to $\mathbf{S}_B$ and $\mathbf{S}_B$ to $\mathbf{S}_C$. Surely $\mathbf{S}_A$ should also be preferred to $\mathbf{S}_C$, in which case $\mathbf{S}_A$, $\mathbf{S}_B$, and $\mathbf{S}_C$ create a transitive relation. But several studies, possibly Dashiell's and Shipley's among them, produced failures of transitivity. These results suggest that something more than stimulus difference is needed in the psychophysics of human and animal judgment.

In order to explain how failures of transitivity may occur C. H. Coombs (1950) created a new concept to augment the Law of Comparative Judgment. In Thurstonian theory, once the stimuli are ordered according to their mean psychological values, *Strong Stochastic Transitivity* (SST) must be observed. That is, letting >P> represent the relation "more preferred than,"

If $\quad Pr\{\mathbf{S}_A > P > \mathbf{S}_B\} \geq 0.50$
and $\quad Pr\{\mathbf{S}_B > P > \mathbf{S}_C\} \geq 0.50$
then,

$$Pr\{\mathbf{S}_A > P > \mathbf{S}_C\} \geq \mathrm{Max}[Pr\{\mathbf{S}_A > P > \mathbf{S}_B\}, Pr\{\mathbf{S}_B > P > \mathbf{S}_C\}].$$

This formulation makes precise the meaning of transitivity for response probabilities. Essentially SST postulates that the probability of choosing the most- over the least-preferred stimulus is greater than or equal to the larger of the pairwise probabilities of choosing $\mathbf{S}_A$ over $\mathbf{S}_B$ or choosing $\mathbf{S}_B$ over $\mathbf{S}_C$. Many possible theories may require the empirical existence of SST. Therefore, SST provides a method of distinguishing between the class of theories for which SST must be true, and those for which it must be false or irrelevant.

SST can easily fail, as is shown in the following example. Four stimuli $\mathbf{S}_A$, $\mathbf{S}_B$, $\mathbf{S}_C$, and $\mathbf{S}_D$ representing lights of increasing luminance, are used to create 16 pairs of stimuli. The subject must choose between the lights on the basis of

TABLE 6.1
Choice Proportions From A Hypothetical Brightness-Preference Experiment

	S_A	S_B	S_C	S_D
S_A	.50	.75	.85	.50
S_B	.25	.50	.65	.25
S_C	.15	.35	.50	.15
S_D	.50	.75	.85	.50

Note 1. Stimuli ordered according to increasing brightness.

Note 2. Entries are probabilities that the column stimulus is more preferred than the row stimulus.

preferred brightness. The choice proportions produced by a hypothetical subject are found in the paired-comparison matrix of Table 6.1.

When presented with the pair ($\mathbf{S}_A$, $\mathbf{S}_B$) the subject tends to choose the brighter stimulus, $\mathbf{S}_B$. When presented $\mathbf{S}_A$ and $\mathbf{S}_C$ the subject most often chooses the brighter stimulus. When presented with $\mathbf{S}_B$ and $\mathbf{S}_C$ again the brighter stimulus is chosen most often. These choices are consistent with stimuli ordered on a Psychological Continuum of brightness. However, when offered the choice between $\mathbf{S}_D$ and $\mathbf{S}_C$, rather than choosing $\mathbf{S}_D$, the perverse subject chooses $\mathbf{S}_C$ with probability 0.85! To further complicate matters the experimenter offers the subject a choice between $\mathbf{S}_A$ and $\mathbf{S}_D$ and discovers the subject to be indifferent??! Is there something peculiar about stimulus $\mathbf{S}_D$ or is there something peculiar about the theoretical prediction?

Indeed, routine Thurstonian (1927b, Case V) analysis of the response proportions in Table 6.1 generates an odd ordering of scale values. Specifically, $\mathbf{S}_A = -0.425$, $\mathbf{S}_B = 0.236$, $\mathbf{S}_C = 0.614$, and $\mathbf{S}_D = -0.425$. Thus, the physical stimuli, which are known to increase in intensity, appear to be ordered,

$$\mathbf{S}_A < \mathbf{S}_B < \mathbf{S}_C$$

but

$$\mathbf{S}_A = \mathbf{S}_D!$$

To the delight of later psychophysicists, Coombs suggested that the stimuli are not peculiar at all. Rather, the model of judgment is incorrect because lurking within the subject's method of choice is an *Ideal* brightness placed somewhere on the psychological brightness continuum. When choosing between a pair of brightnesses the subject prefers the stimulus whose brightness is nearest the Ideal. For example, when the subject's Ideal lies at position I_1 in Fig. 6.3, and the subject must choose between $\mathbf{S}_A$ and $\mathbf{S}_B$, the subject chooses the brighter

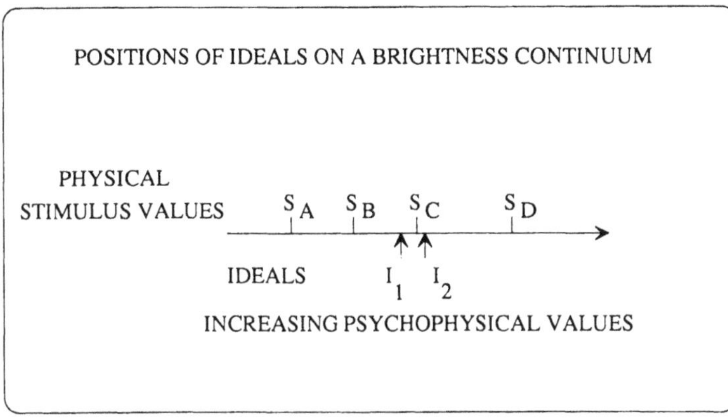

FIG. 6.3. The Psychological Continuum and the Ideals placed at the positions I_1 and I_2.

stimulus, S_B, because it is nearer in brightness to the Ideal brightness, I_1. Examined in this way the subject's earlier choice between S_A and S_D appears consistent with an Ideal at position I_1 on the psychological brightness continuum.

Furthermore, a second subject with an Ideal brightness at position I_2 prefers S_B to S_A, S_C to S_A, S_D to S_A, and prefers S_C to S_B, but prefers S_B to S_D! Plenty confusing, just like real life. Coombs recognized the important point—that when the stimulus representations are fixed, a variety of choice outcomes may occur without capriciousness on the part of the individual. If these choices are averaged across subjects with different preferences, a common method of treating large amounts of choice data, the underlying stimulus structure and the existence of the Ideal may be hidden within the average.

Imagine that for each subject the psychological values created by the stimuli are scattered across a Psychological Continuum. Furthermore, the subjects' Ideals are also located on the same continuum. Some subjects, due to cultural similarities may exhibit similarly positioned Ideals whereas other subjects' Ideals show greater deviations. The dispersion of Ideals creates a probability distribution of Ideals. Because there is a unique Ideal for each subject, the distribution of Ideals also represents the distribution of subjects on the Psychological Continuum. Thus, the location of the subjects with respect to the psychological attribute under investigation, for example, brightness, happiness, or the political appeal of presidential candidates, is determined by the value of the subject's Ideal.

Extraordinary consequences follow from this remarkably insightful view of the relations between the positions on the Psychological Continuum of stimuli and subjects. Certainly, the new joint distribution of subjects and stimuli, on the same dimension, deepens the complexity of theoretical analysis. The first question is whether the rather scanty data of the choice experiment provide sufficient information to locate both the stimuli and the subject on the Psychological Continuum.

Coombs (1954) described new experimental methods for identifying at least the order of stimuli on the Psychological Continuum. By folding the Psychological Continuum at the point of the subject's Ideal the stimuli are ordered according to increasing distance from the Ideal. Perhaps only a folded continuum is necessary for the subject to judge the distance of stimuli from the Ideal. In Coombs' view the Psychological Continuum is revealed by using choice (preference) data to unfold the dimension on which the subject orders the stimuli.

Somewhat unexpectedly, but consistent with the Midwestern behaviorist empiricism that gripped North American psychology for so long, Coombs retreated from suggesting a mechanism for how the subject makes the judgment of whether one stimulus is nearer the Ideal than another. Indeed, by treating a subject as just another point on the stimulus dimension, the problem of sorting out Ideals and stimuli becomes one of mathematical and experimental technique, somehow unrelated to the process giving rise to the judgments.

Other than asserting that the distance of stimuli from the Ideal determines the choice of one stimulus over another, no decision process is postulated. Thus no predictions concerning response time are possible. Of course one could speculate that response times are also a function of distance from an Ideal. But this hardly amounts to a psychological theory that adequately characterizes those features of the psychophysical process that so many experiments on affective value, for example, have shown to exist. Thus the "psychology" of the choice situation goes missing, to be replaced by a concept that Coombs (1960, 1964) very appropriately named, *A theory of data*.

Later Coombs (1958) suggested yet another remarkable result following from the Theory of Ideals. Stimuli located on the same side of the Ideal, such as $\mathbf{S}_A$ and $\mathbf{S}_B$ in Fig. 6.3, are ordered with respect to distance from the Ideal. Therefore, even when such stimuli have dispersions about them, due to the error of the sensory system or other within-subject variability, the probability of choosing a stimulus on the average nearer to the Ideal will exceed the probability of choosing a stimulus farther away. Comparisons between stimuli located on the same side of the Ideal are comparisons of "unilateral pairs."

In contrast, imagine two stimuli, such as $\mathbf{S}_A$ and $\mathbf{S}_D$ in Fig. 6.3, which are located on opposite sides of the Ideal. Let such a "bilateral pair" consist of two stimuli equally distant from an Ideal such as I_1. In this case the probability of choosing one stimulus over another will be 0.50 because they are equally distant from the Ideal. This probability can never occur for ordered unilateral stimuli unless the stimuli are indistinguishable from each other.

This simple observation leads to a powerful consequence. When the Ideal is located near the middle of the stimulus set, and the stimuli are ordered according to increasing brightness, as in Fig. 6.3, stimuli equally distant from the Ideal will generate choice probabilities that fail to satisfy stochastic transitivity. Off-diagonal cells of the paired-comparison matrix will be filled with inexplicable values of 0.50 whenever the two stimuli, obviously quite separate on the stimulus continuum, are equally distant from the Ideal. Inconsistency of response is a consequence of the placement of an Ideal, not a consequence of noncompliant or insensitive subjects.

In order to obtain a large number of stimulus comparisons in a short amount of time Coombs (1958) investigated this prediction by presenting 12 gray chips, 4 at a time. The subjects rank-ordered the grays from most- to least-preferred. Subjects were not required actually to choose a member of a stimulus pair, thus choice probabilities were not available. However, Coombs derived a measure of choice probability through a somewhat complicated procedure. Then, after excluding the two stimuli nearest a subject's Ideal, various forms of bi- and unilateral triples were shown to produce intransitivities that supported the existence of a single Ideal of gray.

Later Greenberg (1962) and Petrusic (1966) pursued this analysis in the context of response times to paired-comparisons drawn from six shades of gray. In order to create four different positions for an Ideal subjects were instructed to choose the darker of the two or to choose the shade that best represented a good medium-gray, or a good light-gray, or dark-gray. Petrusic (1966) defined transitivity of choice for response times and showed that the laterality effects predicted to exist for choice proportions also existed for mean response times. Because response time is a more sensitive measure of performance than response choice, the new techniques held forth the promise that response time may permit tests of transitivity when the many replications of paired-comparisons needed to determine choice proportions are not possible.

Coombs' many theoretical and empirical discoveries provide convincing evidence that subjects compare psychological values of stimuli against Ideals. Kellogg and Cartwright both suggested that the nearer the stimulus comes to the perfect example of the response category type the faster response time becomes. Is this speculation also true for stimuli close to the subject's Ideal?

The results of Petrusic and Greenberg suggest that once stimuli are ordered with respect to distance from the Ideal, the difference between stimuli influences response time. Thus, even in these more complicated judgments involving preferential choice, stimulus difference is still an important element in determining response time. But the important question of whether the comparison of two stimuli equally distant from each other generates a constant mean response time regardless of their distance from the Ideal, remained unanswered.

Perhaps subjects also judged colors against an Ideal in the experiments of Dashiell and Shipley. However, given the reduction of the choice data suggested by Guilford for the determination of a choice scale, there is no apparent way to recover the original choice probabilities from the choice scale or ranking data. Nevertheless, the data analysis yields an ordering of stimuli for each subject. If the true ordering of the stimuli depends upon the distance from the Ideal, then the mean response time will be largest when a standard is near the Ideal and comparison stimuli are clustered near the standard. When the Ideal and standard are near the middle of the range of physical stimuli the average stimulus difference will be small and the response time large. When the standard is extreme the average stimulus difference is large and the mean response time fast. These changes in response time are shown by both Dashiell (Fig. 5.7) and Shipley (Fig. 6.1).

A feature of the data not accounted for by stimulus difference depends upon the instructions given the subject to choose either the most- or least-preferred stimulus. In Fig. 5.7 response times decrease as the difference between stimuli increases. Thus stimulus difference is an important determinant of response time. Yet, as Shipley et al. (1945, 1946) recognized and demonstrated, when subjects chose the most (least) preferred of a pair of colors, response times for choices involving more (least) preferred stimuli were faster than to least (most) preferred stimuli (Fig. 6.1). This result suggests that in addition to distance from the Ideal some other factor must also be a determinant of response time.

7 Discrimination and Response Bias

GEORG VON BÉKÉSY AND STANLEY S. STEVENS

Neural Quantum Theory

Arguments advanced first by von Békésy (1930) in developing a quantum theory of discrimination led to the wholesale adoption of von Békésy's theory by Stevens, Morgan, and Volkmann (1941). As seen by von Békésy, a background stimulus excites a certain number of cells leading to a sensation that is far above any physiological threshold. Nevertheless, whatever level of sensation exists requires an increment beyond the current quantum before a difference in sensation is discriminated. These just noticeably different levels of sensation correspond to the *just noticeable differences* of Fechner and Weber. Blackwell (1963), formalizing and extending this theory, wrote:

> The sensory effect E produced by a background intensity I varies from moment to moment over a considerable range. At one stage in the auditory system the precise value of E is preserved, but at another, the only information available is whether E exceeds one or another criterion value, designated neural quanta. The frequency with which various values of E occur is described by a normal distribution function (*sic*: what is meant is the Gaussian probability density function) as depicted in Fig. 1 [7.1]. Values of Q, the increments in E separating the quantum criteria, are equal. The absolute quantum level achieved by the background intensity I at any moment has no effect on the form of the threshold probability data. (p. 130)

At any moment, the level of E produced by I defines a value of the excitation residual r where

$$r_i = E_i - Q. \tag{1}$$

A stimulus increment ΔI_i produces an increment in sensory effect ΔE_i. There is no qualitative difference between the ΔE produced by ΔI and the E produced by I and

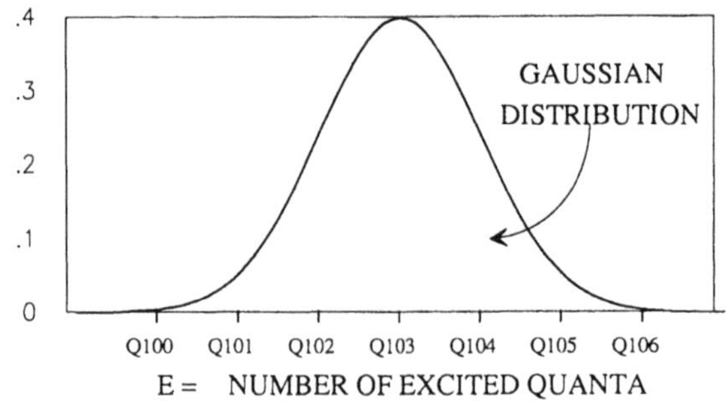

FIG. 7.1. Theoretical construction of von Békésy's neural quantum theory (adapted from Blackwell, 1963, Fig. 1.).

> the quantities ΔE and E may be added. Thus a value of E produced by I plus a value of ΔE produced by ΔI results in a value $E + \Delta E$ which is identical with a value E' produced by a value I'. Furthermore, at least over the range of E produced by I,
>
> $$\Delta E_i = c\Delta I, \quad (2)$$
>
> where c is an arbitrary constant. (pp. 130–131)
>
> In the simplest case, the auditory system discriminates the presence of a stimulus increment ΔI_i whenever
>
> $$\Delta E_i + r_i \geq E_{Qi+1} - E_{Qi}. \quad (3)$$
>
> this represents what is called the one quantum criterion, since the discrimination occurs whenever the value of E is raised above the quantum step next above the quantum level reached by the background intensity alone. (p. 131)
>
> The theory specifies that there is a rectilinear frequency distribution of values of r from 0 to $(E_{Qi+1} - E_{Qi})$. This will be approximated if the distribution of values of E is normal only if values of E corresponding to a background intensity I cover a fairly large number of quantum steps. (p. 131)

The continuous Gaussian density function, $g(Q)$, placed above the discrete *Quanta* (Q) indicates that the separations between the Quanta, the values of Q on the abscissa, are reached with a probability that depends upon the variability of the sensation evoked by background stimulation. By definition the probability that all Quanta up to and including Q_i are stimulated equals $G(Q_i)$, the integral of $g(Q)$ up to Q_i. Any excess stimulation beyond Q_i, up to but not including Q_{i+1} is irrelevant. Thus quantum values less than, say Q_i plus some small excess q, are reached with probability $G(Q_i)$. The probability that the background gives rise to sensory values between two adjacent Quanta equals $G(Q_{i+1}) - G(Q_i)$. Because there is no probability of a background intensity evoking any particular value of Q, only the probability that the background stimulation gives rise to a range of stimulation can be determined.

The theory is beset by vagaries leading to mathematical errors. To calculate the probability that an increment in intensity is discriminated, the decision rule of the subject and the way in which an increment in sensation adds to the sensation produced by the background must be specified. According to Blackwell, the background gives rise to a particular quantum level, say Q_i, plus some excess r_i. Blackwell argued that the excess ranges from 0 to $(Q_{i+1} - Q_i)$, with a rectilinear (uniform) probability frequency (density) function. Unless some unknown process is postulated to give rise to rectilinear error, the excess must be distributed as a conditional Gaussian random variable constrained to fall between Q_{i+1} and Q_i. The residual r_i has, therefore, a probability density function equal to

$$g_R(r_i) = g(r_i)/[G(Q_{i+1}) - G(Q_i)] \qquad (7.1)$$
$$\text{where } 0<r_i<Q = Q_{i+1} - Q_i,$$

which is hardly rectilinear because it is simply a value of the Gaussian density conditioned on the area of the Gaussian distribution truncated at Q_i and Q_{i+1}.

When an increment in intensity, ΔE_i, adds to the background the subject must determine whether the increment produces an increase in sensation sufficient for the total stimulation $E + \Delta E_i + r_i$ to exceed the next quantum boundary, Q_{i+1}. Apparently the increment ΔE_i is not a random variable for the author also states that $\Delta E = c\Delta I$ where c is a constant. Here there is room only for r_i's variability.

For detection to occur the sum $\Delta E_i + r_i = c\Delta I + r_i$ must exceed the quantum beyond Q_i. Adding the constant ΔE_i to a value of r_i increases the mean value of the decision statistic but the variability does not change. Thus if $r_i + c\Delta I$ exceeds $Q_{i+1} - Q_i$ then an increment is detected. The probability distribution of $\Delta E_i + r_i = c\Delta I + r_i$ is identical to r_i except for a shift of mean. Thus

$$\begin{aligned} Pr\{\Delta E_i + r_i + Q_i > Q_{i+1}\} &= Pr\{r_i > Q_{i+1} - Q_i - \Delta E_i\} \qquad (7.2) \\ &= Pr\{r_i > Q - \Delta E_i\} \\ &= 1 - G_R(Q - \Delta E_i). \end{aligned}$$

Even accepting the idea that the increment in sensation is a constant, this probability depends upon the quantum value excited by the background stimulus. As ΔE_i increases from 0 to a value greater than Q the probability of detecting the increment in sensation increases in a nonlinear fashion that depends upon the truncated Gaussian distribution.

The puzzle of Neural Quantum Theory is that the data, particularly those collected by Stevens (1961, 1975), conform well to the incorrectly derived prediction that the Psychometric Function for detecting a difference greater than the background level of stimulation should be linearly increasing. Numerous studies demonstrate the effect, but as currently formulated, Neural Quantum Theory cannot predict the result. The research provides an excellent example of data in search of a theory.

H. RICHARD BLACKWELL

High Threshold Theory

During World War II sensitive new electronic listening devices were developed that taxed the human sensory system's ability to detect small differences in sensation. To best match the human system to the new machines' capabilities the United States Office of Naval Research became an avid consumer of psychological research into factors affecting human performance. In response to problems faced by military personnel, such as vigilance at a radar or sonar screen, the Engineering Research Institute (ERI) at the University of Michigan launched a broad investigation of sensory thresholds and their measurement. The research brought to psychophysics new experimental procedures and concepts that broadened psychophysical methodology. As Blackwell (1953), Director of the Vision Research Laboratory in ERI, put it:

> In the present experiments, the subject was required to discriminate the presence of a difference in visual brightness. A relatively large field of brightness, B_0, was presented to the subject continuously, which served as the background brightness. From time to time a small increment of brightness, ΔB, was added to B_0 over a small portion of the field, for a brief exposure. The subject was always required to indicate whether or not the presence of the brightness increment, ΔB, was discriminated. One or the other of two responses was employed to indicate discrimination: phenomenal report, or forced choice. (p. 9)
>
> When phenomenal report was employed as the indicator response, the subject assessed his sensory experience directly as to whether or not the presence of ΔB was discriminated. The affirmative and negative responses were described to the subjects in terms of "Yes" and "No," as follows: respond "Yes" if you were aware that a stimulus increment was added to the field, and "No" if you were not aware of the addition of a stimulus increment. (p. 9)
>
> When forced choice was employed as the indicator response, the subject was not required to assess his sensory experience directly. Instead, he was required to indicate which of several stimulus conditions was most likely to have contained a stimulus increment. In one general procedural variant, the stimulus increment could occur in one of several spatial locations, at a known moment in time. In a second general procedure, the stimulus increment could appear in one of several temporal intervals at a known spatial location. The number of spatial locations and temporal intervals could be varied. (p. 9)

Blackwell applied to this psychophysical setting the often used correction for guessing of multiple choice test questions. According to Blackwell (1953)

> In the experiments involving phenomenal report, the proportion of "Yes" responses for each stimulus magnitude and for the catch stimuli was recorded. If "Yes"

> responses were made on the catch stimuli, the proportions of "Yes" responses made at all other stimulus magnitudes were corrected by the relation:
>
> $$p = \frac{p' - C}{1 - C} \qquad (1)$$
>
> where p = corrected proportion
> p′ = raw proportion
> C = proportion of "Yes" responses on catch stimuli.
>
> This correction may be shown to depend upon the assumption that "Yes" responses based upon sensory differences, and other "Yes" responses, based, for example, upon expectations, are statistically independent. (p. 16)

In a later, and more formal presentation of the theory, Blackwell (1963) elaborates this thesis by postulating that the total available sensory stimulation produces a sensory effect E. Values of E are assumed to yield a Gaussian distribution. If E exceeds a high threshold, or criterion, E_C, then the subject will respond *Yes*, something more than just background stimulation is present. The background stimulation is assumed to differ so significantly from E_C that the probability of responding *Yes* to background stimulation is vanishingly small.

The subject must always respond either *Yes* or *No*. Thus, when the sensation is smaller than E_C the subject must still decide whether to respond *Yes* or *No*. The probability of responding *Yes* when there is no apparent sensory effect, when the amount of stimulation is below the subject's high threshold at E_C, is manipulable and depends upon task demands, or expectations on the part of the subject. When Catch trials are administered, that is, trials where there is no stimulation at all, the stimulation cannot exceed E_C. The probability of responding *Yes* when no sensory stimulation can exceed E_C must be the same as in the case where stimulation produces a value less than E_C. That is, the probability of guessing, and *Yes* cannot depend upon the stimulus presented. Therefore,

$$Pr\{Yes \mid \text{Catch trial}\} = C \qquad (7.3)$$

$$Pr\{No \mid \text{Catch trial}\} = 1 - C \qquad (7.4)$$

The probability of responding *Yes* when a sensory increment is present derives from two sources. As shown in the finite-state diagram in Fig. 7.2 with probability p the sensory effect exceeds E_C and the detection response *Yes* must occur. With probability $1 - p$ the value of E is smaller than E_C and a *Yes* response still occurs as if no increment were present, that is, with probability C. Therefore, on those trials when an increment is presented,

$$Pr\{Yes \mid \text{Increment trial}\} = p + (1 - p)C \qquad (7.5)$$

and,

$$Pr\{No \mid \text{Increment trial}\} = (1 - p)(1 - C). \qquad (7.6)$$

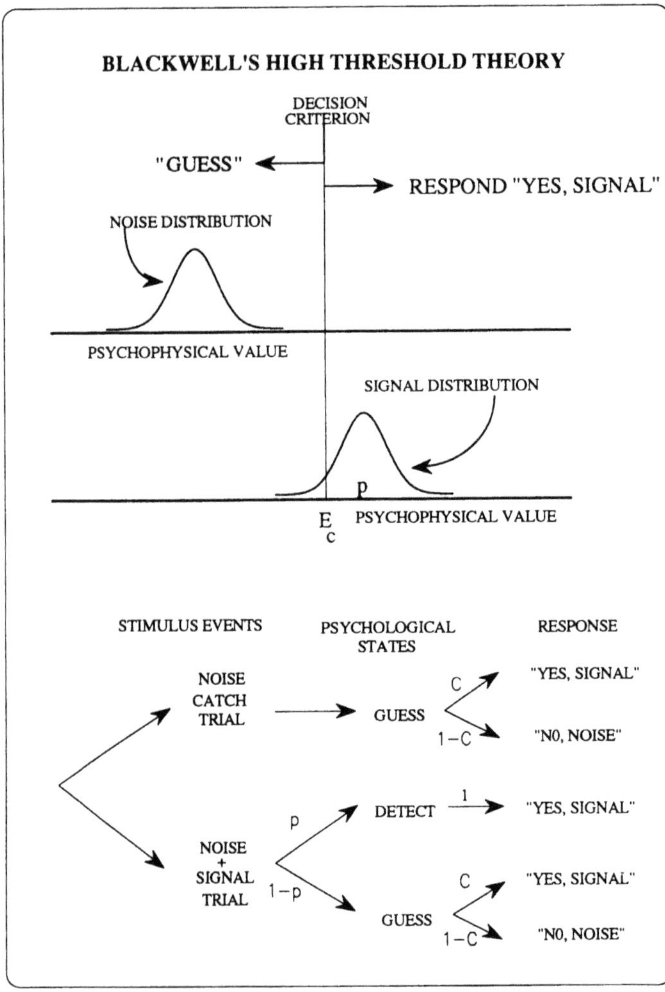

FIG. 7.2. Blackwell's (1963) theory of sensory detection proposes a mixture of two different ways in which errors occur.

Because the unknown probability C is estimated from Catch trials, the unknown probability p can be calculated by substitution of the estimate of C obtained from Equation 7.3 into Equation 7.5. As Blackwell (1953, 1963) observed this yields,

$$p = \frac{Pr\{Yes|\text{Increment trial}\} - Pr\{Yes|\text{Catch trial}\}}{Pr\{No|\text{Catch trial}\}}. \tag{7.7}$$

The value of C is controlled by the design of the experiment. Like the assumptions in correcting a multiple choice test item for guessing, if there are n possible alternatives and one is chosen at random, the probability of choosing the single correct answer equals $1/n$. There is no indication here that the subject manipulates this guessing probability according to any other principle. Continuing his description, Blackwell (1953) remarked,

> The raw data obtained with the forced choice as an indicator response were also corrected by means of equation (1). In this case, C represents the proportion of chance successes and is specified by the number of spatial locations or temporal intervals. (p. 16)
>
> The basic data, therefore, are sets of values of stimulus magnitudes, expressed as brightness contrast, and proportions of correct response after allowance for chance. (p. 16)

These judgments might be influenced by factors inoperable in the simple decision of whether a line length was longer or shorter than a standard. The observer was to *judge*, somehow, whether a certain state of nature, stimulus increment-present or absent, existed. Such decisions might very well determine the fate of a naval vessel if the radar operator at a critical moment missed the appearance of a radar blip signalling the presence of an unseen, hostile enemy.

From these considerations the task faced by the observer and the type of response became two sources of variation in performance. Reviewing some 6 years of research involving a total of 73 subjects who generated 183,000 observations Blackwell's (1953) report concludes with echoes of previous psychophysical findings:

> that optimal psychophysical procedures include the following: (a) the use of forced choice rather than phenomenal report to indicate discrimination; (b) stimulus variation in time rather than in space to provide the basis for forced choice; (c) stimulus magnitudes to be blocked rather than randomized in forced choice; (d) the use of subjects with reasonably extensive experience; and (e) provision for immediate knowledge of the correctness of responses to be given the subjects. (p. 205)
>
> Other topics considered are: (a) systematic differences in threshold data related to data collection procedures; (b) the neural quantum theory of sensory discrimination and other theories which predict the form of threshold data; and (c) the time order of threshold variability. (p. 205)

The fact that sensory thresholds varied as a function of the experimental method confirmed previous suspicions in a new context. The fact that the question asked of the subject produced different measurements of the sensory threshold echoed the earlier findings of Brown and Fernberger, among others, regarding the Interval of Uncertainty. What was new, what infused new vigor into psychophysics and fired the imagination of a generation of psychological researchers, was how these new experimental preparations influenced the unsuspecting subject's judgments under differing motivational instructions.

Other scientists in the Vision Research Laboratory sought to determine how measures of performance changed as a function of the subject's task. In a revolutionary step forward, a step evoking the dormant memory of G. H. Thomson's famous theory of three-choice decisions, these researchers decided that the subject's control over a response must also be part of psychophysical theory. Their contribution brought new life to psychophysics because they focused on how the subject met the task demands imposed by the experimenter-demands that influenced the subject's attitude. Moreover, instead of measuring only discriminability or the difference between stimuli à la Fechner or Thurstone, they included in their model of the decision process an easily measured parameter, controlled by the subject, that changed performance while discriminability remained constant.

The theory sought a convenient method of illustrating how performance varied as the subject's attitude changed. The device constructed to register these changes owes much to Abraham Wald's (1947) famous theory of sequential hypothesis testing and Wald's creation of the *Operating Characteristic*. In psychology a similar device, the *Receiver Operating Characteristic* (ROC), showed the proportion of *Yes* responses when a Signal occurred versus the proportion of *Yes* responses when the Signal did not occur. The relation between these proportions appears in Fig. 7.3. The upper line shows hypothetical proportions from three experimental conditions with possible additional outcomes represented by the straight line passing through the hypothetical experimental results. According to Equations 7.3 and 7.5,

$$Pr\{Yes \mid \text{Increment trial}\} = p + (1 - p)Pr\{Yes \mid \text{Catch trial}\}. \quad (7.8)$$

The intercept value equals two-thirds, which is the unknown value p. The straight line connecting the intercept to the upper right-hand corner must have Slope $1 - p$. The various proportions along the ROC depend upon the subject's

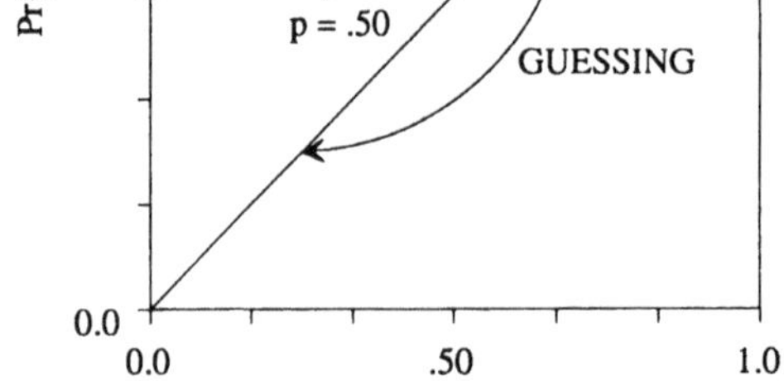

FIG. 7.3. The Receiver Operating Characteristic for Blackwell's High Threshold Theory.

propensity to guess. As p increases the ROC approaches the upper boundary of perfect performance where guessing never occurs. As p decreases the intercept approaches zero where all responses are guesses. Data points on the guessing diagonal, shown in Fig. 7.3, depend only on the subject's propensity to respond *Yes* when guessing.

W. W. PETERSON, T. G. BIRDSALL, AND W. C. FOX

Signal Detection Theory

Signal Detection Theory, as developed by Peterson and Birdsall (1953) and Peterson, Birdsall, and Fox (1954) and explicated in the psychophysical context by Tanner and Swets (1954), postulates a decision mechanism for deciding whether an essentially noisy background also contains an identifiable Signal. According to the theory, a subject faced with only noisy background stimulation will sample a stimulus intensity that is distributed like a Gaussian random variable. The random variation of the intensity of the background noise suggests that intensities are simply a manifestation of noise without any identifiable property. Yet even noise generates a mean. In this case the mean of the Gaussian distribution for the intensity of the background noise depends upon the energy of the background. The greater the intensity the greater the mean.

When a constant Signal is added to the background of Noise the sensory samples derived from the combination of Noise + Signal are again Gaussian, but with a mean equal to the mean of the sum of Noise + Signal intensities. The presentation of a Signal in the context of noise is an extension of the 19th century view of a stimulus as containing a pure stimulus plus a Gaussian distributed amount of error. In accord with the 19th century view the Noise and Noise + Signal Gaussian distributions are separated on an intensity continuum by an amount equal to the difference between their mean intensities.

During an experiment the subject never knows whether the stimulus consists of the background only (Noise alone) or the elusive Signal embedded in background Noise. Tanner and Swets (1954) remarked:

> Thus, if an observer is asked if a signal exists, he is assumed to base his judgment on the quantity of neural activity. He makes an observation and then attempts to decide whether this observation is more representative of N or S + N. His task is, in fact, the task of testing a statistical hypothesis. (p. 403)

But Peterson et al. (1954) noted: "The nature of the 'noise' and 'signal plus noise' must be known to some extent by the observer" (p. 171).

The two Gaussian distributions representing magnitudes of sensation extracted from a sensory sample are illustrated in Fig. 7.4. For convenience the distributions are allowed equal variance. Any measure of neural activity which exceeds a predefined response criterion, C, is accepted as a Signal, and any value

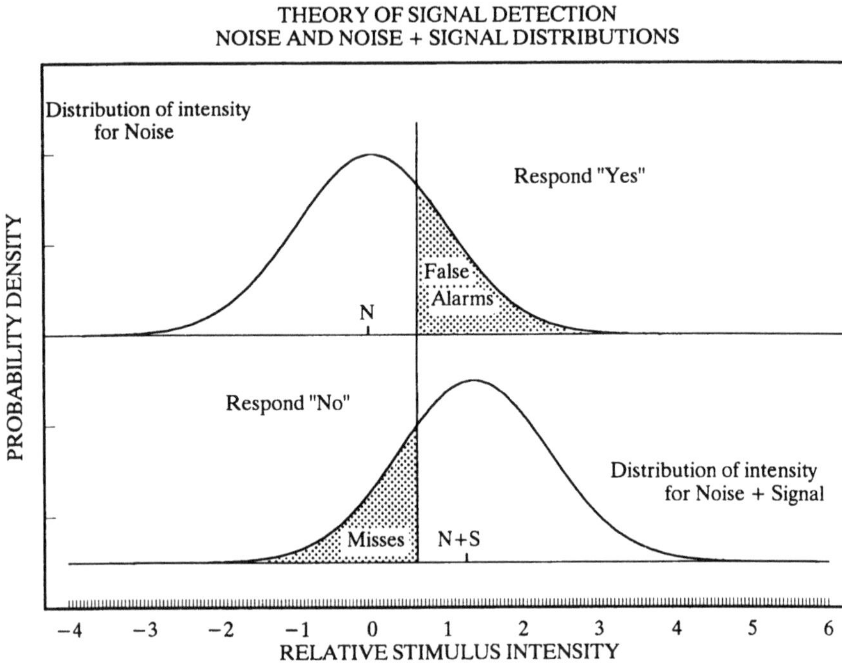

FIG. 7.4. Two Gaussian probability distributions characterize the sensory effects of a background stimulus and the background plus an energy increment.

less than the criterion is accepted as background Noise alone. If the criterion is midway between the mean of the Gaussian distributions, the theory is indistinguishable from Fechner's. The theory, thus far, contains the essential elements of psychophysical discrimination advanced by Fechner so many years earlier.

Unlike Fechner and Thurstone, however, the Theory of Signal Detectability permits the subject to move the criterion to meet the experimenter's performance demands. When the criterion falls midway between the means for Noise and Noise + Signal, the case assumed by Fechner to apply to sensory discrimination, the probability of an error is identical for both stimuli. As the criterion slides to the left, the probability of detecting the presence of the Signal in the Noise background steadily increases. A less intensity increment is required for a Signal to be reported.

This increase in the probability of a *Hit* is at the expense of a larger number of errors in deciding that the Signal is present when only the Noise background occurs. In the bodacious parlance of detection theory this error is called a *False Alarm*. Other response probabilities measure the occurrence of *Misses* and *Correct Rejections*. A Miss occurs when the subject fails to detect the Signal when it is present. When the subject correctly decides that only the Noise background occurred a Correct Rejection is recorded.

Because the four probabilities come from disjoint areas of probability distributions, the probabilities are related in the following way:

$$Pr\{\text{Hit}\} = 1 - Pr\{\text{Miss}\} \tag{7.9}$$

$$Pr\{\text{False Alarm}\} = 1 - Pr\{\text{Correct Rejection}\}. \tag{7.10}$$

Therefore, only two probabilities, say $Pr\{\text{Hit}\}$ and $Pr\{\text{False Alarm}\}$, are necessary to characterize performance for a given Signal and a fixed criterion.

To analyze changes in performance due to attitudinal factors, Tanner and Swets adopted the definition by Peterson and Birdsall (1953) of the Receiver Operating Characteristic (ROC) as the set of points representing these joint probabilities for a fixed difference between the means of the two Gaussian distributions. Tanner and Swets defined d' to be a (standardized) measure of the difference in means divided by the standard deviation of the Gaussian Noise distribution. To each d' there corresponds a unique ROC. That is the ROC is the set of points $\{P_{\text{HIT}}, P_{\text{FA}} \mid d'\}$, where P_{HIT} is the probability of a Hit and P_{FA} is the probability of a False Alarm.

Figure 7.5 presents three ROCs for values of d' equal to zero, 1.0 and 2.0. In the case of no difference between the background Noise and background Noise + Signal, like Catch trials, the probability of a Hit equals the probability of a False Alarm regardless of the position of the criterion. When d' increases discrimina-

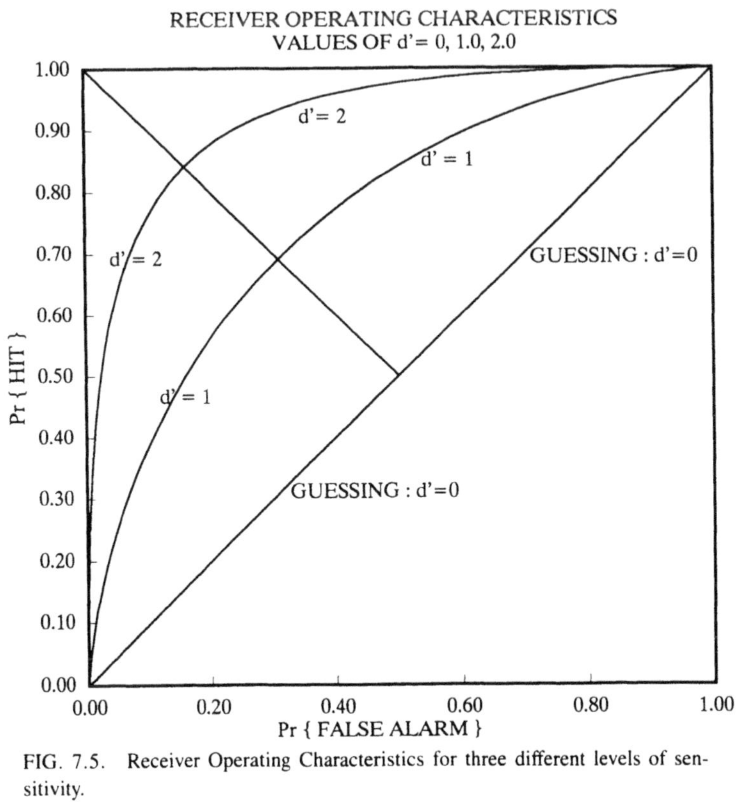

FIG. 7.5. Receiver Operating Characteristics for three different levels of sensitivity.

tion improves. For example, when the value of d' equals 1.0 the discrimination is fairly accurate. A value of d' equal to 2 yields even better discrimination between background Noise (alone) and background Noise + Signal.

As a means of comparing predicted changes in Hit and False Alarm probabilities for Signal Detection Theory and Blackwell's High Threshold Theory, Tanner and Swets showed how the two theories generate distinct ROC predictions. The simplest version of SDT, shown in Fig. 7.4, generates ROCs symmetric with respect to the diagonal defined by the relation, $P_{HIT} = 1 - P_{FA}$. For High Threshold Theory, however, the ROC is a straight line, as in Fig. 7.3, with an intercept equal to the probability that a sensory sample exceeds the subject's high threshold. Such dramatic differences might be expected to reveal which theory accounts more satisfactorily for detection data. Surprisingly the data support both points of view. Blackwell easily found circumstances that produced linear ROCs. Swets, Tanner, Birdsall, Peterson, and many others found curved ROCs to support the predictions of SDT.

W. P. TANNER, J. A. SWETS, AND D. M. GREEN

For example, Tanner, Swets, and Green (1956) reported a classic study of a single subject who discriminated between Noise + Signal (**N** + **S**), and Noise (**N**), under a variety of inducements to bias responses. In Phase I the probabilities of presenting either **N** + **S** or **N** varied from session to session. The lowest probability of presenting **N** + **S** was 0.10 and thus the probability of presenting **N** in that block of trials was 0.90. Figure 7.6 shows the resulting Hit and False Alarm probabilities of 0.335 and 0.09 for this condition.

Other experimental conditions increased the probability of presenting **N** + **S** to values of 0.30, 0.50, 0.70, and 0.90. The resulting probabilities for Hits and False Alarms appear in Fig. 7.6. The response probabilities yielded estimates of d' and C shown in Table 7.1. The average value of d' equals 0.85 and the decision criterion varies from −0.50 to 0.134. An important feature of the theory is that as the response criterion varies from negative to positive values the locus of operating characteristic values forms a smooth curve as shown by the line in Fig. 7.6 for the case of $d' = 0.85$.

In Phase II the authors maintained constant probabilities of presenting either **N** + **S** or **N** alone. To illustrate that motivational factors also influenced the bias toward producing Hits and False Alarms, differing incentives were placed on the production of *Yes* or *No* responses across experimental sessions. The incentive values, ranging from −9 to +9 points, encouraged correct *Yes* or *No* responses and penalized Misses and False Alarms.

These incentive values and response proportions appear in Table 7.1. The average value of d' equals 0.75, somewhat less than for the previous experimental conditions. The decision criteria change to reap benefits available from

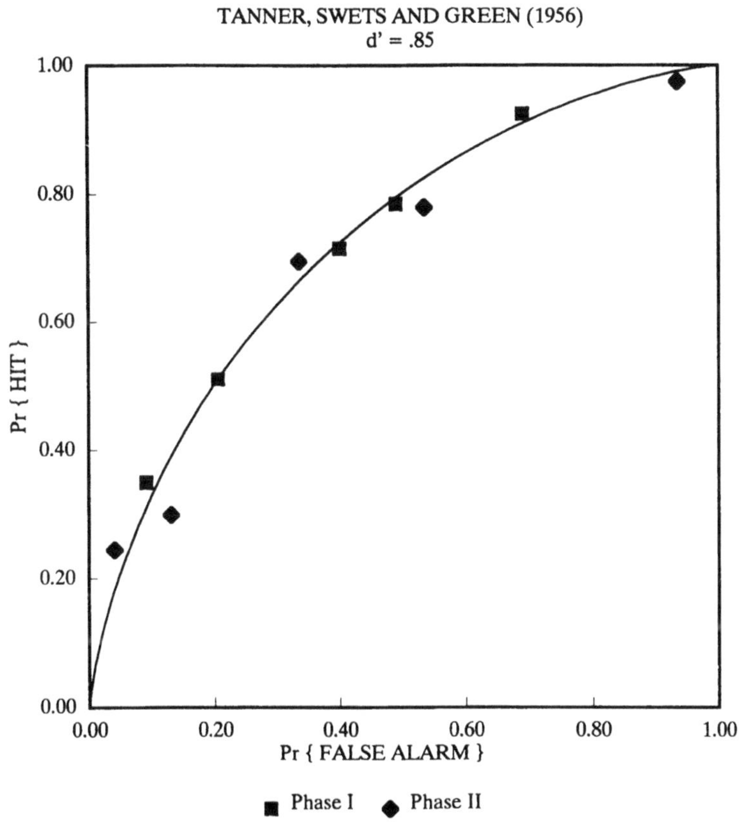

FIG. 7.6. A single ROC provides an excellent account of performance in two detection experiments using fixed stimuli but varying motivational conditions.

changing rewards for Hits and False Alarms. These variations in performance result from the steady change in the response bias reflected in the changes of Hit and False Alarm probabilities shown in Fig. 7.6. In Phase II the data seem in less agreement with the prediction that d', the measure of the difference between the Gaussian distributions, is constant but the influence of payoffs is evident. The ROC for Phase I, which is based on a d' of 0.85 gives a fair account of the Hit and False Alarms probabilities of Phase II.

The gulf between traditional psychophysical methods and Signal Detection Theory widened as the theory found new applications in psychological research. Some remarkable extensions of the theory to recognition (Tanner, 1956, 1961), to auditory frequency analysis (Green, 1960, 1961; Swets, 1963; Swets, Green, & Tanner, 1962; Tanner, 1961), to speech communication (Egan & Clarke, 1956) and to response confidence (Clarke, 1960) motivated a surge of interest in applications of detection theory. When applied appropriately the theory provides

TABLE 7.1
Results and Parameter Estimates

	Values of Payoffs							
	No		*Yes*					
Pr{S}	*CR*	*M*	*H*	*FA*	*Pr{FA}*	*Pr{H}*	*d'*	*C*
				PHASE I				
.10	1	−1	1	−1	.090	.335	0.87	1.34
.30	1	−1	1	−1	.205	.510	0.85	0.82
.50	1	−1	1	−1	.400	.715	0.82	0.25
.70	1	−1	1	−1	.490	.785	0.82	0.03
.90	1	−1	1	−1	.690	.925	0.91	−0.50
				PHASE II				
.50	9	−9	1	−1	.040	.245	1.06	1.75
.50	2	−2	1	−1	.130	.300	0.60	1.13
.50	1	−1	1	−1	.335	.695	0.94	0.35
.50	1	−1	2	−2	.535	.780	0.68	−0.09
.50	1	−1	9	−9	.935	.975	0.45	−1.51

striking measures of a subject's response to an experimenter's performance demands.

This phenomenal success spawned numerous extensions of the theory to new, more cognitive, domains far removed from traditional psychophysics. In some instances the applications became an odd mixture of an overzealous pursuit of a new technique for data analysis and a solid misunderstanding of how the mathematical model applies to the psychological problem under investigation. After the initial successes of SDT, these new applications appear as poor substitutes for the development of new mathematical models designed to meet new needs.

Not every experiment resulted in wholesome support for the expectations of detection theorists. In some experiments the value of the response criterion changed for no apparent reason, as if the estimate of the d' and the criterion changed due to an unknown influence on both parameters. The speed of response and/or accuracy demands placed upon the subject also produced dramatic changes in the value of d' and C. The most troubling results came from experiments that measured both response frequency and response time. But, with no mechanism to explain how the subject's response time could covary with response proportion, and many new experiments based on response proportions leading experimenters into new applications, there was little interest in disturbing the emerging picture of the influence of response bias on response probability by introducing the bugaboo of response time.

8 Discrimination and Response Strategy

R. DUNCAN LUCE

Low Threshold Theory

Signal Detection Theory gathered added support as increasing numbers of experiments generated the predicted curved shape of the Receiver Operating Characteristic. This success, at the obvious expense of Blackwell's High Threshold Theory ROC predictions, suggested that there was no sensory threshold at all, only a response criterion that the subject manipulated to meet the performance demands. The collapse of threshold based theories appeared immanent.

To demonstrate further failures of threshold based theory Swets, Tanner, and Birdsall (1955) noted that in a forced-choice experiment, when sensory excitation is below threshold, a subject forced to respond must guess among the possible response alternatives. If the subject is given a second opportunity to guess among the remaining response alternatives the second guess should also show no discrimination. If the second response proportions show some deviation from chance guessing then the threshold theory cannot be correct because some suprathreshold experience is necessary to guide the second choice.

Drawing on data from Swets, Tanner, and Birdsall (1955), Swets (1961) noted that

> In the forced-choice procedure, four temporal intervals were defined on each trial, exactly one of which contained the signal. The signal was a small spot of light projected briefly on a large, uniformly illuminated background. Ordinarily, the observer simply chooses the interval he believes most likely to have contained the signal. In this experiment the observer made a second choice as well as a first. (p. 177)

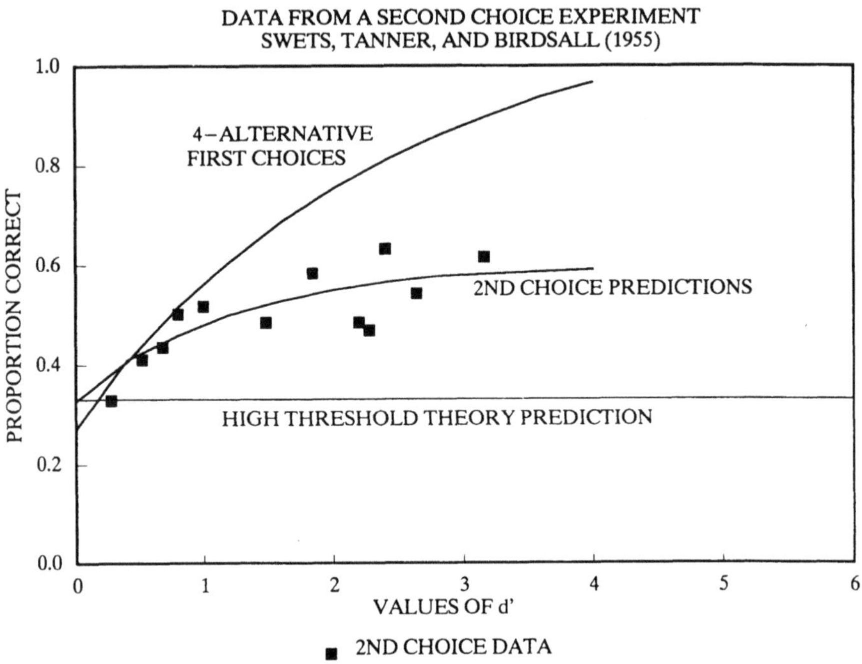

FIG. 8.1. The results obtained in a second-choice experiment, shown with the prediction from detection theory. Data from Swets, Tanner, and Birdsall (1955).

The results are shown in Fig. 2 [Fig. 8.1]. The top curve is the theoretical function relating the proportion of correct first choices to d′; the lower curve is the theoretical relation of the proportion of correct second choices to d′. The points on the graph represent the proportions of correct second choices obtained by experiment. They are plotted at the value of d′ corresponding to the observed proportion of correct first choices. (p. 171)

It may be seen that the data points are fitted well by the theoretical curve. The rather considerable variability can be attributed to the fact that each point is based on less than 100 observations. In spite of the variability, it is clear that the points deviate significantly from the horizontal line. The horizontal line may be taken as a baseline; it assumes a sensory threshold such that it is exceeded only on a negligible proportion of the trials when noise alone is presented. Should such a threshold exist, the second choice would be correct only by chance. The data indicate that the observer is capable of ordering values of sensory excitation well below this point. Two sensory thresholds are shown in the insert at the lower right in Fig 2 [Fig. 8.1]. The threshold on the right, at three standard deviations from the mean of the noise distribution, corresponds to the horizontal dashed line in the upper part of the figure. The data indicate that, were a threshold to exist, it would have to be at least as low as the left hand threshold, at approximately the mean of the noise distribution. (p. 171)

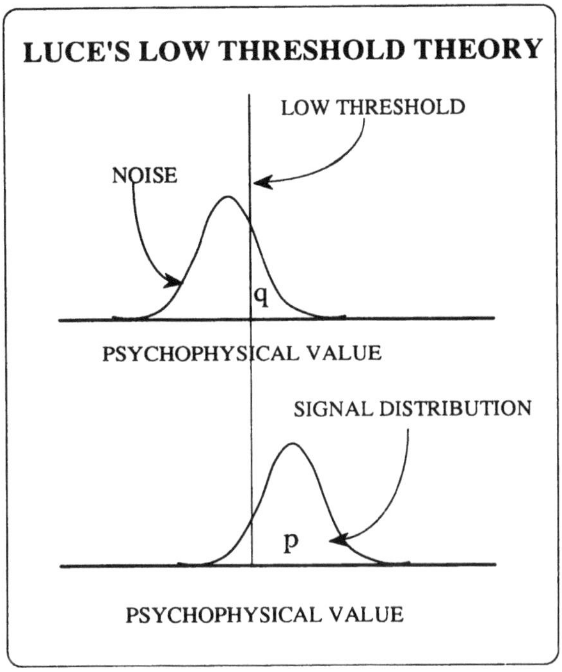

FIG. 8.2. The hypothesized probability distributions for sensory events given a low threshold as postulated by Luce (1960).

Although the results degrade High Threshold Theory they are not nearly so damaging to the existence of a sensory threshold. Luce (1960, 1963), for example, supposed that Noise sometimes produces an above-threshold sensation. This "low threshold" is exceeded by the Noise stimulus with probability q and by the Signal with probability $p>q$, as shown in Fig. 8.2. To complete the theory Luce also assumed that subjects adopted response strategies; sometimes biased to avoid, and sometimes biased to risk, False Alarms. Because the response strategy in no way depends upon the stimulus this assumption characterizes a true form of response bias independent of the psychophysical stimuli. The theory is complicated by needing four parameters to define various stimulus-dependent response probabilities. Nevertheless, Luce ingeniously simplified this complex set of relationships by relating Hit and False Alarm probabilities to each other in terms of observable response proportions.

The equations for Hit and False Alarm probabilities depend on the response strategy adopted by the subject to meet the experimenter's demands. Figure 8.3 shows the stimulus-response probability tree when the subject is biased to avoid False Alarms. The response *No* is always given to a below-threshold event, thereby eliminating the possibility of False Alarms. An above-threshold event,

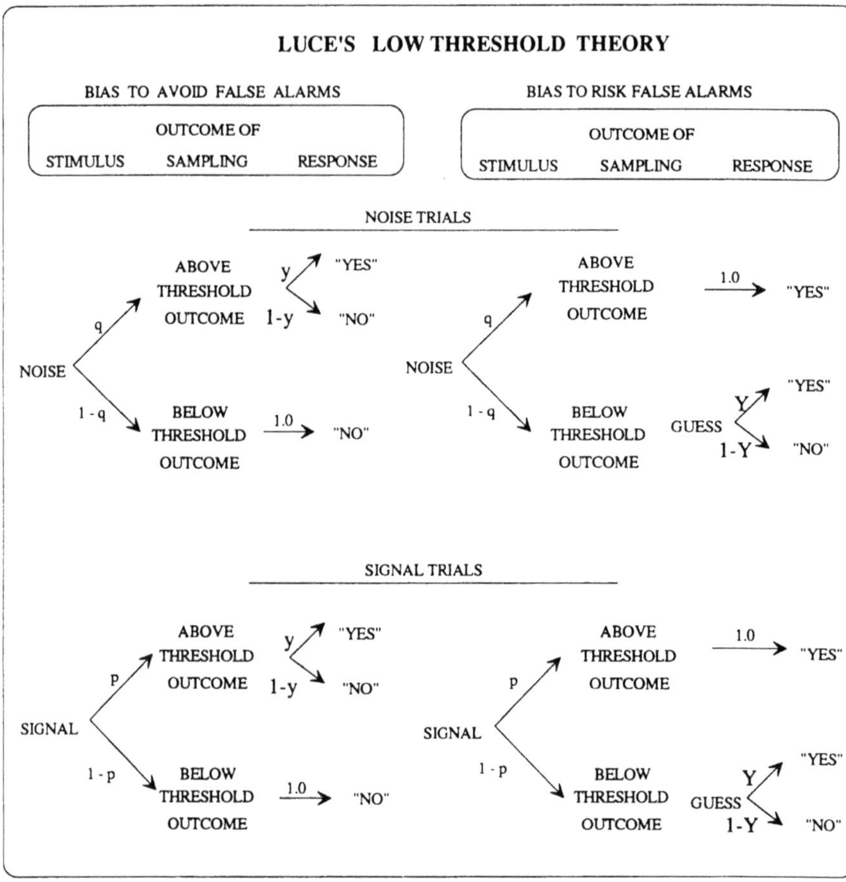

FIG. 8.3. Two modes of bias generate distinct probability trees. At the left a bias to avoid False Alarms leads to the response *No* (no Signal present) whenever a subthreshold event occurs. At the right a bias to risk False Alarms leads to a probability of responding *Yes* (Signal is present) whenever a suprathreshold event occurs.

generated either by the Signal or Noise stimuli, produces a *No* response with probability $1 - y$ and, a *Yes* response with probability y. The probability cannot depend upon the stimulus, for the subject cannot know which stimulus produced the above-threshold sensations.

The probability of responding *Yes* (Signal present) equals the probability that the Signal is above threshold, that is p times the probability of guessing the interval correctly. Therefore, the probability of a Hit is,

$$Pr\{\text{Hit}\} = py \tag{8.1}$$

where $y = 1/n$ is the probability of guessing correctly given n possible response alternatives.

The probability of a False Alarm equals the probability that the Noise exceeds the low threshold times the probability of correctly guessing the temporal interval containing the Signal. That is,

$$Pr\{\text{False Alarm}\} = qy. \tag{8.2}$$

The Hit probability is seen to be a function of the False Alarm probability,

$$Pr\{\text{Hit}\} = \left(\frac{p}{q}\right) Pr\{\text{False Alarm}\}. \tag{8.3}$$

Thus, under the influence of instructions motivating the avoidance of False Alarms the relation between Hits and False Alarms is linear with slope $(p/q) > 1$. This *Isosensitivity Function* is shown in Fig. 8.4 as the lower segment of the Receiver Operating Characteristic.

The upper ROC segment derives from instances when the subject, through instructions, payoffs, or whims, risks False Alarms. Because the stimuli are constant the probabilities that Signal or Noise stimuli exceed the low threshold remain unchanged at p and q, respectively. However, the probability of responding *Yes* when a below-threshold event occurs (regardless of whether the event is caused by Noise or Signal) now increases to Y from zero. Furthermore, consistent with the idea of a bias toward risking False Alarms, the subject always responds *Yes* to an above-threshold event regardless of whether Noise or Signal generate it. These various probabilities are shown in the probability tree of Fig. 8.3. For this form of bias the probability of a Hit is

$$Pr\{\text{Hit}\} = p + (1 - p)Y, \tag{8.4}$$

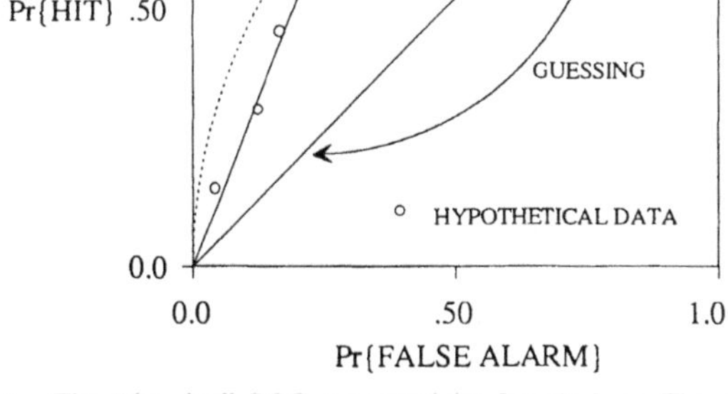

FIG. 8.4. The "dog leg" ROC generated by Luce's Low Threshold Theory represents performance drawn from two different response strategies.

and the probability of a False Alarm is

$$Pr\{\text{False Alarm}\} = q + (1 - q)Y. \tag{8.5}$$

Again the two probabilities are related through the common parameter Y. Removing this parameter shows that

$$Pr\{\text{Hit}\} = \frac{1 - p}{1 - q} Pr\{\text{False Alarm}\} + \frac{p - q}{1 - q}. \tag{8.6}$$

Thus, under a bias to risk False Alarms the Hit and False Alarm probabilities are linearly related with a slope equal to one minus the intercept. This function is the upper portion of the Receiver Operating Characteristic shown in Fig. 8.4. This linear function intersects the ROC for avoiding False Alarms at the common point, (q, p), thereby facilitating the estimation of the unknown probabilities of exceeding the low threshold. Finally, when $q = 0$ this ROC is identical to that of High Threshold Theory.

In forced-choice experiments using n intervals, the values of y and Y are equal and fixed by the choice of the number of temporal intervals. However, these relations make obvious the fact that when the bias to produce Signal responses is manipulated by payoffs or other instructions to the subjects, the values of y and Y may change due to situational factors evaluated by the subject. In such experiments the Signal and Noise remain constant, thus q and p remain constant, but Hit and False Alarm probabilities are expected to slide along the two branches of the ROC. The hypothetical data in Fig. 8.4 are characteristic of data reported by Luce as evidence that linear, rather than curved, ROCs best describe psychophysical performance. The dotted curved line in Fig. 8.4 illustrates a prediction of Signal Detection Theory assuming that the estimate of d' obtained from unbiased performance should apply to all other predictions. The two theories produce quite distinct predictions.

RICHARD C. ATKINSON

Variable Sensitivity Theory

A quite psychological theory occurred when Atkinson (1963) united the principles of Stimulus Sampling Learning Theory (Estes, 1950) with the ideas of psychophysical discrimination. For Atkinson abstract stimulus elements create a sensory pattern characteristic of each stimulus, but some patterns form a set, S*, associated only with background stimulation. When a particular stimulus is presented the subject samples a pattern. "On every trial a single pattern is sampled from the background set S* and simultaneously one of the sensory patterns may or may not be activated" (Atkinson, 1963, p. 93). If a stimulus pattern is sampled then the stimulus is identified correctly. However, if only a background pattern is sampled the subject guesses about which stimulus oc-

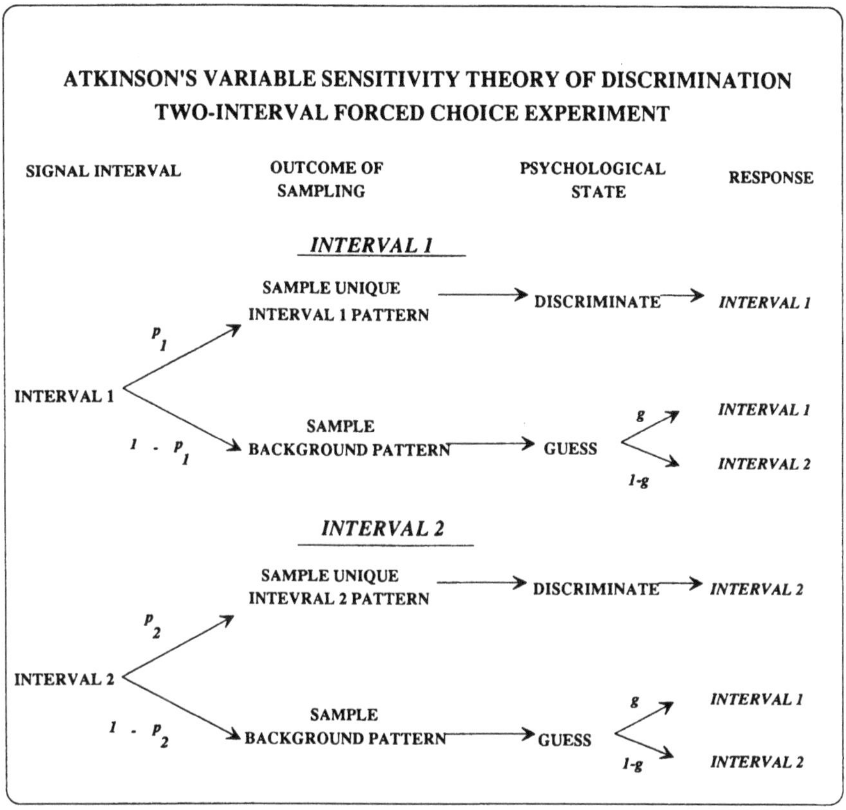

FIG. 8.5. The probability trees for Variable Sensitivity Theory show the events in the stimulus to response path.

curred. The theory extends the sampling theory proposed first by Thomson (1920) by adding the assumption that there are some neutral (i.e., background) elements (LaBerge, 1959) common to all stimuli.

The theory also focused attention on the role of response reinforcement principles in the development of a subject's response bias. In particular, the theory postulated that if a sensory pattern was not activated then the response that occurred was conditioned to the background pattern sampled from S*. When this background pattern was again sampled the guessed response would be based on the previous conditioning. The model for conditioning responses to the background patterns closely followed the assumptions of Bush and Mosteller's (1955) linear learning model. Thus the theory introduced into psychological discrimination sequential dependencies (Bertelson, 1961; Atkinson, Carterette, & Kinchla, 1962) already shown to exert strong influences on performance. The theory of trial by trial changes in performance generated analyses of sequences of stimulus-response events more typical of learning experiments.

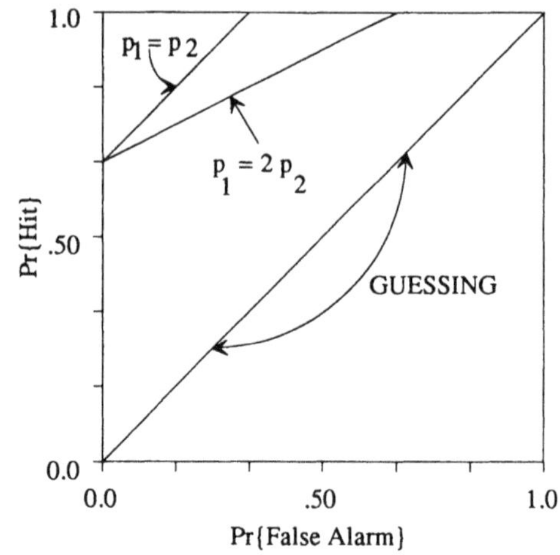

FIG. 8.6. The ROC space shows two unique linear functions characteristic of Variable Sensitivity Theory.

For the two-interval detection experiment modelled by Atkinson (1963), the Hit and False Alarm probabilities derive from the probability tree shown in Fig. 8.5. The probability of a Hit equals the sum of the probabilities at the ends of branches leading to a correct response, that is, when the Signal is identified in the correct Interval. The probability of a Hit given that the Signal occurs in Interval 1 equals p_1, the probability that a pattern unique to the occurrence of a Signal in Interval 1 is sampled, plus the probability that a background element is sampled, $(1 - p_1)$, times g, the probability of guessing that a Signal occurred in Interval 1. That is,

$$Pr\{\text{Hit} \mid \text{Signal in Interval 1}\} = p_1 + (1 - p_1)g. \tag{8.7}$$

If the guessing probability g is known then this function may be plotted in a ROC-like space with the Hit probability on the ordinate and the value of g on the abscissa. The result will appear similar to the ROC of Blackwell's High Threshold Theory. Usually the guessing probability g is not known.

The False Alarm probability equals the probability of sampling only a background pattern, $1 - p_2$, when the Signal occurs in Interval 2, times the probability, g, of guessing that the Signal occurred in Interval 1. Thus the probability of a False Alarm is

$$Pr\{\text{False Alarm} \mid \text{Signal in Interval 2}\} = (1 - p_2)g. \tag{8.8}$$

Atkinson (1963) showed that the theory produces a unique linear ROC. When the subject is biased to always guess Interval 1 the value of g equals 1.0. The Hit probability equals 1.0 and the False Alarm probability equals $1 - p_2$. This point $(1 - p_2, 1)$ is shown at the top of the ROC space of Fig. 8.6. At the other extreme, when the subject never guesses Interval 1 the value of g equals 0, the

Hit probability equals p_1, and the False Alarm probability equals 0. This outcome, $(0, p_1)$, is shown at the left edge of the ROC space of Fig. 8.6. Relating the Hit and False Alarm probabilities through the common parameter g shows that

$$Pr\{\text{Hit}|\text{Signal in Interval 1}\} = p_1 + \left[\frac{1 - p_1}{1 - p_2}\right] Pr\{\text{False Alarm}|\text{Signal in Interval 2}\}. \quad (8.9)$$

If $p_1 = p_2 = p$ then a Signal in either Interval 1 or 2 is just as likely to produce unique stimulus elements. In this case the ROC is a linear function with Slope 1.0 and Intercept p. Atkinson and Kinchla (1965) tested this rather unlikely possibility in a uniquely designed two-interval auditory detection experiment in which the stimuli were created by adding, either in Interval 1 or 2, a sinusoidal Signal to a continuously present band-limited noise background. The subject's task was to identify which interval contained the Signal. A total of eight subjects showed Slopes of the ROC function nearly equal to 1.0. Thus the two different types of stimulus presentation, Signal in Interval 1 or Signal in Interval 2, appeared to produce equal sampling probability values, i.e., $p_1 = p_2 = p$.

The second phase of the experiment contained the unusual feature of presenting no signal in either Interval on half the experimental trials. On a No Signal trial the subject must only guess; but to keep the subject from suspecting that no signals were presented the experimenters misinformed the subject about a nonexistent signal that the subject should have detected.

In this way the experimenters created four unique situations. After a response on a No Signal trial the subject was informed that Interval 1 contained the (nonexistent) Signal with probability $\pi = 0.25$ or $\pi = 0.75$, values that remained fixed during an experimental session. On the Signal trials a Signal did occur in one or another interval and the subject was accurately informed about which interval contained the Signal. The probability that a particular interval contained the Signal changed across sessions from $\gamma = 0.25$ to $\gamma = 0.75$. The experimenters arranged that in two types of experimental session the probability of presenting a Signal in Interval 1 agreed with the misinformation delivered on No-Signal trials, ($\pi = 0.25$, $\gamma = 0.25$, and $\pi = 0.75$, $\gamma = 0.75$). In two other types of experimental session the presentation probabilities and the misinformation conflicted, ($\pi = 0.25$, $\gamma = 0.75$, and $\pi = 0.75$, $\gamma = 0.25$). In all, then, there were four types of experimental sessions and in each session half the trials were No Signal trials.

The first phase of the experiment led to the hypothesis that the values of p_1 and p_2 were equal. To test this hypothesis the data for 14 subjects were combined. The trials from each of the four experimental conditions on which a Signal actually occurred produce estimates of four Hit and False Alarm probabilities. As Fig. 8.7 illustrates, the four pairs of data points fall along an ROC with Slope 1 and Intercept 0.57. These results not only confirm, again, the unlikely ROC with

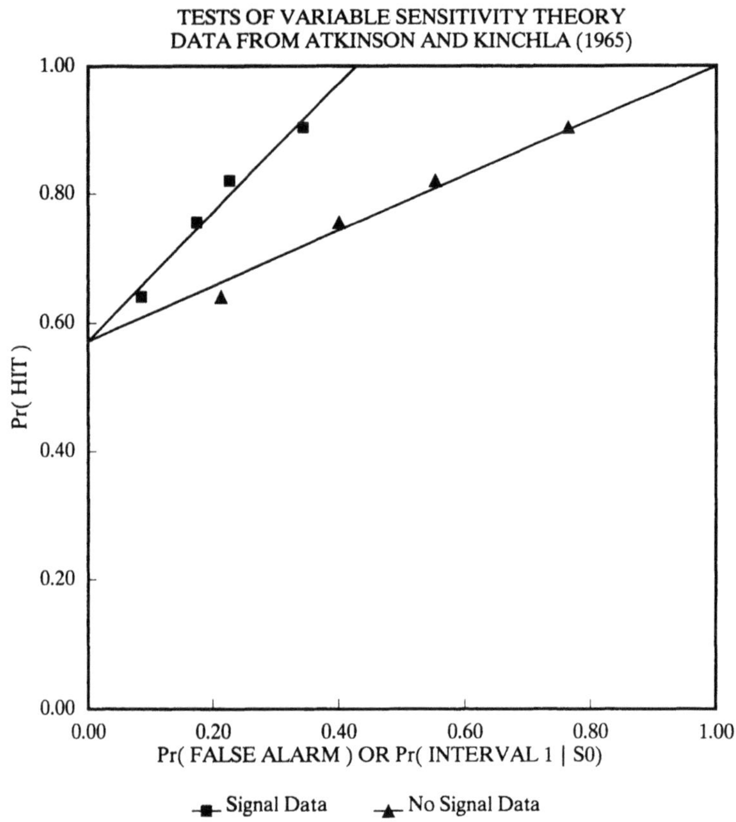

FIG. 8.7. The probability of a Hit as a function of the probability of a False Alarm, and as a function of guessing the Signal occurred in Interval 1 when No Signal occurred.

Slope 1.0 but also agree well with more detailed theoretical predictions that depend upon the theory's assumptions about the effects of response reinforcement.

A unique feature of this experiment is the method of measuring the unknown guessing probability, g. On 50% of the trials a Signal never occurred. However the subject did respond "Signal in Interval 1" even when no signal occurred in either Interval. Although this is not a False Alarm, it must be a guess. Therefore the previously unobservable probability of guessing that a Signal occurred in Interval 1 can be estimated from trials on which no signal occurred at all. Let this response probability be $Pr\{Interval\ 1 \mid \mathrm{S0}\}$. The relation between Hits and the probability of responding Signal in *Interval 1* when no signal was presented at all is,

$$Pr\{\mathrm{Hit}\} = p_1 + (1 - p_1)Pr\{Interval\ 1 \mid \mathrm{S0}\}. \tag{8.10}$$

Atkinson and Kinchla's four experimental conditions were designed to promote changes in guessing probability. The success of this experimental manipulation is evident in Fig. 8.7. The data points, expected to spread along the predicted ROC as a consequence of the effects of reinforcement, cover a broad range of guessing values ranging from 0.21 to 0.77. The data from Signal trials used in the preceding analysis predict the Intercept, p_1, to equal 0.57. Thus the Slope in Equation 8.10, $1 - p_1$, must equal $1 - 0.57$. The four pairs of Hit and guessing proportions agree closely with this parameter-free prediction. All four data points are quite near the predicted operating characteristic.

The diverse set of discriminations in Chaps. 7 and 8 derive from studies of Neural Quantum Theory, High Threshold Theory, Signal Detection Theory, Low Threshold Theory, and Variable Sensitivity Theory. The results come from conscientious subjects responding to similar stimuli in similar experiments conducted by dedicated, capable scientists. The theories prompted by the data are more complex than Fechner's theory, bringing into the psychophysical pantheon response bias, response strategies, and stimulus elements. But few scientists measured both response proportions and response times. The relations between these two common measures of performance remained largely unexplored until Carterette, Friedman, and Cosmides (1965) provided data and Thomas (1971) provided a method for investigating a Signal Detection Theory based ordering of Hit, Miss, False Alarm, and Correct Rejection median response times.

SUMMARY

Swift progress in understanding mental phenomena occurred when the measurement of sensory events became the focus of psychophysical inquiry. Fechner's theory of discrimination created the first bridge between mental theory and the measurement of subjective experience. Yet Fechner's theory stumbled at the threshold to measuring the inner, hidden psychological values that are of primary interest in psychology. Not until the extraordinary shift in view introduced by Thurstone did the focus of measurement turn from sensitivity to stimulus differences. The change in emphasis provided psychologists with a practical tool for the measurement of psychological values—a tool that applied to stimuli without physically measurable characteristics.

This new measurement tool came head to head with the formidable chore of measuring elusive affective values. Although Fechner initiated experimental aesthetics as a psychophysical approach to the measurement of these difficult psychological quantities, little progress occurred until the Law of Comparative Judgment suggested a method for measuring differences between these obscure feelings. Later, experimental results at variance with the Law of Comparative Judgment pushed the Psychological Ideal into the psychophysical lexicon. Further research revealed the influence of motivation. Then Response Bias enlarged the psychophysical vocabulary.

New experimental methods complemented the development of psychophysical theories. No routine application of dubious models of chance performance brought better understanding of the subject's role in the process of discrimination. Instead, new mathematical models of the judgment process struggled with new demands. The student familiar with classical statistical methods will see in Fechner's theory the hypothesis-testing ideas common to later linear statistical models. But the mind is not a garden where the linear models of plant growth flourish. The problems of measuring mental phenomena are quite unique to psychology. The hidden elements of the mind are best exposed by crafting mathematical tools to measure psychological phenomena.

Data supporting the many theories housed in the psychophysical pantheon and the experiments suggested by these newly developed theories form a remarkable sequence of scientific advances. The various corelations between the three response measures, proportion, latency, and confidence, suggest that a single mechanism may provide the basis for these many empirical phenomena. Part II introduces a theory that encompasses these fascinating data and the diverse theories that form the basis for understanding mental abilities.

II

WAVE THEORY

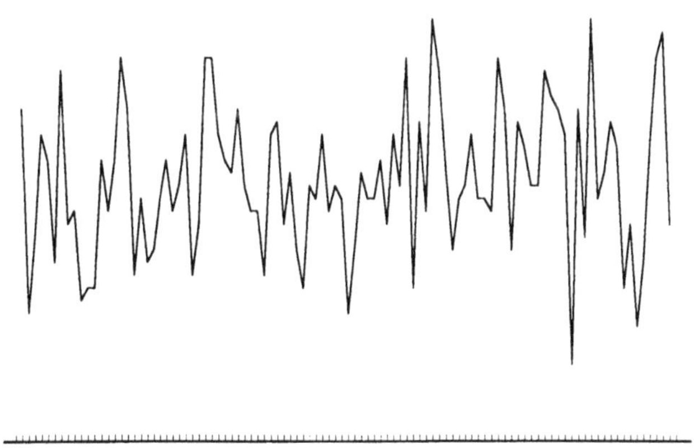

Introduction

The first psychophysical theory portrayed the brain as a primitive measuring device. The device labored to measure a pure stimulus but, in the process, accumulated error through a reactive sequence of thoughtless events. New experimental findings pressed this theoretical view for improvements until a succession of new theories fostered accounts of an ever broadening range of empirical discoveries. Theory and experiment stimulated each other's development, each acting as a springboard for the advancement of the other. Through these advances the ideas of threshold, discrimination, sensitivity, comparative difference, response bias, and ideal entered the psychophysical lexicon. These psychophysical concepts guide the development of a new theory of psychological judgment that exposes the role of these factors in experiments on mental abilities.

The new theory, Wave Theory, is composed of two parts. First, Wave Discrimination Theory develops the theory of comparative bipolar judgments such as *greater-lesser*, *larger-smaller*, *heavier-lighter*, or in general, *more* or *less*. The second component, Wave Similarity Theory, describes the other form of comparative judgment discovered during the preceding 150 years of psychological research—judgments based on a particular example, or a matching to a sample or ideal—such judgments as *same or different*.

Each theory shows how a stimulus is turned into a response. Each theory postulates that sequential comparisons between electrical waveforms are the genesis of mental judgments. Each theory relates properties of these electrical signals to key elements of previous psychophysical theories. In an important departure from the assumptions about error that shrouded earlier theories in a veil of probability density, new methods of analysis no longer require probability density functions as a basis for discriminative judgments.

Wave Discrimination Theory introduces a new within-trial architecture for discriminative judgments. In this regard, the distinction between input and output sides of the stimulus-response path remains essential, but the processes that intervene between input and output constitute much of the theory. Analysis of these processes shows how stimulus difference affects both response probabilities and response times. Because both measures of performance derive from a single theory, it is possible to derive testable predictions relating response time to probability. Thus, early suggestions of a relationship between these two primary measures of performance receive a formal treatment. As several investigators suspected, more than just stimulus difference must enter the theory of performance. The subject's response bias and responsiveness, that is, the willingness to generate responses, dramatically alter response time and probability. Wave Theory analyzes the role of these forces on performance and provides methods for their measurement.

Further analysis shows how Weber's, Fechner's, and Stevens' Laws arise from fundamental quantum forces, thereby removing some of the mystery surrounding their frequent appearance. Weber's Constant proves to be a necessary consequence of wave comparison, and new light illuminates much of Thurstone's work. Logistic Response Functions arise from the mechanism that uses wave differences to control behavior. These theoretical discoveries, supported by experimental data described in Part I and extended by newer research, create a single structure to house the apparently contradictory theories of Weber, Fechner, Thurstone, Helson, Stevens, and Swets and the finite state theories of Atkinson, Blackwell, and Luce.

Resolving old conflicts between theories facilitates new theoretical progress. Wave Discrimination Theory predicts response probabilities, response times, response confidences, and the existence of empirical laws that relate these measures of performance. The Psychological Function that relates feeling to sensation is created. Extraordinarily simple methods for measuring feeling remove a major impediment to the study of those unobservable psychological objects whose magnitudes are the energy of mental life.

Wave Similarity Theory develops a comparison mechanism for the only other form of judgment to be discovered in nearly 150 years of psychophysical experimentation, equality judgments. Urban's (1908) behavioristic approach to equal judgments sought to capitalize on Fechner's observation that as a comparison stimulus neared a standard an interval of uncertainty appeared. Later von Békésy (1930) created Neural Quantum Theory to account for judgments of equal loudness. Eventually S. S. Stevens, Volkmann, and Morgan adopted von Békésy's theory. These and other theories of equality treat the mechanism for judgments of *sameness* in a manner akin to bipolar judgments such as *larger* and *smaller*. As Fernberger (1930), and Bindra, Donderi, and Nashisato (1968) suggested, judgments of *sameness* are not just the flip side of judgments of *difference* but are fundamentally different. Wave Similarity Theory provides a formal theory that explains the results of Cartwright (1941b), predicts new phenomena, and shows how Stevens' sometimes puzzling studies of equality and Coombs' formulation of Ideals are each facets of a single jewel.

9 Wave Discrimination Theory

STIMULUS REPRESENTATION

The art of Caravaggio differs from the art of Titian. To some Caravaggio's *Amor Vincit Omnia* may appear more luminous than Titian's *Venus with a Mirror*, or newer, or more beautiful, or more pleasing. These judgments are of complicated stimuli, yet performed as routinely as judgments about simpler stimuli such as size or weight. The frequent occurrence of such aesthetic judgments and their similarity to judgments of more easily manipulated laboratory stimuli suggests that both forms of judgment derive from a single mechanism; only the input to the judgment process changes. The judgment about the luminance of a Caravaggio masterpiece differs from the judgment of the brightness of a light in the selection of data from the stimuli, not in the mechanism for judgment.

The stimuli are neither the paintings nor lights themselves. Instead, they are the physical paintings or lights as "seen" by the mind making the comparison. The process of seeing involves the neural transduction of the *proximal* stimulus into a *central*, unobservable, psychological stimulus (Brunswik, 1956). Wave Theory posits that within the body all stimuli are electrical signals—the physical manifestation of messages transmitted through the body. Messages travel through the body, but over time. Therefore, signals are components of a complex time-amplitude electrical waveform. Many such signals may be combined and transmitted within a single waveform. Thus, many signals may be present at the same time. From this potentially chaotic ensemble of signals those signals of interest can be filtered from the ensemble any time, at will. That is, the mechanism of

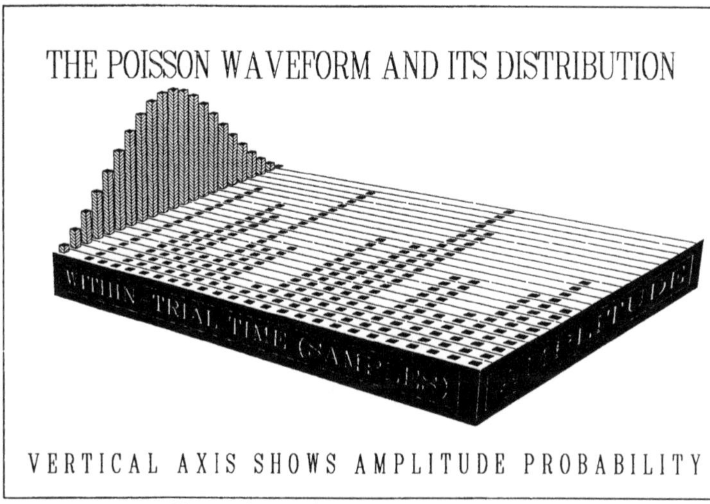

FIG. 9.1. Wave Theory characterizes stimuli as real valued functions of time. In this example the amplitude of a signal is plotted against time. The probability of a particular amplitude derives from a stationary signal of infinite length.

attention can select components of the waveform, *qua stimulus*, at any time.[1]

Successive within-trial values that define a wave for stimulus $\mathbf{S}_A$ are illustrated in Fig. 9.1. These within-trial, sequentially sampled, values of stimulus $\mathbf{S}_A$ are defined as $W_A(\Delta t)$. The amplitudes result from the summation of quantum pulses during a small unit of time. The amplitude value varies over time and defines a sensory string that is a composite of a spatial and temporal dimension. From sample to sample differences arise. The peaks and valleys of the wave illustrate that some wide deviations from the average value of $W_A(\Delta t)$ occur. However, the probability distribution for amplitudes, shown on the ordinate, suggests that these large deviations are quite unlikely.

This figure makes apparent an important difference between the mental process presented here and that of previous psychophysical theories. Here the components of a wave, the quanta giving amplitude to the wave in Fig. 9.1, are the most fundamental elements of analysis. They are direct descendents of Leibnitz's "*petites perceptions*." The smallest sample of the stimulus may contain a number of such elements. These numbers are the amplitudes that define a wave's envelope.[2]

---

[1]"Any time" for the physiological system may be anytime, but the real-time clock of psychological experiments ticks at a fixed rate. Therefore the signals, and the consequences of operations upon them, are only visible at those instants when the experimental clock ticks. It is important to show, as is done by Martingale analysis, that the conditions revealed by a fixed-duration clock are also revealed by a clock that provides time-outs leading to variable durations.

[2]Waves are real variable functions of time. Rather than use the analytical advantages found in their complex variable representation, the mathematical development is also greatly simplified by the application of Martingale Theory (Levy, 1937; Doob, 1940). This elegant method of analysis provides simple solutions to previously intractable problems.

The 19th-century Fechnerian view supposed that within the sensory system Gaussian error caused fluctuations in the perceived value of the stimulus. Such changes were the source of failures to discriminate between signals. This view persisted in one form or another in subsequent theories such as The Law of Comparative Judgment (Thurstone,1927a) and Signal Detectability Theory (Peterson et al., 1954). Thus probability distributions came to represent stimuli, and such properties of the probability distribution as the mean became a measure of the stimulus.

ACCUMULATING WAVE AMPLITUDES

Wave Theory focuses on the sequence of amplitudes passing through the body and, as a consequence, on the statistical properties of the sequence of amplitudes characterizing signals. In this regard, the probabilities on the ordinate of Fig. 9.1 represent the relative frequency of amplitudes for a single signal of infinite duration. At the level of probability distributions this view is indistinguishable from the 19th-century view proposed by Fechner. However, much information exists in a sequence of amplitudes that does not exist in the probability distribution itself. The existence of this added information permits the analysis of mental mechanisms through the application of stochastic processes (Feller, 1966).

Two important properties of stationary random signals tie the fluctuations of a signal in time to the probability distribution for the signal amplitude (Cooper & McGillem, 1986; Gikhman & Skorokhod, 1969; Mason & Zimmerman, 1960). First, the expected value of the signal, that is, amplitude, equals

$$E\,\{\mathbf{W}_A(\Delta t)\} = \Sigma Pr\{\mathbf{W}_A = \mathrm{w}_A\}\mathrm{w}_A.$$

This mean amplitude is also known as the time average of the signal,

$$= \lim_{T\to\infty} \frac{1}{2T} \int_{-T}^{T} \mathbf{W}_A(\Delta t)\Delta t. \tag{9.1}$$

Second, the mean power of the signal (amplitude) equals the variance of the amplitudes plus the mean amplitude squared. That is,

$$\lim_{T\to\infty} \frac{1}{2T} \int_{-T}^{T} \{\mathbf{W}_A(\Delta t)\}^2\Delta t = Var\{\mathbf{W}_A(\Delta t)\} + [E\{\mathbf{W}_A(\Delta t)\}]^2. \tag{9.2}$$

These equations relate the stimulus probability distribution common to previous psychological theories to the stochastic theory of the variations of signals in time that are fundamental to electrical signal analysis. Although the analysis of the signals, per se, is important in its own right, in the present context the analysis of mental processes is facilitated by the use of probability theory rather than electrical signal analysis, although the methods are interchangeable. The method used here leads to the most fundamental results the most simply. As will be seen some phenomena are independent of the exact form of the signal. Other important results on Weber's, Fechner's, and Stevens' Laws appear to depend upon the precise description of waves. In this regard, chap. 10 develops some important ideas concerning a stimulus representation and the signals that define a wave.

Signals are manifestations of the electrochemical processes that bring a mental event into existence. Considering mental events as signals puts signals from peripheral stimuli on a common ground with internal signals that are also waves. This common basis opens to psychophysical analysis mental processes more fundamental than those previously associated with the analysis of psychological stimuli—in particular, the analysis of such concepts as heaviness, tiredness, seriousness, plainness, loveliness, healthiness, sameness, difference, and number (Link, 1990a). These concepts are represented as waves. They vary in amplitude over time. To Wundt they are elements of thought. To Thurstone they are dimensions on which to order stimuli. Here they are considered as waves in their own right.

Not all stimuli result from external stimulation. Ample evidence exists that externally transduced stimuli are often, perhaps always, compared against an internal referent (Wever & Zener, 1928; Link & Heath, 1975). The values of a *referent*, $\boldsymbol{W}_{R}(\Delta t)$, are represented in Fig. 9.2 by the string labelled +. This image extends Fig. 9.1 by showing the amplitudes of both a stimulus and a referent against which it is compared. Although obvious extremes occur, the sequentially sampled values of $\boldsymbol{W}_{A}(\Delta t)$ and $\boldsymbol{W}_{R}(\Delta t)$ are difficult to compare by eye. Judging each wave separately to determine which wave has the largest average amplitude is at best difficult. A more useful comparison results from calculating the momentary difference between values of $\boldsymbol{W}_{A}(\Delta t)$ and $\boldsymbol{W}_{R}(\Delta t)$, sample by sample.

In this regard, an important component of theories of discrimination is comparative difference. For Fechner a comparative difference occurred when a sensation value was judged against a referent equal to the average sensation of the two compared stimuli. For Thurstone stimulus values were compared against each other. In Signal Detection Theory a sensory value was compared against a criterion that varied to meet certain experimenter-imposed task demands. Adaptation Level Theory (Helson, 1959; Restle & Greeno, 1970) proposed that stimuli were compared against a geometric mean of the ensemble of experimental stimuli, whereas Coombs (1950, 1964) postulated the existence of a psychological Ideal against which the psychophysical values of stimuli were compared.

Previous psychophysical theories are distinguished by how a comparative difference contributes to the subject's eventual response. Fechner (1860/1966) assumed one response to occur when a sensation exceeded the average sensation, another when it was less. For Thomson (1920), on the other hand, a response resulted from comparing a sum of activated sensory elements to each of two response thresholds. Thurstone (1927b), in his monumental Law of Comparative Judgment, postulated that a response depended upon comparing a stimulus differ-

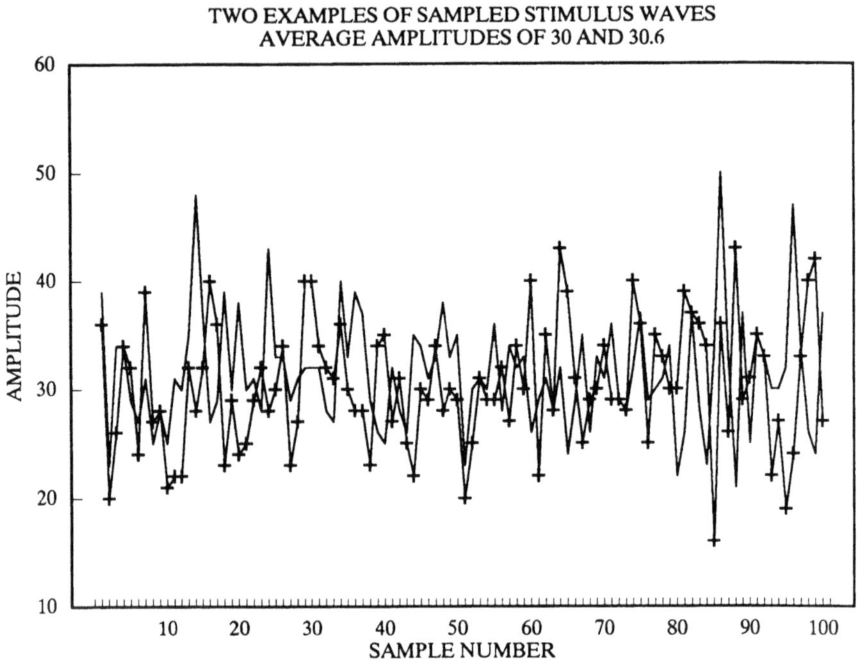

FIG. 9.2. Two different signals with mean amplitudes of 30 and 30.6 generated these waves.

ence against a referent with a value of zero—a theory we now know, from Part I, to be identical to the Fechnerian theory. For Coombs a response occurred by comparing each of two stimulus values against an Ideal and then comparing the difference between the resulting discrepancies to another referent equal to zero. Blackwell (1963) posited the existence of a high threshold that determined whether the subjects guessed or discriminated. In Signal Detection Theory, Tanner and Swets (1954) let a response depend upon the comparison of a sensory value, or a function thereof, to a referent called the *decision criterion*.

In these earlier theories the referent acted in a dual capacity. First as a standard against which a sensory value was compared, and second, as a response threshold that tied the two response alternatives together in an either/or relation. This dual role is particularly evident in Signal Detection Theory where the "decision criterion" is, at once, a standard against which a sensory value is judged and a response threshold. This dual role of a psychophysical referent is not a requirement of psychophysical theory. Instead the two roles can be disambiguated by specifying a separate process for each. In Wave Theory the referent acts only as a standard wave against which the stimulus wave is compared.

Waves are compared by subtraction. At an instant in time the result is a difference between the two momentary amplitudes that represent the stimulus and the referent against which it is compared. This discrepancy oscillates about

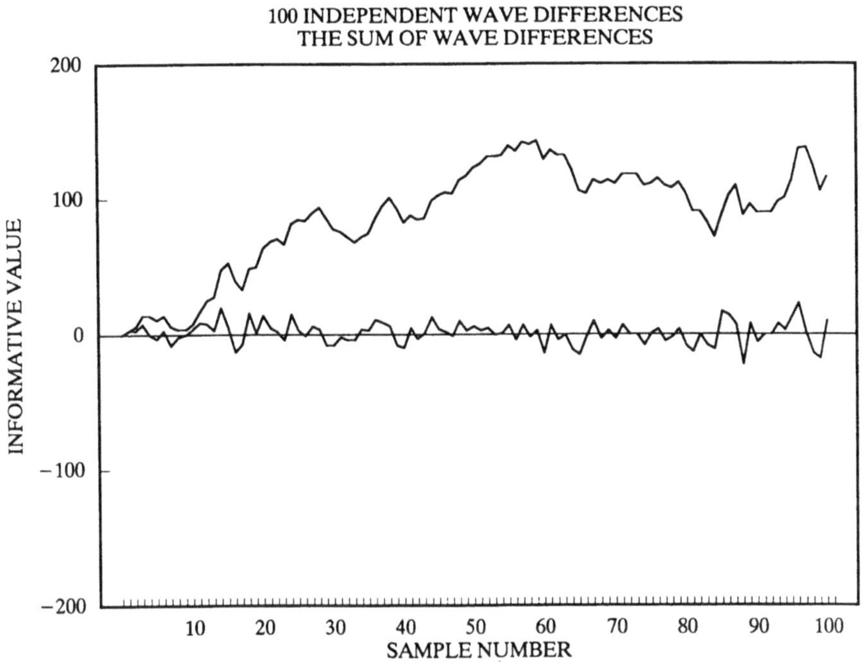

FIG. 9.3. A wave representing the difference between two waves fluctuates about the abscissa. Its cumulative value is a fundamental component of the decision process.

the abscissa of Fig. 9.3. In contrast to some previous psychophysical theories, and as an extension of the notion of comparative difference, these differences occur sequentially during an experimental trial to create a comparative wave. The difference, $\boldsymbol{D}_A(\Delta t) = \boldsymbol{W}_A(\Delta t) - \boldsymbol{W}_R(\Delta t)$, is called an *Informative Value* because the difference contains some, but not complete, information about which response is more likely to occur.

The psychophysical problem of measuring sensitivity is, in part, the problem of how to characterize Informative Values. Both the expected value and variance measure features of the Informative Values. However, the probability distribution for $\boldsymbol{D}_A(\Delta t)$ is dependent for its existence on the sampled values of the waves for $\mathbf{S}_A$ and the referent, $\boldsymbol{W}_R(\Delta t)$. Thus the distribution of Informative Values also depends on the process that samples the signal. To the extent that the sampling process preserves the statistical characteristics of the signal the probability distribution for amplitude difference characterizes stimulus difference.

The time during which wave amplitude differences such as $\boldsymbol{D}_A(\Delta t)$ occur is treated as a time quantum of small value. The very large quanta investigated by Stroud (1955) are too large to fit here. The smaller duration suggested by Geissler (1990) may be more directly relevant. In any case the development of Wave Theory will maintain the notation for Δt whenever new ideas are introduced. Otherwise the time quantum notation will be suppressed but always assumed.

MODELING RESPONSIVENESS

Wave Theory postulates that the process of sampling Informative Values and adding the newly sampled values to the sum continues until the sum exceeds a subject-controlled response threshold. That is, the amplitude of the sum exceeds the subject's preset resistance to respond. For the typical two-response discrimination experiment this resistance is positive, with value A for response R_A, and negative with value $-A$ for response R_{-A}. The total resistance equals $[A-(-A)]$ or $2A$. The ordinate of Fig. 9.4 shows values of $A = 100$ and $-A = -100$.

The tendency to produce responses, called the responsiveness of the subject, is measured by $1/A$. When the value of A is small a response is likely to occur more quickly than when A is large. The subject's responsiveness is high when a small value of A is necessary for a decision to occur. Alternatively, when A is large, a large sum of Informative Values must accrue before a response occurs. In this case the subject's responsiveness is low. When A is small and the rate of sampling the signals is constant the average time to a decision is also small.

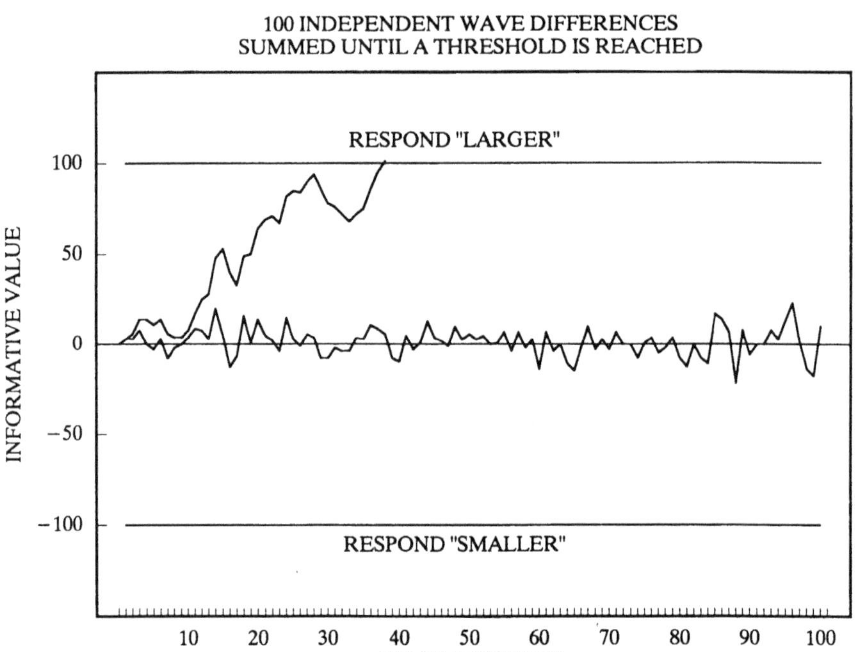

FIG. 9.4. Comparative differences between two random variables are Informative Values that are added sequentially to form a stochastic path to the response threshold at A from the starting position at 0.

By controlling the resistance to respond a subject indirectly controls response time. Experiments discussed later show this ability to be both accurate and selective. Subjects required to respond quickly reduce the value of A, thereby responding in less time, but increasing the probability of an erroneous response. Increases in A increase response time and increase correct response probability. Thus the changes, or tradeoffs, between response time and error probability are mediated by response resistance (Link, 1978b). These surprising changes provide insights into human performance that unify many seemingly disparate empirical results that were previously accounted for by various psychophysical theories.

The major features of this theory are evident in Fig. 9.4. The response threshold values A and $-A$ are shown at the left. The axis on which these values are defined is Informative Value and can be thought of as a dimension of psychological difference. In some previous psychophysical theories, for example Thurstone's Law of Comparative Judgment, comparative difference was the dimension on which stimuli were defined. The time dimension is represented on the abscissa by equal spacings of time units from a clock that uniformly measures the passage of time. Time is considered to be discrete rather than continuous, homogeneous, and linear, so that the discrete clock measurements of time made during an experiment will correspond to the mathematical derivations on which the theory is based.

To summarize, a sampled value of a stimulus wave is compared to a sampled value of the referent wave. The result is an Informative Value. These Informative Values are created during a trial and accumulated over time (Fig. 9.3). The stochastic nature of this accumulation of Informative Values leads to stochastic paths such as the one illustrated in Fig. 9.4.

THE RANDOM WALK

This theory is characterized as a "random walk" for two reasons. First, the process of sampling momentary differences results in a value on the line connecting A and $-A$. The sampling of differences produces momentary changes that appear to be haphazard—a random movement or walk on the line with an average difference that remains constant over time. Second, the accumulation of these differences produces a stochastic path which also projects values onto the dimension of Informative Value at the figure's left. The accumulated value will appear to walk randomly along that axis with a drift equal to the average value of the momentary differences. Viewed on the dimension of Informative Value alone, either process generates changes in Informative Value that increase or decrease erratically, although at every accumulation there is a tendency for the current value to change by the average difference.

In earlier and simpler random walks (Feller, 1957, 1968; Uspensky, 1939) there were no thresholds, and the analogy of an imaginary coin tossed to determine whether a unit step would occur in the positive (heads) or negative (tails) direction provided an ideal example of a random walk. The difference between the number of heads and tails produced by a coin with a tendency toward positive outcomes might generate a path similar to that in Fig. 9.4. In psychology, Estes (1959) introduced a seminal random walk theory of choice behavior in his classic article, "A random walk analysis of two-choice discrimination." For probabilists events in continuous time defined the time axis, but in psychology and statistics the time increments are points at which the process might be seen. To the extent that sampling occurs regularly in time the number of samples is a good approximation of duration.

The accumulation of amplitudes is a sum of random variables. Let $\boldsymbol{D}_{\mathrm{Ai}}$ be the i-th sampled difference between the amplitudes of $\boldsymbol{W}_{\mathrm{Ai}}$ and $\boldsymbol{W}_{\mathrm{Ri}}$. Thus $\boldsymbol{D}_{\mathrm{Ai}}$ is a measure of the difference between signals during the i-th time interval. Let the Informative Value accumulated at the end of the **N**-th time interval be the sum,

$$\boldsymbol{V}_{\mathrm{N}} = \sum_{i=1}^{\mathbf{N}} \boldsymbol{D}_{\mathrm{Ai}} \qquad \mathbf{N} = 1,2,3, \ldots ; i = 1, \ldots, \mathbf{N} \tag{9.3}$$

The value $\boldsymbol{V}_{\mathrm{N}}$ increases and decreases over time depending upon whether the contributions of the $\boldsymbol{D}_{\mathrm{Ai}}$ are positive or negative. The process stops when $\boldsymbol{V}_{\mathrm{N}}$ first equals or exceeds the response threshold at either A or $-A$.

An often overlooked, but important, theoretical point is that when a response occurs the sum of random variables, $\boldsymbol{V}_{\mathrm{N}}$, giving rise to a response is a **binary random variable** constrained to only two possible values. Because the response threshold is reached by computing a sum it is occasionally argued (on the basis of the normal convergence theorem) that the sum will be Gaussian distributed. But for any theory that proposes the existence of response thresholds this is clearly impossible, for the sum must equal one or the other of the two response threshold values. Thus the sum has only two possible values and not the large number needed to satisfy the demands of Gaussian error theory.

In previous theories response proportions from discrimination experiments were a consequence of Gaussian error. Error played a significant role by providing a unit of measure, the standard deviation, for determining how much the standard and comparison differed. **In Wave Theory error is not a basis for failures to discriminate.** Rather, variability of a signal is an inherent part of the way in which signals are transmitted through the nervous system. A network based on the accumulation of differences is, if anything, a system that reduces the influence of error. Indeed, a neurological system operating on the basis of accumulation up to threshold values reduces the influence of randomness on the output of the system by yielding only a binary output—an extreme form of signal shaping.

A second point of interest is that previous theories, such as Thurstone's Law of Comparative Judgment and Signal Detection Theory, postulate the existence of random variables with known probability distributions, typically Gaussian. In Wave Theory a distribution-free approach is taken so that fundamental predictions do not depend upon unknown probability distributions. This distribution-free approach proves a valuable aid in showing how similar psychological phenomena occur across a variety of quite different experimental stimuli and procedures. The fact that virtually all of psychophysical theory can be derived in this distribution-free way says much for the powerful advances in modern probability theory and signal analysis. The fact that certain major results depend upon a specific distribution says much about how the system operates.

Nevertheless, two parameters of the stochastic process of accumulating wave differences play a fundamental role in the analysis and interpretation of the process of comparison. These two parameters measure characteristics of the difference between waves. One parameter, μ equals the average difference between waves and is related to stimulus difference, Δs. This parameter measures the rate of drift of the accumulation of differences toward a response threshold. A positive value of μ suggests that the processes will tend to stop at the positive-(upper) response threshold. A negative value produces differences in the opposite direction that will tend to cause the process to stop at the negative-(lower) response threshold. The following formal analysis shows that the reciprocal of μ, $1/\mu$, is an important determinant of decision time. Since μ depends on Δs the decision time depends on $1/\Delta s$.

The second parameter affects mainly response probability. The parameter θ^* together with the value of the response thresholds A and $-A$ determine the probabilities of exceeding the response thresholds. Typically, as the value of Δs increases, θ^* also increases and the probability of stopping at the upper-response threshold (responding *larger*) increases. This parameter relates the mean, μ, and higher moments of the probability distribution of differences. For example, if the distribution of differences is Gaussian with mean μ and variance σ^2 the value of θ^* is $2\mu/\sigma^2$. Thus, for fixed variance, the value of θ^* increases with increases in μ. For a two-valued random variable, X, for which two values $+1$ and -1 occur with probabilities p and $1 - p$, respectively, the value of $\theta^* = ln[p/(1 - p)]$. The expected value and variance of X are p and $p(1 - p)$, respectively. Letting $\mu = E\{X\}$ and $\sigma^2 = Var\{X\}$ and applying some algebra shows that $\theta^* = 2ln(\mu/\sigma)$. The monotonic increase in θ^* as a function of μ is sufficiently general to characterize many probability distributions of differences. In general, the sign of μ is the same as the sign of θ^*, and when μ equals 0, θ^* also equals 0 (Appendix A).

A consequence of these considerations is that a reciprocal relation exists between response probability and decision time. Generally, when μ increases θ^* increases to yield an increase in the probability of responding *larger*. As Δs increases μ increases but $1/\mu$ decreases to yield a decrease in decision time. Thus, under general conditions that do not depend upon the specific probability distribution for differences, the relation between μ, θ^*, and Δs provides for the often observed reciprocal relation between response probability and response time. These various relations are illustrated by application of Wave Theory to the data of Part I.

10 Response Probabilities and Response Times

LOGISTIC RESPONSE PROBABILITIES

Wave Discrimination Theory postulates that the accumulation of wave differences ceases when the value A or $-A$ is first equalled or exceeded. At moment in time N the accumulated value, V_N, equals the value of a response threshold, either A or $-A$. The analysis in Appendix A shows that the stopped random variable V_N is related to the parameters of the distribution of differences through the expected value of $e^{-\theta^* V_N}$ which must equal 1.0. The fact that there are only two possible values for V_N, either A or $-A$, provides a solution to the problem of determining the probability that either the threshold at A or the threshold at $-A$ will be exceeded first.

When the process of sampling stops, V_N equals either A or $-A$. Let the probability that $V_N = A$ be P_A and the probability that $V_N = -A$ equal $1 - P_A$. Because there are only two values for V_N the expected value, $E\{e^{(-\theta^* V_N)}\}$, expands into only two terms; either $e^{(-\theta^* A)}$ or $e^{(-\theta^*(-A))}$ which occur with probabilities P_A and $(1 - P_A)$, respectively. The sum of these two terms equals 1.0. Putting these various facts together yields,

$$
\begin{aligned}
E\{e^{(-\theta^* V_N)}\} &= (1 - P_A)e^{[-\theta^*(V_N=-A)]} + P_A e^{[-\theta^*(V_N=A)]} \\
&= (1 - P_A)e^{[-\theta^*(-A)]} + P_A e^{[-\theta^* A]} \\
&= P_A(e^{-\theta^* A} - e^{\theta^* A}) + e^{\theta^* A} \\
&= 1.
\end{aligned}
$$

The unknown probability, P_A, that the accumulation of amplitude differences first reaches A, resulting in the response *larger*, follows by solving directly for P_A:

$$P_A = \frac{e^{\theta^* A} - 1}{e^{\theta^* A} - e^{-\theta^* A}}.$$

Some algebra shows that,

$$P_A = \frac{1}{1 + e^{-\theta^* A}}. \tag{10.1}$$

This famous response-probability relation defines a *Logistic Response Function* in the parameter θ^* for a fixed value of A. Because the parameter θ^* increases with increases in stimulus difference this response probability, when plotted against stimulus difference, will appear to trace out a smoothly increasing Logistic Response Function as shown in Fig. 10.1. The derivation of this response probability (Link, 1975, 1978b) is distribution-free. Thus, the particular waves that characterize the signals are not critical to the appearance of the function that is often assumed (Anscombe, 1956; Berkson 1944; Feller, 1939;

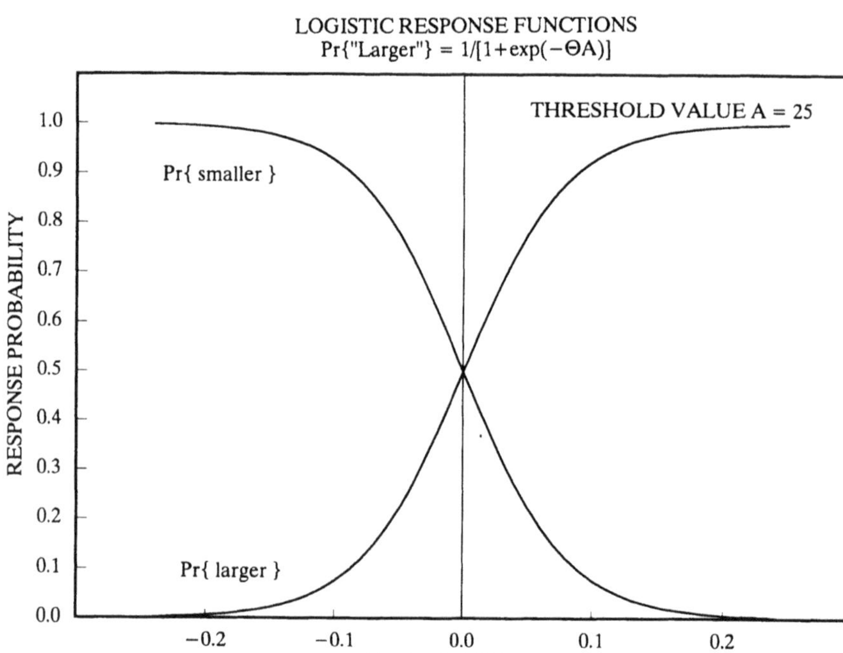

FIG. 10.1. A Psychometric Function with the form of a Logistic distribution function appears similar to a smoothly increasing Gaussian distribution function. For the purpose of illustration the resistance to respond, A, equals 25.

Luce, 1959; Rasch, 1960) to model response performance in psychological and biometric experiments. The frequent appearance of the Logistic Response Function is an empirical clue to how judgments arise from this simple theory of wave comparison. Furthermore, this derivation of the Logistic Response Function is applicable in general, not only to the mental phenomena that are of immediate concern, but to such areas as toxicology where the Logistic Response Function plays a vital role.

The probability of responding *smaller* is the probability that the alternative threshold is reached first. It equals,

$$1 - P_A = \frac{1 - e^{-\theta^* A}}{e^{\theta^* A} - e^{-\theta^* A}}$$

$$= \frac{1}{1 + e^{\theta^* A}}. \qquad (10.2)$$

Figure 10.1 illustrates the shapes of these two remarkable functions. The abscissa increases in value with increases in θ^*, which generally increases with the average difference, μ, between stimulus waves. Thus the greater the difference between two stimuli the larger the values of μ and θ^*. When θ^* equals zero the value of P_A equals 0.50, as may be seen by expanding the exponential part of the equation as a Taylor's series and then letting θ^* approach zero. That is, from Equation 10.1,

$$P_A = \frac{1}{1 + e^{-\theta^* A}}.$$

Expanding the exponential as a Taylor's series yields,

$$= \frac{1}{1 + [1 - (\theta^* A) + (\theta^* A)^2/2! - (\theta^* A)^3/3! + \ldots]}.$$

Because all terms involving θ^* approach zero as θ^* approaches zero, the final result is,

$$P_A = \frac{1}{2}. \qquad (10.3)$$

For a particular value of θ^* the probability of responding *smaller* plus the probability of responding *larger* must sum to one. To express this in another form, let the logistic function for *larger* responses be $L(\theta^*)$. Similarly, let the logistic function for *smaller* responses be $S(\theta^*)$. Then for any value of θ^*, $L(\theta^*) + S(\theta^*) = 1$. Figure 10.1 shows both functions. The probability of responding *larger* appears to be a cumulative probability distribution, but these are not cumulative probabilities at all. Rather, they are choice probabilities that depend upon the comparison stimuli.

ESTIMATING θ^*A

The values of θ^* and A usually are not known in advance, yet their estimation is of special importance because values of θ^* and A provide measures of internal psychological events. As Fechner proposed, the response probabilities may be determined through experimentation by using as an estimate the proportion of times a particular response occurs. Let R_A be the response, such as *larger*, that occurs when the response threshold at A is first exceeded. Let $\mathbf{S}_A$ be a stimulus generating an average difference μ greater than zero so that θ^* is also greater than zero. Then P_A is the probability that response R_A occurs given a presentation of $\mathbf{S}_A$. Estimates of θ^*A come from the relation:

$$ln\ [P_A/(1 - P_A)] = \theta^*A. \tag{10.4}$$

Let R_{-A} be the response, such as *smaller* which occurs when the response threshold at $-A$ is first exceeded. Let $\mathbf{S}_{-A}$ be a stimulus generating an average difference $\mu<0$ so that θ^* is also less than zero, and let P_{-A} be the probability that a presentation of $\mathbf{S}_{-A}$ evokes response R_{-A}. Then an estimate of the unknown value of θ^*A is obtained from the relation,

$$ln\ [P_{-A}/(1 - P_{-A})] = \theta^*A.$$

Finally, to separate θ^* from A, the logarithm $ln(\theta^*A)$ equals $ln(\theta^*) + ln(A)$. This additive model for stimulus controlled and subject controlled effects will come to play an important part in the measurement and analysis of psychological experiments.

The natural logarithm of the ratio of correct-to-error response proportions yields an estimate of the unknown value of θ^*A. For example, Peirce and Jastrow (1884) used as stimuli pressures exerted on the pad of the index finger. The subject judged whether a pattern with an increase, or a decrease, in pressure occurred. The experimental conditions (Part I, chap. 2) suggest that changes in θ^* result from changes in pressure, whereas A, the resistance to respond, remains constant.

Values of θ^*A estimated from the experiment by Peirce and Jastrow (1884) are shown in Fig. 10.2. Quite clearly the data conform well to the hypothesis that the physical stimuli produce linear increases in the values of the discrimination parameter θ^*. What remains unexplained is the "Constant Error" so often discovered in studies of the Psychometric Function. Quite clearly, the value of θ^* reaches zero before the difference from the standard, Δs reaches zero.

RESPONSE TIMES

Having computed the previously unknown response probabilities it is now possible to compute the other important unknown—how many samples of differences

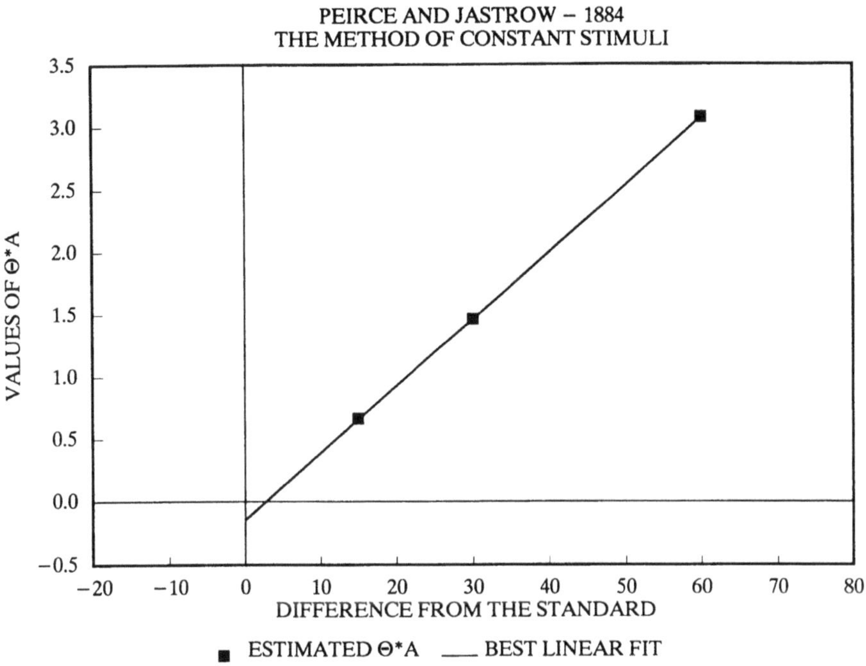

FIG. 10.2. Results from Peirce and Jastrow (1884) and their Method of Constant Stimuli produce estimates of θ^*A that are a linear function of Δs.

need to accumulate before a threshold is exceeded? The number of differences is a random variable, $\boldsymbol{N}$, with an expected value derived in Appendix A. The expected number equals the average response threshold value divided by the average stimulus difference, $E\,\{\boldsymbol{D}_A\}$. That is, letting $\boldsymbol{N}$ be the time to reach a response threshold, the expected value of $\boldsymbol{N}$ is,

$$E\{\boldsymbol{N}\} = \frac{E\{\boldsymbol{V}_N\}}{E\{\boldsymbol{D}_A\}}. \tag{10.5}$$

The numerator of this equation is the average stopping position. The average value depends upon the values generated by the accumulation process when a response threshold is first reached or exceeded. The response threshold acts as a trigger, releasing a fixed response once the threshold is crossed. The crossing may lead to an accumulation somewhat in excess of A, that is, there may be a certain amount of overshooting, as occurs when squeezing a trigger beyond the point necessary to fire a gun. The critical value for the response process is A. With probability P_A the value A is reached before $-A$. With probability $(1 - P_A)$ the value $-A$ is reached first. Thus, the average response threshold is somewhere between $-A$ and A. The average threshold is a probability mixture of stopping at $-A$ or stopping at A,

$$E\{V_N\} = P_A(A) + (1 - P_A)(-A).$$

Therefore the expected stopping time is,

$$= \frac{P_A(A) + (1 - P_A)(-A)}{E\{\boldsymbol{D}_A\}},$$

$$= \frac{A(2P_A - 1)}{E\{\boldsymbol{D}_A\}}.$$

The average Informative Value, $E\{\boldsymbol{D}_A\}$, is μ_A. Thus,

$$E\{N\} = \frac{A(2P_A - 1)}{\mu_A}. \qquad (10.6)$$

These changes in response time as a function of μ_A are illustrated in Fig. 10.3. The abscissa is defined by values of μ which vary from -1.0 to 1.0. As the value of μ approaches zero the average number of samples increases. As the value of μ increases past zero the number of samples steadily drops. Notice that the drop is quite noticeable for deviations from zero that lead to little discernible difference in response probability. As with the calculation of response probabilities

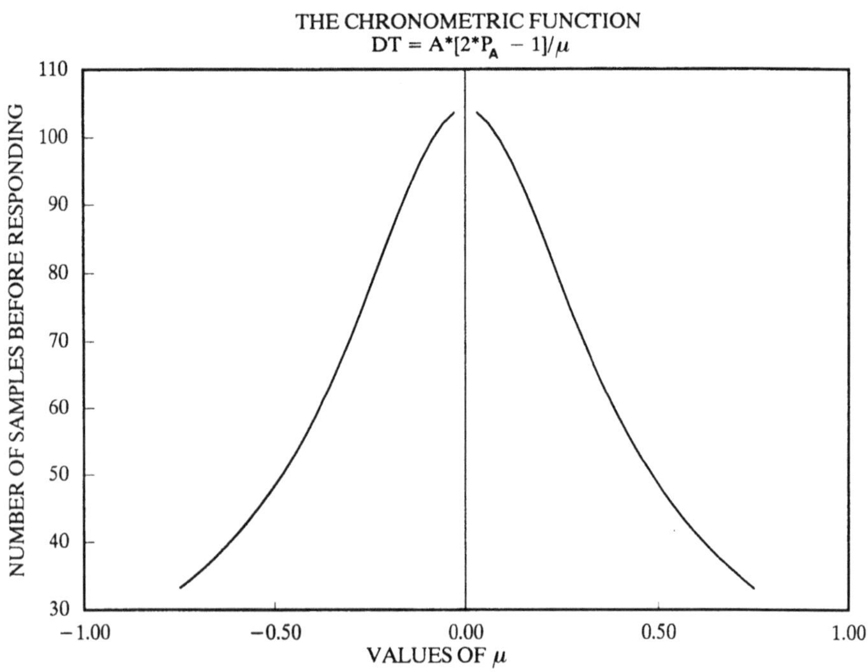

FIG. 10.3. The number of samples needed to exceed a response threshold at the value $A = 25$ is a function of the three parameters θ^*, μ, and A. Values of μ tend to increase as a function of θ^*. The decision time equals the number of samples times a constant equal to the sampling duration.

when $\theta^* = 0$, the case of $\mu = 0$ requires special attention. From Appendix B, and the fact that the average difference between waves is zero, comes the expected number of samples,

$$E\{N\} = \frac{A^2}{Var\{\boldsymbol{D}_A\}} \tag{10.7}$$

where $Var\ \{\boldsymbol{D}_A\}$ is the variance of the difference between waves. Many possibilities exist for selecting a variance to fill in the function in Fig. 10.3. However, without any specified distribution of wave differences the choice is entirely arbitrary. Therefore this point on the function is omitted.

The exact relation between μ_A and θ^* derives from the difference between waves characterizing the stimuli undergoing comparison. Here, for the purpose of illustration the relation is assumed to be linear. The values of P_A are those from Fig. B.1. For response probabilities near 0.50 the decision time is large, reflecting the fact that the average Informative Value, μ_A, and θ^* are near zero. A small value in the numerator of Equations 10.6 or 10.7 generates a long expected stopping time. As the values of μ_A and θ^* deviate from zero the average number of samples decreases. The number of samples is never negative because negative values of μ_A and θ^* occur for probabilities of P_A less than 0.50. Thus the sign of the numerator always matches the sign of the denominator. The result is similar to many Chronometric Functions obtained from psychophysical studies reported in Part I.

The average number of samples to a decision depends upon three parameters, A, μ_A, and θ^*. Though complicated, Equation 10.6 has a simple interpretation. As θ^* increases the probability of responding *larger* also increases and the value of $2P_A - 1$ approaches 1.0. Under fairly regular conditions the values of μ_A tend to increase monotonically as a function of θ^*. Thus, as θ^* increases, μ_A also increases. Because μ_A appears in the denominator of Equation 10.6, increases in μ_A cause decreases in decision time. The decrease is particularly evident for response probabilities near 1.0. Increases in θ^* make little difference in probability but, as many early psychological experimenters discovered, response time still changes dramatically.

For many years psychologists realized that some relation between response probability and response time related the two measures of performance (Welford, 1980). A very simple, testable, distribution-free prediction can be derived from Equations 10.1 and 10.6. Multiplying the numerator and denominator of Equation 10.6 by θ^* provides an empirical law relating the number of accumulated differences and the probability of response R_A. That is,

$$E\{N\} = \frac{(\theta^* A)(2P_A - 1)}{\theta^* \mu_A}, \tag{10.8}$$

$$E\{N\} = \frac{\{ln[P_A/(1 - P_A)]\}(2P_A - 1)}{\theta^* \mu_A}. \tag{10.9}$$

The numerator is estimated from experimentally determined values of P_A, whereas the left-hand side represents the average number of samples when the process stops.

The fixed time to generate and compare two samples of stimulus waves is Δt. For one experimental trial the average number of samples multiplied by the constant unit of time, Δt, defines the average decision time. The decision time is just one component of an observable response time that is a sum of the decision time and all other nondecision components of the observed response time. Labelling the nondecision time $\boldsymbol{K}$ gives the measurable response time as

$$\boldsymbol{RT} = \boldsymbol{N} \times \Delta t + \boldsymbol{K}.$$

To simplify the analysis let Δt be a unit of temporal measurement so that it may be set equal to 1.0. Then the expected response time is,

$$E\{\boldsymbol{RT}\} = \frac{\{ln[P_A/(1 - P_A)](2P_A - 1)\}}{\theta^* \mu_A} + E\{\boldsymbol{K}\}. \tag{10.10}$$

Labelling the term in brackets, { }, as $\boldsymbol{Z}$ makes apparent the linearity of the relation:

$$E\{\boldsymbol{RT}\} = \left\{\frac{1}{\theta^* \mu_A}\right\} \boldsymbol{Z} + E\{\boldsymbol{K}\}. \tag{10.11}$$

This basic relation between response time and response probability, called the ***RT*** versus ***Z*** relation (Link, 1975; Link, 1978a; Link, 1990a; Link & Heath, 1975), relates response time and response choice based on each trial of an experimental session. For a fixed experimental stimulus only changes in the threshold values at A and $-A$ can cause joint changes in response time and response probability. However, for fixed experimental conditions designed to maintain a constant value for A, differences between stimuli do produce changes in θ^* and μ. If the psychophysical function is known then Equation 10.10 is completely determined except for the nondecision component $E\{\boldsymbol{K}\}$. When μ is known but θ^* is not then Equation 10.7 is determined except for A and $E\{\boldsymbol{K}\}$. These relations define empirical laws which relate response time and response probability. Subsequent chapters show them to exist across a variety of stimuli and experimental conditions.

THE ANALYSIS OF KELLOGG'S DATA

These various relations occur in empirical results from carefully run psychophysical experiments. Among the many such experiments Kellogg's (1931) study of brightness discrimination stands out as exceptional for the following reasons. First, the subjects received extensive practice by making some 4,000 comparisons before the data described in Figs. 5.5 and 5.6 were gathered. Second, the experimenter gathered both response probabilities and response times. Third,

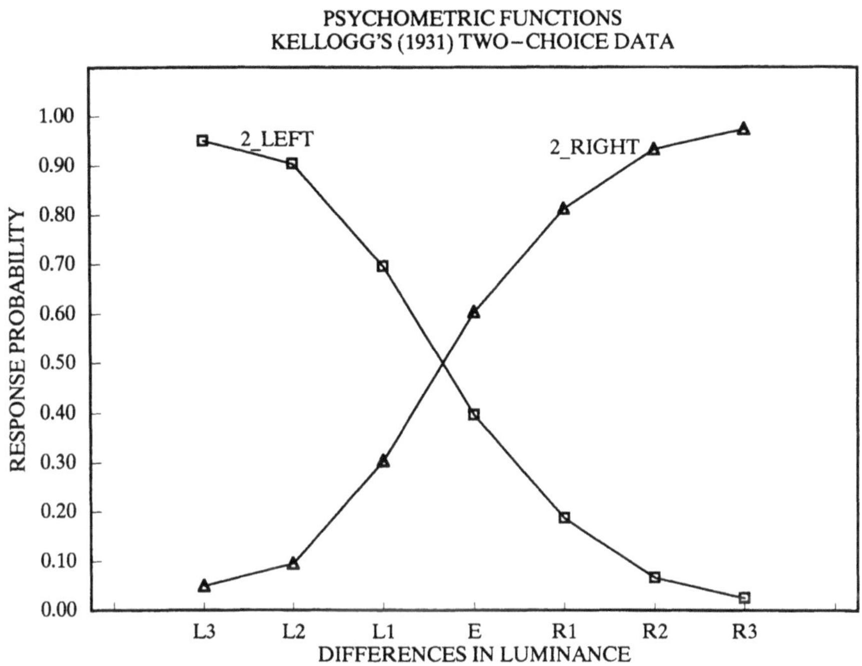

FIG. 10.4. The Psychometric Functions for Kellogg's study of brightness discrimination agree well with theoretical expectations shown in Fig. 10.1.

the experimenter constructed four stimuli, ($\mathbf{S}_1$, $\mathbf{S}_2$, $\mathbf{S}_3$, and $\mathbf{S}_4$), to appear to have equal increments in brightness. Then, symmetric stimuli were created by combining these stimuli to form seven pairs ($\mathbf{S}_1$, $\mathbf{S}_4$), ($\mathbf{S}_2$, $\mathbf{S}_4$), ($\mathbf{S}_3$, $\mathbf{S}_4$), ($\mathbf{S}_4$, $\mathbf{S}_4$), ($\mathbf{S}_4$, $\mathbf{S}_3$), ($\mathbf{S}_4$, $\mathbf{S}_2$), ($\mathbf{S}_4$, $\mathbf{S}_1$). The subject's task was to choose which side of a split disk containing ($\mathbf{S}_i$, $\mathbf{S}_j$) appeared brightest.

Because Kellogg chose stimuli that defined equal brightness intervals, the response proportions can be viewed as a function of steadily increasing stimulus differences as shown on the abscissa of Fig. 10.4. The abscissa indicates the apparent brightness difference (1, 2, 3 or Equal) for the less intense side (Left or Right). The response proportions are taken from Fig. 5.5. The response proportions for *right side brighter* increase steadily to 0.99 from 0.05. The other Psychometric Function defines the corresponding proportions for the response *left side brighter*. When the two brightest stimuli are presented for comparison, ($\mathbf{S}_4$, $\mathbf{S}_4$), the stimuli are apparently not seen as equal. Rather, the right-hand side is judged to be brighter with probability 0.59. Still, the general form of these Psychometric Functions agrees well with Fig. 10.1.

Using Equation 10.4 to estimate values of θ^*A yields the estimates shown in Fig. 10.5. The instructions to subjects and the design of Kellogg's experiment emphasized stability in performance. Under these conditions the response thresh-

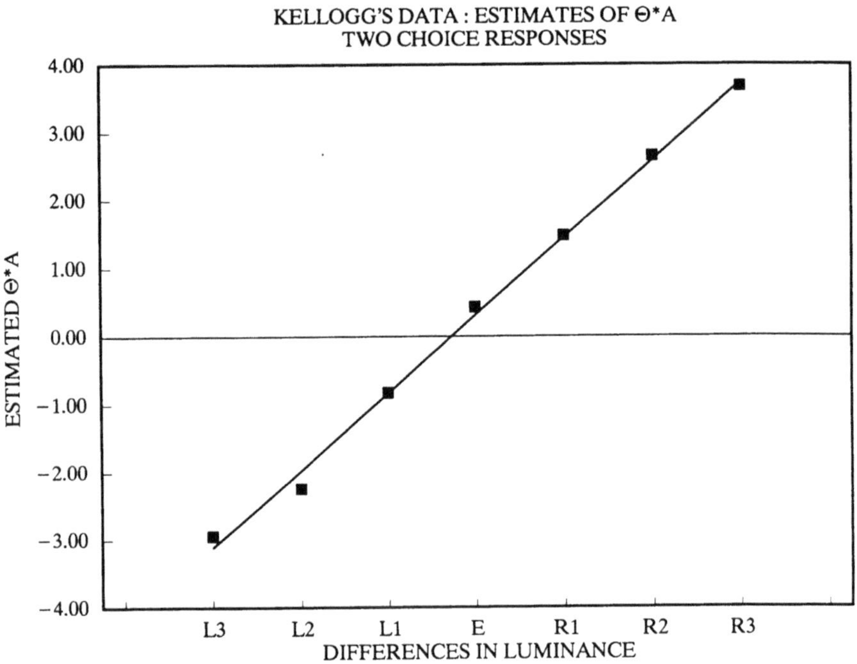

FIG. 10.5. Estimates of θ^*A vary linearly with increases between the left- and right-hand luminances of a split disk.

olds at A and $-A$ may be taken as constant, whereas changes in stimulus luminance produce differences between stimuli that, in turn, produce changes in θ^*. Positive values of θ^*A occur when the probability of responding *right side brighter* exceeds 0.50. Negative values of θ^*A occur for probabilities that are less than 0.50. Estimated values of θ^*A show a linear increase as a function of the luminance difference between the right- and left-hand sides of the split disk. The linear increase in parameter estimates confirms the subjective reports of equal increases in brightness.

Assuming the abscissa to range from -3 to $+3$ luminance units provides a basis for a least squares fit of a linear function to the estimated values of θ^*A. The best fit yields Slope and Intercept values of 0.31 and 1.14, respectively. The best fitting linear function crosses the abscissa at -0.273, well below the point of objective equality, where stimuli of equal intensity are compared against each other. There appears to be a Constant Error. This shift in the Psychometric Function away from the brighter end of the stimulus scale may result from response bias, as will be demonstrated in the next chapter.

Values of mean response time, that is, correct and error times averaged together are plotted in Fig. 10.6. The response times are quite fast for comparisons between quite different brightnesses, but become slower as the stimuli under

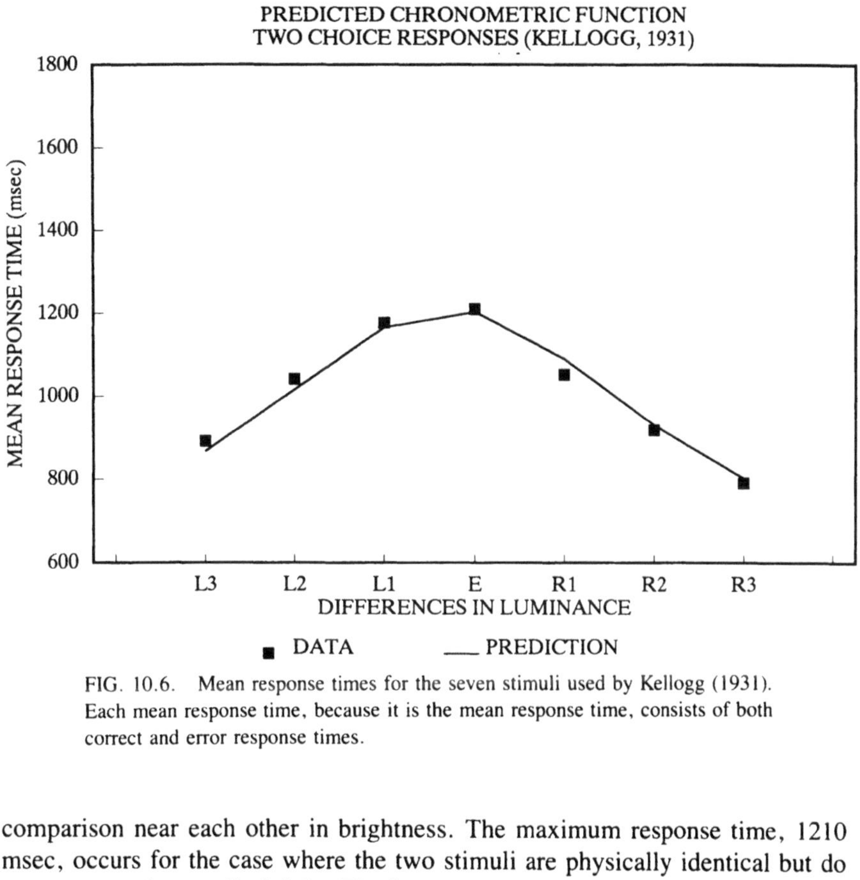

FIG. 10.6. Mean response times for the seven stimuli used by Kellogg (1931). Each mean response time, because it is the mean response time, consists of both correct and error response times.

comparison near each other in brightness. The maximum response time, 1210 msec, occurs for the case where the two stimuli are physically identical but do not appear to be equally bright. The fastest response times occur at the extremes where the brightness differences are greatest.

The distribution-free analysis of these data continues by creating an empirical law relating response time and response proportion. From Equation 10.6,

$$E\{N\} = \frac{A(2P_A - 1)}{\mu}.$$

The actual values for μ are unknown, but the linear increase in brightness differences supported by the relation between θ^*A and stimulus difference suggests that μ increases linearly as a function of luminance differences ranging to +3 from −3 units. Thus, μ is a linear function of stimulus differencc, Δs, and may be written as $\mu = \alpha(\Delta s) + \beta$. The mean response time then equals,

$$E\{\boldsymbol{RT}\} = \frac{A(2P_A - 1)}{\alpha(\Delta s) + \beta} + E\{\boldsymbol{K}\}. \tag{10.12}$$

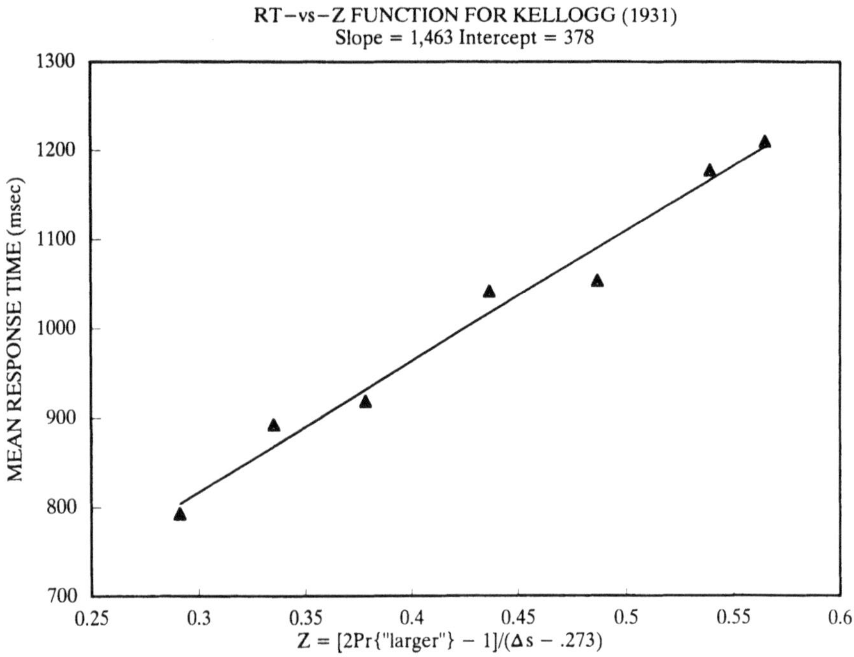

FIG. 10.7. The existence of an empirical law relating mean response time and response probability is confirmed by the results shown in this figure. The linear function derives from Equation 10.12.

This linear function of $(2P_A - 1)$ has slope A and intercept $E\{\boldsymbol{K}\}$. Searching for values of α and β that minimize the squared deviations between the predicted and observed results leads to the finding shown in Fig. 10.7. The data points span a linear function with Intercept 378 msec, which is an estimate of the nondecision time components of this discriminative task, $E\{\boldsymbol{K}\}$. The Slope of 1463 is a measure of the threshold value A, scaled in the same units as μ.

The results support the existence of a law relating response time and response probability. The linear function crosses the abscissa near the actual intercept value determined by estimates of θ^*A. The value of Δs for which $\mu = \alpha(\Delta s) + \beta = 0$ equals $-\beta/\alpha$ or -0.258. This value compares favorably with the estimated value using values of θ^*A in Fig. 10.5 of -0.273. Finally, using these parameter estimates generates the predicted mean response times shown in Fig. 10.6. These predicted response times are quite close to the mean response times reported by Kellogg (1931).

The noteworthy features of this analysis are:

1. The estimates of the parameters θ^*A increase linearly with stimulus difference.
2. The mean response time data conform to the theoretical expectation.
3. The predicted linear relation between mean response time and the function of response probability, labelled $\mathbf{Z}$, is confirmed.
4. Two estimates of the zero crossing value, the point of subjective equality, agree that the value is not zero but slightly negative, about -0.26. Thus, both response probability and response time data are affected by what appears to be a Constant Error.

THE EFFECTS OF RESPONSIVENESS

The Psychometric Function swept out by early experimentalists required changes in stimulus difference to secure changes in response probability. Given the idée fixe at the time, that error in the sensory system's measurement of stimuli produces erroneous responses, the experimentalist's decision to vary the difference in amount of stimulation appears inevitable. Yet response probabilities, according to Wave Difference Theory, depend upon both stimulus difference and the subject's resistance to respond. Furthermore, changes in resistance also produce changes in response time. Subjects willing to reduce resistance produce quick responses at the expense of correct responses. Alternatively, increasing the threshold value A increases response time and increases the number of correct responses. By controlling the resistance to respond the subject controls *responsiveness*, $1/A$, and indirectly decision time. In this way the subject maintains voluntary control over response time and response accuracy. These simultaneous effects are illustrated through the discussion and figures that follow.

Figure 10.8 illustrates changes in Psychometric Functions as a function of changes in responsiveness. For the most responsive case, A assumes the smallest of three values, $A = 20$, and the Psychometric Function exhibits a shallow slope. The 75% threshold determined from this Psychometric Function is about 52.8. The function attains the value 0.50 at exactly the magnitude of the standard, which is 50. An increase in resistance to respond to $A = 30$ reduces responsiveness but increases the probability of correct responses. For comparisons that are less than the standard, the probability of an error is smaller than for the case of $A = 20$. For comparisons larger than the standard, the probability of a correct response is greater than for the case of $A = 20$. Increasing the resistance to respond even higher, to $A = 40$, produces a more steeply sloped Psychometric Function as indicated by another increase in the number of correct responses.

As shown in Table 10.1 the measured value of the 75% and 50% thresholds depends on the value of A, which is controlled by the subject. Yet these threshold estimates are deceiving because the discriminability of this hypothetical subject

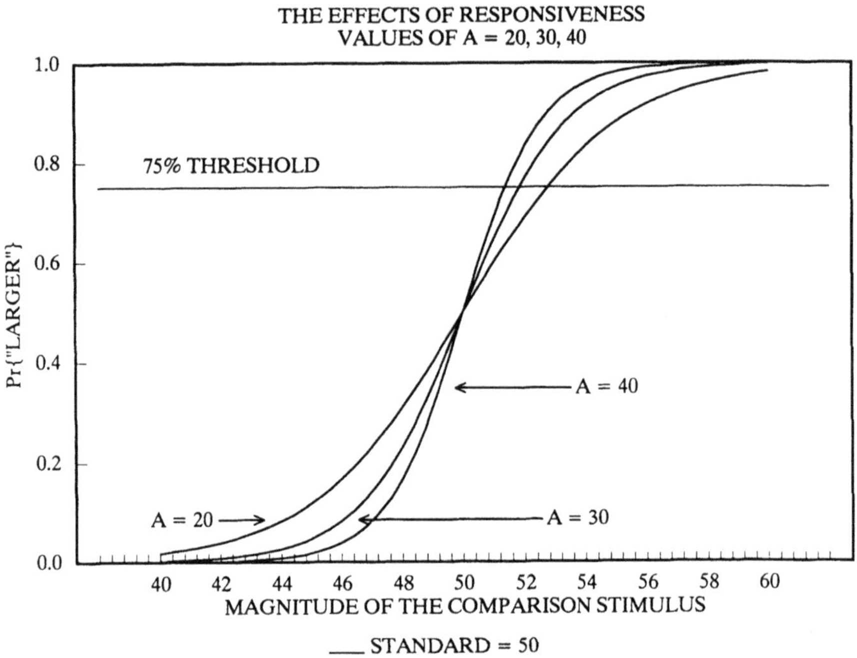

FIG. 10.8. These three Psychometric Functions appear to show changes in sensitivity. They are generated by changes in responsiveness, a subject controlled part of the decision process.

actually remains constant. The changes in the Psychometric Functions are not due to a change in the subject's sensitivity. Rather the subject's sensitivity remains constant but responsiveness changes. The more steeply sloped Psychometric Function derives from multiplying the resistance to respond for the intermediate Psychometric Function by a factor of $^4/_3$. Not knowing that responsiveness changed, the experimenter estimated the 75% threshold to be the smallest of the three. The deception continues in the form of apparently reduced sensitivity for the less steeply sloped Psychometric Function. These probabilities occur by multiplying the resistance to respond of the intermediate Psychometric Function by $^2/_3$. The result produces the largest of the three 75% thresholds.

The various threshold measurements, all based on a fixed sensitivity, show that the measurement of a threshold is contaminated by subjective factors that influence the estimated threshold value. The threshold value depends on at least two parameters, one of which is directly under the control of the subject. By suitable inducements the experimenter may obtain nearly any desired result. Any number of unknown factors such as the experimenter's demeanor, the subject's

TABLE 10.1
The 75% Thresholds Determined from Three Psychometric Functions

Value of A	*50% Threshold*	*75% Threshold*
20	50	52.8
30	50	51.8
40	50	51.4

dislike for the closeness of a darkened experimental room, the rush to complete the experimental session as soon as possible, encouragement to respond as accurately as possible, all influence the subject's responsiveness. Changes in responsiveness, in spite of constant sensitivity, result in changes in the estimated 75% threshold. Thus the estimated threshold requires correction for responsiveness before its value can be directly related to discriminability.

It follows that the Psychometric Function obscures features of performance. Subjects may meet performance demands through the strategic manipulation of the value of A, thereby substantially improving or degrading performance. Nearly every experimenter knows from personal experience that subjects who begin a 1-hour experimental session late are quite capable of completing a lengthy series of judgments before rushing to their next class. Their performance is generally poor due to the reduction in resistance to respond which yields quick, rather inaccurate, judgments but reduces the length of each experimental trial, and consequently the duration of the experimental session.

Data suggesting that a change in A does lead to different response time performances appear in an overlooked monograph by D. M. Johnson (1939). The experimenter required subjects to discriminate differences between black horizontal line segments inked onto a white card. Vertical uprights at the ends of the line and one placed 50 mm from the left-hand side defined the two line segments. The 50 mm long standard always occurred on the left side of the stimulus card. The comparison varied from 40 mm to 60 mm. Subjects were to make judgments constrained by certain response time considerations. In the "usual" case no mention was made of the importance of response time. Subsequently subjects were instructed either that response time was important and they should respond as quickly as possible, or told to take time and make no errors.

Three subjects each made 15 judgments of each stimulus difference in each response time condition. The small numbers of responses per stimulus difference precludes discrimination between Psychometric Functions that should change as A changes. However, the mean response times define three quite distinct Chronometric Functions shown in Fig. 10.9. Quite clearly there is a very major influence of the response time instructions on the subjects's performance.

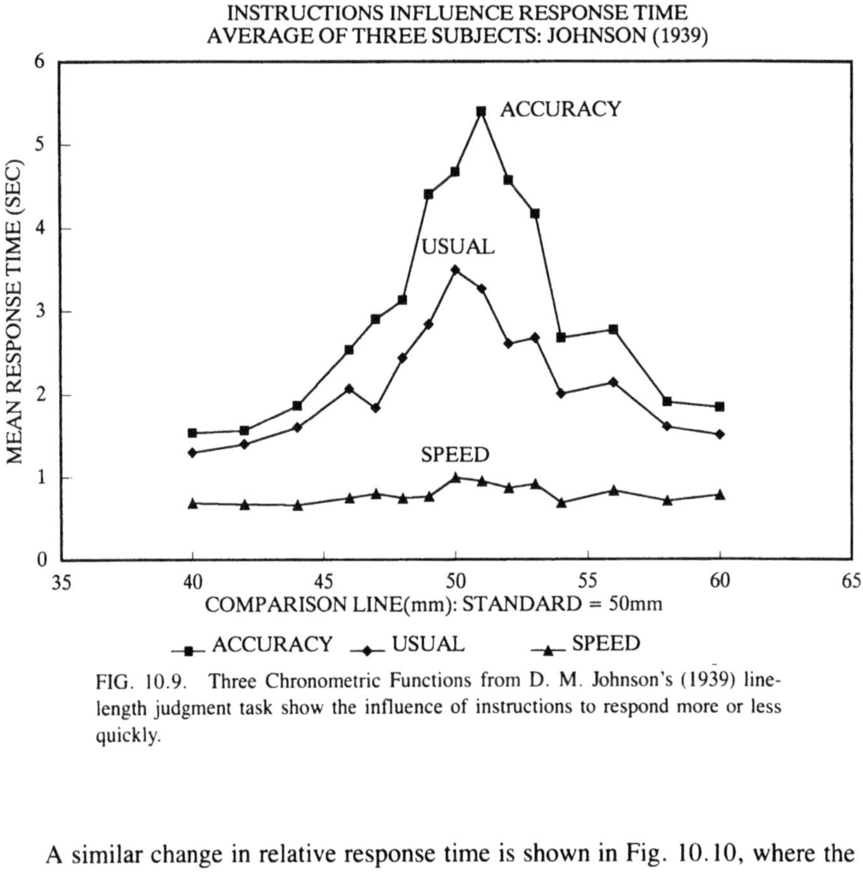

FIG. 10.9. Three Chronometric Functions from D. M. Johnson's (1939) line-length judgment task show the influence of instructions to respond more or less quickly.

A similar change in relative response time is shown in Fig. 10.10, where the abscissa shows comparison stimuli and the ordinate shows mean decision time. Only one parameter, *A*, (20, 30, and 40) is needed for each Chronometric Function. Even without an exact fit the correspondence is quite remarkable. The hypothesis supported by these results is that the subjects controlled their resistance to respond by changing the value of *A*. No ***RT***-versus-***Z*** test of this hypothesis is available from Johnson's experiment because of methodological features that precluded the measurement of response probabilities within each experimental condition, but a similar finding is easily obtained by using the Psychometric Function with values of *A* equal to 20, 30, and 40, like those shown in Fig. 10.8.

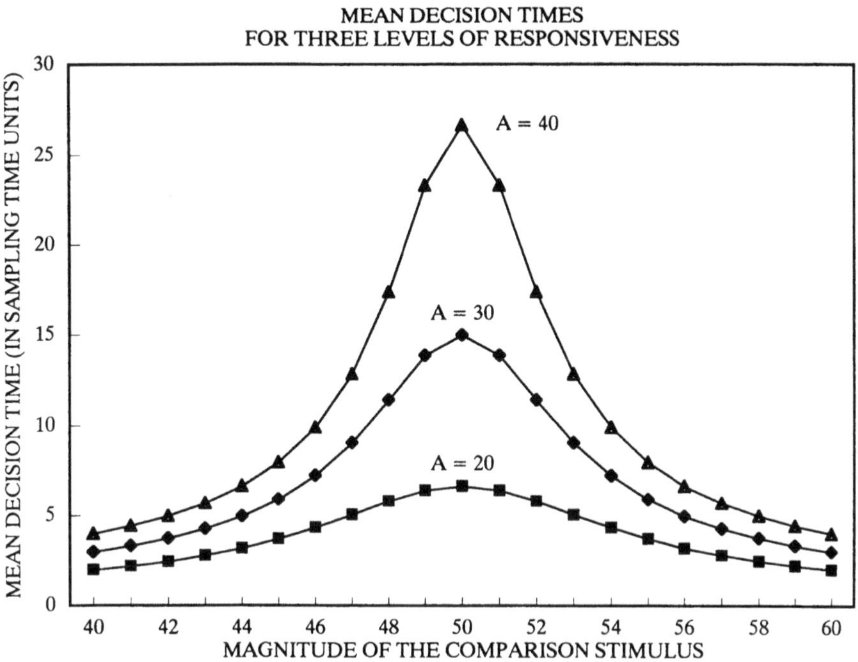

FIG. 10.10. Three Chronometric Functions predicted by the theory are based on only a single parameter. They are remarkably similar to Johnson's results.

11 Response Bias

Response bias is a powerful determinant of performance. Thomson (1920) proposed how bias might account for skewed Psychometric Functions but the discovery of response bias belongs to Blackwell (1953, 1963). The continued interest in response bias is due to the many experimental studies that showed how Signal Detection Theory accounted for changes in performance in terms of response bias. In SDT a bias toward one response simultaneously produces bias away from the alternative response. Equal bias toward *Yes* and *No* is represented by the decision criterion shown in Fig. 7.4 and Fig. 1.1. The decision criterion determines the amount of response bias and also acts as a referent against which a sampled sensory value must be compared. The decision criterion also acts as a response threshold that determines which response will occur—if the sampled sensory value is less than the criterion the response *No* occurs, if greater then the response *Yes* occurs.

In Wave Theory response bias represents a directionality of response (Shipley et al., 1946), or a preparatory set (Bertelson, 1961), toward one or the other of the two response alternatives. In particular when A is the response threshold for response R_A and $-A$ is the response threshold for response R_{-A}, the preparatory response bias B must lie between $-A$ and A. When B equals zero there is no response bias at all and when B is greater than zero the resistance to response R_A, $(A - B)$, is smaller than resistance to response R_{-A}, which equals $(A + B)$. An implication of this representation of bias is that favoring one response is necessarily at the expense of the alternative response.[1]

[1] In previous presentations of Relative Judgment Theory, the parameter C indicated the amount of response bias. However, the change in notation introduced here facilitates association of bias with its parameter, and allows the letter C to indicate response Confidence.

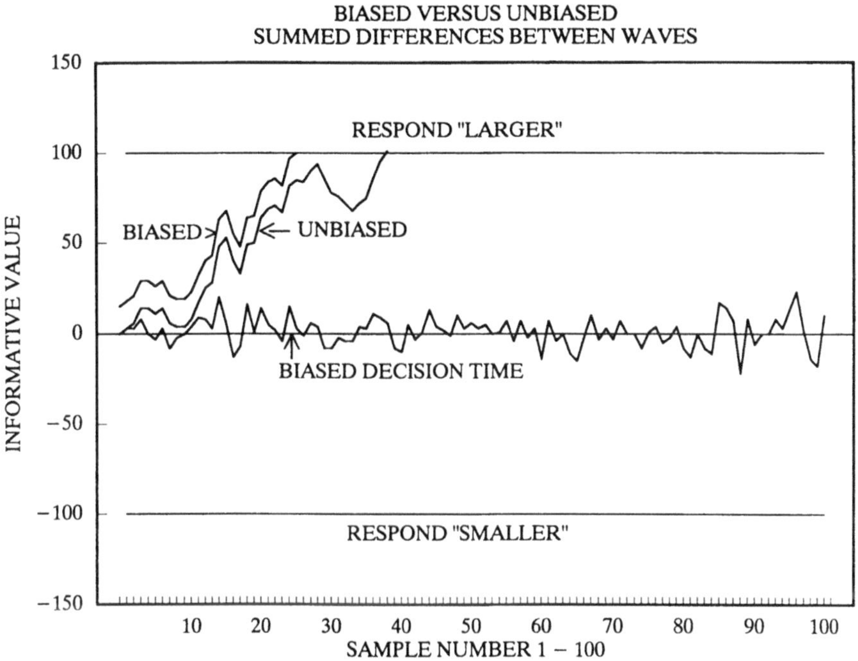

FIG. 11.1. Response bias is a predisposition toward one response threshold. As a consequence of this bias both response probabilities and response times change.

Figure 11.1 illustrates the influence of response bias. Formally, the amount of bias is the value $V_0 = B$, the initial value of the process. The accumulated differences shown in this figure are the same as in Fig. 9.3 but due to the initial bias of 15 units toward A, the accumulation reaches A much earlier than for the unbiased case, also shown. The decision *larger* or *Yes* occurs considerably faster than in the unbiased case. Furthermore the probability of reaching A increases. The process stops at A when the value of V_N accrues an additional amount equal to $(A - B)$, or stops at $-A$ when V_N accrues an added amount equal to $-(A + B)$.

BIASED RESPONSE PROPORTIONS

When a threshold is exceeded the only possible values of V_N are $(A - B)$ or $-(A + B)$. Therefore,

$$V_N = \begin{cases} A - B \text{ with probability } P_A \\ -(A + B) \text{ with probability } 1 - P_A. \end{cases} \tag{11.0}$$

The response probabilities under these biased conditions follow immediately from the computations previously used to derive P_A when $B = 0$. In general the result is,

$$P_A = \frac{e^{\theta^*(A+B)} - 1}{e^{\theta^*(A+B)} - e^{-\theta^*(A-B)}}. \tag{11.1}$$

This probability is easily separated into components that represent the probability that the threshold at A is exceeded given no response bias, $B = 0$, plus another term that depends on the amount of response bias. Rewriting Equation 11.1 to separate out the effect of bias gives,

$$P_A = \frac{e^{\theta^*A} - e^{-\theta^*B}}{e^{\theta^*A} - e^{-\theta^*A}}. \tag{11.2}$$

By adding and subtracting 1 in the numerator the two distinct terms become apparent. In this format (Link, 1978) the probability is,

$$P_A = \frac{e^{\theta^*A} - 1}{e^{\theta^*A} - e^{-\theta^*A}} + \frac{1 - e^{-\theta^*B}}{e^{\theta^*A} - e^{-\theta^*A}}$$

= Probability of R_A given No Bias + Change due to Bias.

The first term is the probability of responding R_A given no bias at all. The second term is a measure of the advantage to responding R_A given a bias of size B. The advantage is positive when the bias is toward response R_A, that is, $B > 0$. When the bias is toward response R_{-A}, B is less than zero, the probability of responding R_A decreases. The corresponding equation for response R_{-A} is,

$$1 - P_A = \frac{1 - e^{-\theta^*(A-B)}}{e^{\theta^*(A+B)} - e^{-\theta^*(A-B)}} \tag{11.3}$$

$$= \frac{1 - e^{-\theta^*A}}{e^{\theta^*A} - e^{-\theta^*A}} + \frac{e^{-\theta^*B} - 1}{e^{\theta^*A} - e^{-\theta^*A}}$$

= Probability of R_{-A} given No Bias + Change due to Bias.

The first term is the probability of responding R_{-A} given no response bias. The second term reduces the probability of responding R_{-A} when B is positive and increases the probability when B is negative. When there is no bias, $B = 0$, these equations reduce to Equations 10.1 and 10.2.

When there is no discriminability the wave differences produce an expected value of zero and the measure of discriminability, θ^*, equals 0. Expanding the numerator and denominator in Equation 11.2 in a Taylor series and letting θ^* approach zero shows that

$$P_A = \frac{A + B}{2A}.$$

$$= \frac{1}{2} + \frac{B}{2A}. \tag{11.4}$$

Thus, the probability of responding when no stimulus difference exists still depends upon the ratio of response bias, B, to the resistance to respond, A. A positive bias, $B > 0$, toward responding R_A generates a value of P_A greater than the chance value of 0.50. Because of such biases, unsuspecting subjects are easily manipulated to indicate a degree of discriminability simply by suggesting that response R_A might be the more appropriate of the two responses. The probability of response R_A increases due to bias but there is no discriminability.

Such suggestions are common in both clinical and laboratory settings where the patient or experimental subject is easily swayed by the manner and tone of the examiner or by the preferred directionality of the response. Such comments as "Tell me whether you can hear this," easily produce a predisposition to the response *Yes*. Threshold values calculated from such data are bound to be inaccurate. The amount of the inaccuracy can be determined by noting that the unknown relative amount of bias, B/A is easily estimated when there is no stimulus difference. From Equation 11.4 it follows that

$$2P_A - 1 = B/A. \tag{11.5}$$

Somewhat surprisingly Equation 11.5 produces the usual correction for guessing, previously derived from quite different sources such as the finite state theories or Blackwell's Low Threshold Theory or under more general considerations (Link, 1982). In previous theories the value B/A simply represented the probability of selecting R_A when guessing. In Wave Discrimination Theory there is no guessing. Instead B/A is determined by the subject in response to task demands.

Theoretical results are often ideals against which actual data are compared. For example, Fig. 11.2 presents three empirical Psychometric Functions, one taken from an actual weight discrimination experiment (Brown, 1910). The others result from adding response bias to the actual data. The experimenter used a standard of 100 g. Which Psychometric Function is described by the experimental data? Which are the result of response biases? An examination of the "Which" data of Fig. 3.3 will provide the reader with a startling conclusion.

The Psychometric Function passing through the value 0.50 when $\Delta s = 0$ illustrates the case of no response bias. For these data the subject yields a probability of responding 0.50 when a comparison stimulus equals the standard of 100 g. The comparison stimulus generating 75% *heavier* judgments equals 102 g. The Psychometric Function displaced to the right appears to represent a bias away from responding *heavier* because the response probabilities for responding *heavier* are generally lower in value for a fixed-comparison stimulus. For example, the 102 g-comparison generates a proportion of *heavier* responses equal to about 0.35, versus 0.75 for the middle Psychometric Function. The extreme left Psychometric Function appears to indicate significant bias toward the response *heavier* because the probability of responding *heavier* is quite high for even small comparison stimulus values. At 100 g, for example, the proportion of *heavier* responses is not 0.50, but nearly 0.85. A disturbing result appears

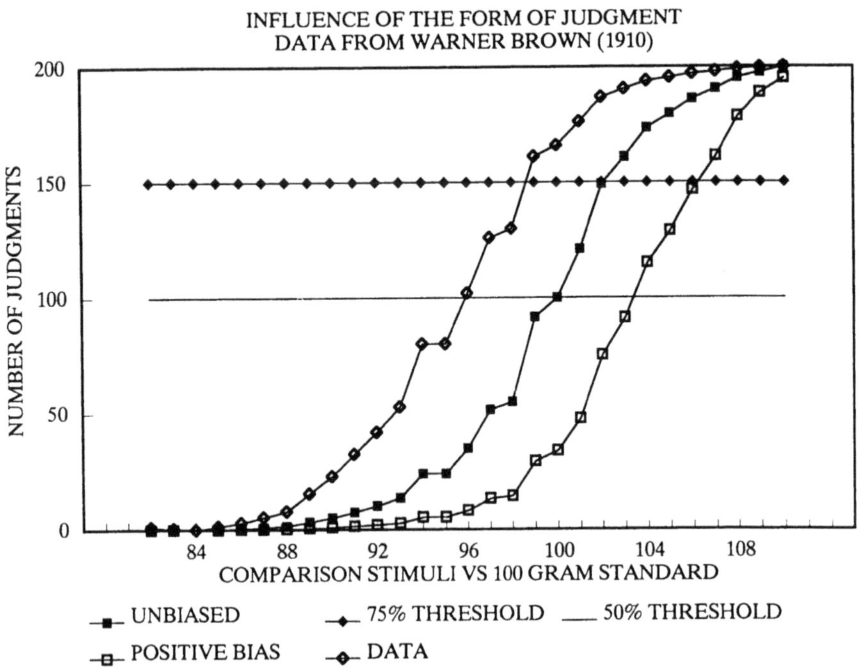

FIG. 11.2. One of these Psychometric Functions is from a Method of Constant Stimuli experiment employing a standard of 100 g. Which Psychometric Function comes from the experiment? Which results from adding response bias?

here because the stimulus value for a 75% threshold is about 99.7 g, a value smaller than the standard itself!

BIASED RESPONSE TIMES

Although response probabilities may offer a bewildering variety of possible results, response bias also causes changes in response times. Figure 11.1 makes evident the fact that response bias toward response R_A leads to a reduction in the time to reach the R_A response threshold, and thus a reduction in the time for these responses. With regard to a particular stimulus, say $\mathbf{S}_A$, the question naturally arises as to whether only R_A response times should be analyzed or whether the error response times to $\mathbf{S}_A$ should also be analyzed, or whether the mean response time to $\mathbf{S}_A$ provides the best insight into the subject's performance.

The answer to the question is to keep the analysis simple enough to determine what the subject does. In some cases an elaborate analysis of the difference between correct and error times leads to useful insights. For a first brush with the simultaneous analysis of response probabilities and response times the most direct approach is to treat the mean response time to a particular stimulus. The mean response time includes correct and error responses. Even this statistic shows such remarkable change as a function of discriminability (θ), resistance to respond (A), and bias (B) as to provide much grist for the theorist's mill.

With response bias, the expected number of samples given stimulus $\mathbf{S}_A$ becomes,

$$E\{N\} = \frac{P_A(A + B) + (1 - P_A)(-(A + B))}{E\{\mathbf{D}_A\}}$$

$$= \frac{A(2P_A - 1) - B}{E\{\mathbf{D}_A\}}. \tag{11.6}$$

Thus the expected response time becomes,

$$E\{\mathbf{RT}\} = \frac{A(2P_A - 1) - B}{\mu_A} + E\{\mathbf{K}\}. \tag{11.7}$$

where $E\{\mathbf{K}\}$ is the average nondecision component of response time.

To illustrate how changes in this mean response time occur as a function of changes in μ some relation between μ_A and θ^* must be specified in order to determine the value of P_A for a given μ_A. In the example shown in Fig. 11.3 the relation derives from considering the difference between stimulus waves to be a difference between Poisson distributions, an assumption presented in greater detail in chap. 12. As can be seen in Fig. 11.3 the effect of response bias is to slide the peak of the Chronometric Function away from $\Delta s = 0$ and toward a value of Δs opposite in value to the response bias.

In Fig. 11.3 the rightmost Chronometric Function occurs with positive bias, that is, a bias toward response R_A as shown in Fig. 11.1. The largest response time occurs for a somewhat negative stimulus difference. The negative value of Δs, which corresponds to a negative value for μ_A, offsets the advantage of positive response bias to result in a large mean response time. When the bias is negative, $B < 0$, the largest mean response time occurs for a positive value of Δs, which corresponds to a positive value for μ_A. Again, the slight positive value for μ_A offsets the amount of response bias toward R_A, resulting in a long mean response time.

To summarize these results, for bias toward the response *larger* the peak of the Chronometric Function occurs at about $-0.1\ \Delta s$. Equal, but opposite, bias toward the alternative response, *smaller*, produces a peak in the Chronometric Function at about $+0.1\ \Delta s$. The two decision-time functions are quite symmetric. By reflecting the mean times for the negative bias condition with respect to $\Delta s = 0$, so that the result for $-\Delta s$ is plotted as $+\Delta s$, both functions are seen to be identical. This is a consequence of equal but opposite response bias.

An example of similar experimental results appears in the study of color preference (Shipley et al., 1945, 1946). Shipley's data are reported in Fig. 6.1

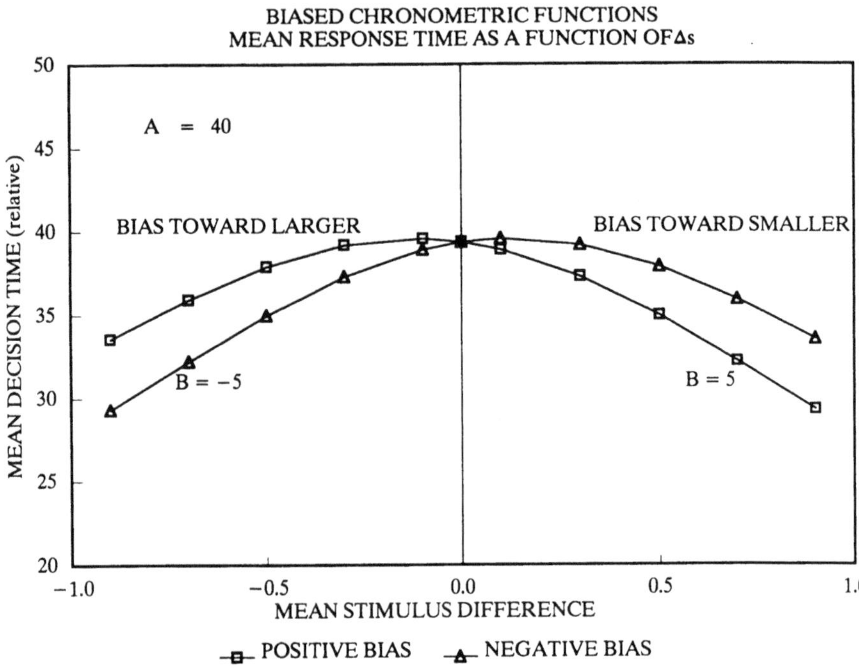

FIG. 11.3. The influence of response bias on the Chronometric Functions is to shift the point at which the maximum response time occurs.

where the abscissa shows stimuli ordered in terms of the least to most preferred color, as if the magnitude of preference increases along the abscissa. In one experiment subjects were to choose the stimulus preferred *most*. The Chronometric Function shows a form characteristic of positive response bias, as in Fig. 11.3. When, in a second experiment, subject's were instructed to choose the *least* preferred color the Chronometric Function changes in a manner consistent with a change in response bias toward the response *least*, as seen in Fig. 11.3. In such studies the experimenter may unknowingly introduce bias toward the one or the other response by emphasizing a particular response.

The amount of bias generated by the experimenter's instructions appears to be equal but opposite in the two Shipley experiments. Given that the stimuli are ordered with respect to preference, the average stimulus represents a stimulus with neither positive nor negative preference, that is 0 preference. Taking the average stimulus to characterize a case of $\Delta s = 0$ and reflecting the response times from the second Shipley experiment shows the remarkable correspondence in response times in the two experiments. The similarity of the two Chronometric Functions shown in Fig. 11.4 is most likely due to the equal but opposite biases generated by the experimenter's instructions.

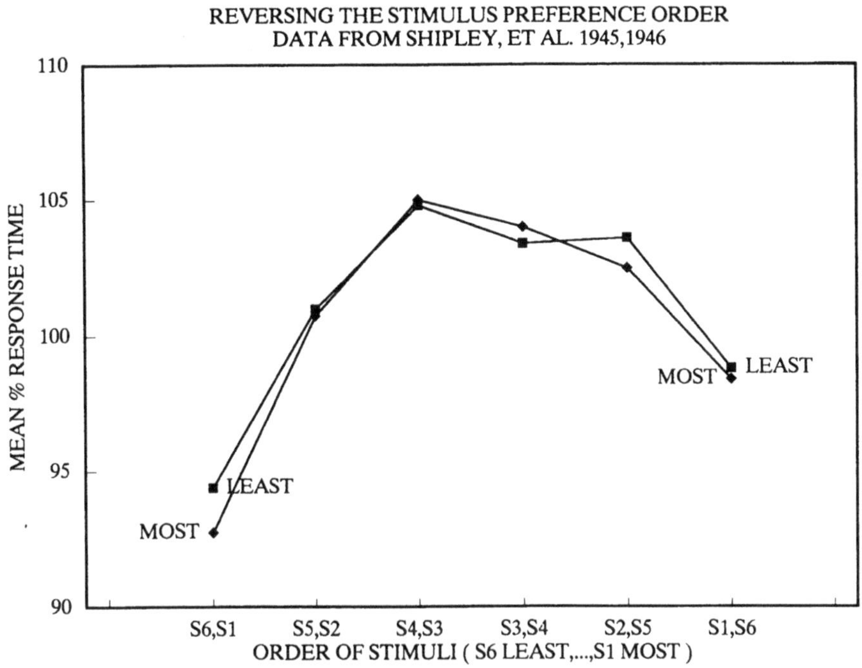

FIG. 11.4. Shipley et al. (1945, 1946) data from two experiments. The data from the second experiment are reflected with respect to the ordering of stimulus preference to show that the amount of response bias in the two experiments is the same but opposite in direction.

These response biases are produced by the experimental procedure. Yet changing the instructions to remove the bias must also change the data. Knowing that response bias is the cause of these results suggests that the mathematical model be used to correct the data for bias, leaving behind the effects due only to differences in preference. For a fixed stimulus difference the separation of Equation 11.7 into two parts shows the influence of bias to be largely a function of the difference between A and B.

$$E\{\boldsymbol{RT}\} = \frac{A(2P_{\mathrm{A}} - 1)}{\mu_{\mathrm{A}}} - \frac{B}{\mu_{\mathrm{A}}} + E\{\boldsymbol{K}\}. \tag{11.8}$$

The value of P_{A} can be separated into two parts, the first free of the effect of response bias. To wit,

$$P_{\mathrm{A}} = \frac{e^{\theta * A} - 1}{e^{\theta * A} - e^{-\theta * A}} + \frac{1 - e^{-\theta * B}}{e^{\theta * A} - e^{-\theta * A}}.$$

Substituting this value into Equation 11.8 produces

$$E\{\boldsymbol{RT}\} = \left[\frac{2A}{\mu_A}\right]\left[\frac{e^{\theta * A} - 1}{e^{\theta * A} - e^{-\theta * A}}\right] - \frac{A}{\mu_A} + \left[\frac{2A}{\mu_A}\right]\left[\frac{1 - e^{-\theta * B}}{e^{\theta * A} - e^{-\theta * A}}\right] - \frac{B}{\mu_A} + E\{\boldsymbol{K}\}. \tag{11.9}$$

The third and fourth terms are determined by the amount of response bias. Isolating them permits the expected response time to be written in terms of the response time when no bias is present plus the effect of bias. That is,

$$E\{\boldsymbol{RT}\} = \left[\frac{2A}{\mu_A}\right]\left[\frac{e^{\theta * A} - 1}{e^{\theta * A} - e^{-\theta * A}}\right] - \frac{A}{\mu_A} + E\{\boldsymbol{K}\} + \left[\frac{2A}{\mu_A}\right]\left[\frac{1 - e^{-\theta * B}}{e^{\theta * A} - e^{-\theta * A}}\right] - \frac{B}{\mu_A}. \tag{11.10}$$

$E\{\boldsymbol{RT}\}$ = Average Response Time without Bias + Effects due to Bias.

The major contributor of response bias to response time is the value of the bias B. Thus for a fixed stimulus comparison the bias will increase or decrease response time from the no bias value depending upon the magnitude and sign of B. For positive values of Δs positive values of B decrease response time and negative values of B increase response time. As illustrated by Fig. 11.5 an unbiased Chronometric Function is symmetric with respect to $\Delta s = 0$. The introduction of positive bias tends to reduce the response time for R_A responses which are correct responses for positive Δs comparisons. The overall effect is to also reduce the average time, correct and error times included.

Although the effect of response bias is additive, the question of whether bias is additive with respect to experimental design parameters used by the experimenter to alter performances is a matter for future study. At the outset, however, there is no a priori reason to suspect that such linear response models as those typically used for analyses of variance should have application to the analysis of choice response times. Instead, the various exponential terms in these equations suggest that the additive models typical of analyses of variance are not directly applicable. In any case, if one knows that performance is influenced by response bias there is no reason to construct a statistical straw man to demonstrate that response bias has no effect. The more important job is to measure the amount of the effect through the development and application of a suitable theory.

RESPONSE BIAS AND THE CONSTANT ERROR

A disturbing finding turned the attention of early psychophysicists toward a ubiquitous deviation of data from theoretical expectations. In studies of discrimination, most clearly in those studies using the Method of Constant Stimuli, the

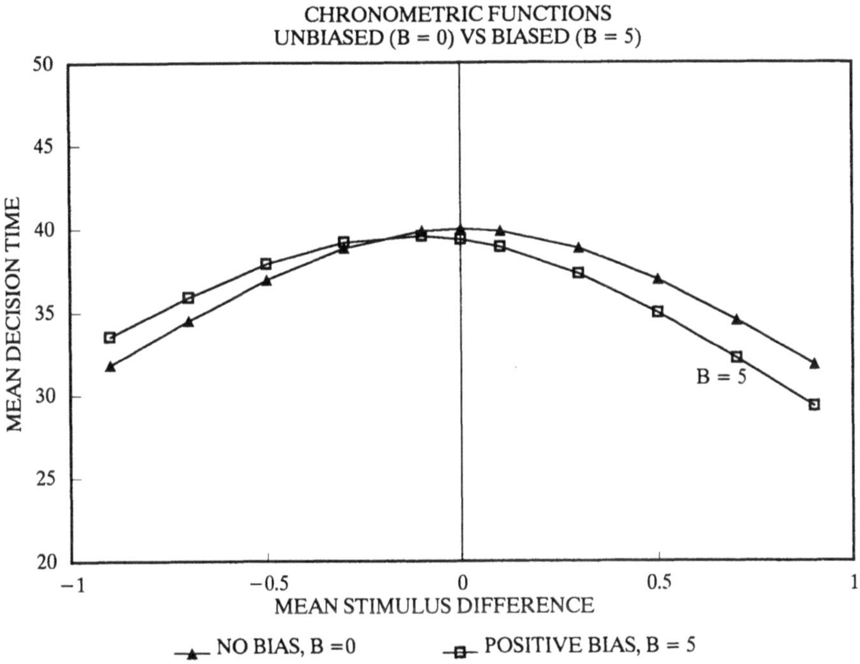

FIG. 11.5. Comparison of unbiased versus positively biased Chronometric Function illustrates the changes in response times attributable to response bias.

Point of Subjective Equality occurred for a comparison stimulus generally smaller than the experimenter's standard. This vexing Constant Error between the stimulus that proved equally likely to be *larger* or *smaller* than the standard and the standard itself persisted, even in studies using the most carefully balanced equiprobable presentations of comparison stimuli (see, for example, Schumann, 1898, 1902).

In studies of the Psychometric Function this Constant Error played havoc with the Lagrange methods introduced by Urban to estimate the Interval of Equality. To combat the influence of the Constant Error, Urban (1908) devised the stratagem of shifting a set of comparison stimuli to the left of the standard, so that the standard no longer corresponded to the midpoint of the range of comparison stimuli. Instead, as discussed in chap. 2, Urban presented a set of comparison weights ranging from 84 g to 108 g against a standard of 100 g. The resulting Psychometric Functions for *larger* and *smaller* judgments appeared to reach the Point of Subjective Equality at about 96 g, the midpoint of the range of comparisons. In spite of Urban's stratagem the Constant Error remained because the Psychometric Functions were not centered with respect to the standard of 100 g.

The Constant Error became a theoretical interest that lost significance as response bias came to play an increasingly important role in theories of discrimination. Yet response bias may provide an explanation for the ubiquitous Constant Error, especially in those experiments where the greatest care is taken to avoid any possible overt influences that might bias the subject's response.

Suppose that the experimenter, aware of the pervasive occurrence of response bias, seeks to balance all influences on the subject's response in order to "average out" the influence of response bias. For example, the stimuli are carefully balanced so that comparisons are as likely to be larger as they are to be smaller than the standard, rewards for correct and error responses are identical regardless of which comparison stimulus occurs, and sequences of stimulus presentations are controlled to provide equal probabilities for each sequence. The experimenter seeks to "average out" various extraneous influences but the Constant Error appears as an inexplicable, vexing contaminant of an otherwise flawless experimental procedure. As the following analysis shows the appearance of the Constant Error is likely to be due to the method of averaging data across unseen response biases—biases introduced by the subject and not directly controlled by the experimenter.

To compute the consequence of averaging across response biases imagine a distribution of response bias contained within the range $-A$ to A, the boundaries for the two response thresholds. For a particular experimental trial let b be the amount of Bias. Let $\boldsymbol{B}$ be the random variable representing response bias and let $f_{\boldsymbol{B}}(\mathrm{b})$ be the biases probability (density) function. The probability of responding *larger* given a comparison stimulus with a value of $\theta^* \neq 0$ is (from Equation 11.2)

$$P_{\mathrm{A|b}} = \frac{e^{\theta^* A} - e^{-\theta^* \mathrm{b}}}{e^{\theta^* A} - e^{-\theta^* A}}. \tag{11.11}$$

When there is no bias the value of b is zero, the right-hand term of the numerator equals 1, and Equation 11.11 reduces to Equation 10.1. On this basis one might suspect that averaging response proportions across trials with an average response bias of zero would produce a response proportion uninfluenced by response bias. This is not true. Instead averaging response proportions generates an apparent diminution of performance that appears in the form of a flattened Psychometric Function.

Averaging across response bias is the same as conditioning on a value $\boldsymbol{B} = \mathrm{b}$ for response bias, multiplying the response probability $P_{\mathrm{A|b}}$ by the probability density for b and integrating across b. That is,

$$P_{\mathrm{A}} = \int_{-A}^{A} P_{\mathrm{A|b}} f_{\boldsymbol{B}}(\mathrm{b})\mathrm{db}. \tag{11.12}$$

$$P_{\mathrm{A}} = \frac{e^{\theta^* A} - M_{\boldsymbol{B}}(\theta^*)}{e^{\theta^* A} - e^{-\theta^* A}} \tag{11.13}$$

where $M_{\boldsymbol{B}}(\theta^*)$ is the *moment-generating function* (mgf)[2] for $\boldsymbol{B}$ evaluated at θ^*.

When bias is centered at zero, a condition often taken as an experimental ideal, $\boldsymbol{B}$ has expected value zero. Therefore, the moment-generating function $M_{\boldsymbol{B}}(\theta^*)$ assumes a minimum value of 1.0 at $\theta^* = 0$ but must be larger than 1.0 for any other value of θ^*. Because $M_{\boldsymbol{B}}(\theta^*)$ is greater than 1.0 it can be represented in an exponential form as

$$M_{\boldsymbol{B}}(\theta^*) = e^{+\theta^*B'} \tag{11.14}$$

where B′ is selected to satisfy the necessary relation, $ln[M_{\boldsymbol{B}}(\theta^*)]/\theta^* = B'$. Substitution into Equation 11.13 gives the marginal probability of the *larger* response as,

$$P_A = \frac{e^{\theta^*A} - e^{+\theta^*B'}}{e^{\theta^*A} - e^{-\theta^*A}}\,. \tag{11.15}$$

This response probability is of the same form as Equation 11.11. However, because the value of the moment-generating function $M_{\boldsymbol{B}}(\theta^*)$ is greater than 1.0 this probability will be smaller, for positive values of θ^*, than the corresponding probability when there is no bias at all. By comparison to the unbiased case there appears to be a bias toward responding *smaller*. For negative values of θ^*, for those comparison stimuli smaller than the standard, the value of P_A will increase as if there is a bias toward responding *larger*.

In general, the effect of averaging across bias is to make the Psychometric Function appear flatter, as if there is a diminution in performance due to a loss of discriminability. An example of the apparent loss of discriminability due to averaging across bias appears in Fig. 11.6. The two Psychometric Functions show the probability of responding *larger* to the same set of comparison stimuli. The more steeply sloped Psychometric Function occurs when there is no bias at all. This logistic function passes through the point 0.50 when $\theta^* = 0$. The less steeply sloped logistic function also passes through the value 0.50 when $\theta^* = 0$, but the slope is less than for the case of unbiased responding. In Urban's parlance, the Interval of Uncertainty widens due to the operation of averaging across response bias.

Previous analyses of sequential effects in psychophysical judgments (Atkinson, Carterette, & Kinchla (1962), Atkinson, Carterette, & Kinchla (1964), and Kinchla (1966) illustrate the pervasive influence of response bias. Averages across these various stimulus-response sequences produce a diminished measure of discriminability by flattening the Psychometric Function and thereby enlarging estimates of 75% thresholds. Thus comparisons across different probability distributions of response bias, unseen and uncontrolled by the experimenter, will produce differences in threshold measurements that appear inexplicable. Vari-

[2]The mgf of $\boldsymbol{B}$ is the expected value of $e^{-\theta^*\boldsymbol{B}}$. Equation 12.2 provides an explanation of the name, moment-generating function.

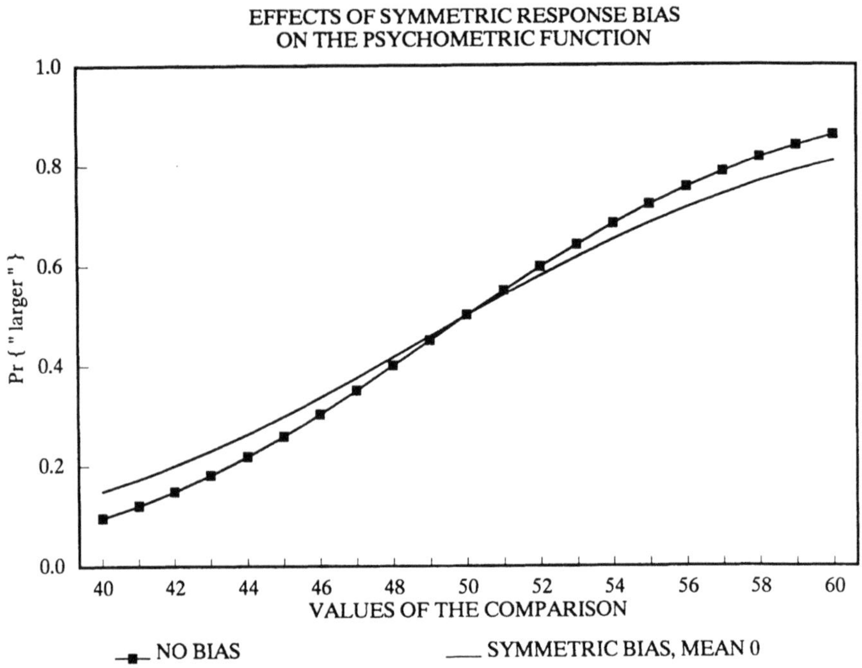

FIG. 11.6. Two Psychometric Functions show the difference between unbiased performance and the effect on response probability of averaging across the distribution of response bias.

ability occurs without cause. Experimental methods of encouraging a particular response bias, such as presenting the biasing information at the beginning of each experimental trial (Hinrichs & Krainz, 1970; Link, 1975; and Noreen, 1979) produce more stable results that permit analysis of the effects of the experimenter-imposed bias and a subsequent correction of thresholds for the amount of response bias.

When the probability distribution for response bias shifts toward one response the influence is equivalent to shifting and flattening an unbiased Psychometric Function. The amount of the shift in the Psychometric Function is related to the average value of bias. Let the average bias equal ν. To view the influence of ν on the Psychometric Function let a new moment-generating function, including the mean bias, be $M_{\boldsymbol{B}+\nu}(\theta^*)$. The new moment-generating function, defined for $\boldsymbol{B} + \nu$, is

$$\begin{aligned} M_{\boldsymbol{B}+\nu}(\theta^*) &= E(e^{-\theta^*(\boldsymbol{B}+\nu)}) \\ &= e^{-\theta^*\nu} M_{\boldsymbol{B}}(\theta^*). \end{aligned} \tag{11.16}$$

Writing the response probability with this value of $M_{\boldsymbol{B}+\nu}(\theta^*)$ yields

$$
\begin{aligned}
P_{\mathrm{A}} &= \frac{e^{\theta^*A} - M_{B+\nu}(\theta^*)}{e^{\theta^*A} - e^{-\theta^*A}} \\
&= \frac{e^{\theta^*A} - e^{-\theta^*\nu}M_B(e^{-\theta^*B})}{e^{\theta^*A} - e^{-\theta^*A}} \\
&= \frac{e^{\theta^*(A+\nu)} - M_B(e^{-\theta^*B})}{e^{\theta^*(A+\nu)} - e^{-\theta^*(A-\nu)}}. \qquad (11.17)
\end{aligned}
$$

This response probability is formally equivalent to a response bias of amount ν toward the response threshold at A plus the flattening of the Psychometric Function due to the distribution of response bias around ν. The entire Psychometric Function is increased toward the response *larger*. The effect, apparent to all experimenters, is a change in the Point of Subjective Equality that suggests a Constant Error. The source of the Constant Error is not in the perception of the subject but in the averaging of response biases. Emphasis on a particular response may inadvertently lead the subject toward a bias in favor of that response. The results of Shipley et al. (1945, 1946) and Brown (1910) show clear evidence of this basis for the Constant Error.

The Chronometric Function also exhibits properties consistent with the spread and shift of the Psychometric Function. When the average response bias equals zero, the effects are due entirely to the distribution of response bias. A shift in response bias away from zero will shift the peak of the Chronometric Function away from a location directly above the standard. Positive response bias generates a negative shift in the position of the maximum response time consistent with the existence of a Constant Error. Kellogg's (1931) brightness discrimination study (see Fig. 10.6) illustrates the negative shift along the abscissa for the Chronometric Function. The effect is not limited to studies of human discrimination. Peak shift also occurs in discrimination studies with animals, (Hanson, 1959). The generalization gradient obtained during extinction (Guttmann, 1965) tends to shift away from the standard used to train the discrimination. This too appears to be a consequence of response bias.

A tendency to favor one response over the other is easily introduced by the experimenter or clinician. With respect to response probability the effect of this bias is to create an apparently more compliant subject who generates the response the experimenter wants to hear. With respect to response time, the subject produces generally faster responses for stimuli more closely identified with the favored response.

OPTIMAL STOPPING

Wolfowitz (1947) proved that Wald's (1947) sequential likelihood-ratio test of a Gaussian-based null hypothesis requires fewer samples than a comparable fixed sample-size test. Against a large class of alternative hypothesis-testing theories

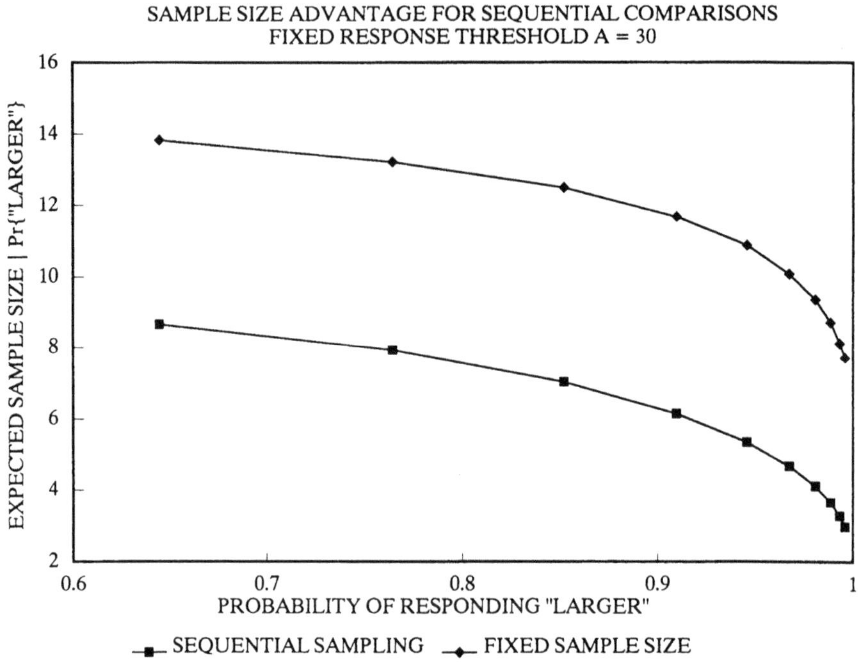

FIG. 11.7. The number of samples required to reach the decision *larger* with probabilities varying from 0.64 to 0.99 is shown for a fixed sample-size model and the sequential theory described in the text. The sequential method is considerably more efficient.

Wald's sequential-analysis model proved to be optimal, generating faster decisions for a fixed probability level. The same efficiency exists in Wave Theory.

Figure 11.7 shows the expected number of samples for Wave Theory versus the corresponding number for a fixed sample size Gaussian-based hypothesis test. The ordinate shows the expected number of samples, which is, in the fixed sample-size case a constant. The abscissa shows the probability that the accumulation of wave differences will first exceed the response threshold at A resulting in the response *larger*. The two curves illustrate that the number of samples required by Wave Theory to determine that a comparison stimulus is larger than a standard is always smaller than the corresponding fixed sample-size test.

These results follow from examining two Poisson signals, such as those described in chap. 12. In this investigation the standard Poisson wave has mean 50 and a variance of 50 because Poisson random variables have equal mean and variance. The comparison Poisson wave increases in expected value, and thus in variance as well, from 51 to 60, thereby creating a succession of expected differences ranging from 1 to 10. It is these differences, together with results from chap. 12, that lead to the response probabilities along the abscissa. For the difference of size one, Wave Theory predicts that for a response threshold $A = 30$, the probability of first exceeding the *larger* response threshold is 0.644, and the expected number of wave differences needed to reach this threshold equals 8.6. For a corresponding Gaussian-based fixed sample-size test of whether the difference between the two waves is greater than zero, the mean difference equals one and the total variance equals $50 + 51 = 101$. The number of observations required to reject the hypothesis of equality with probability 0.644 is 13.8 (14, to be discrete). Other probabilities along the abscissa derive from similar comparisons of calculations based on Wave Theory or a simple Gaussian-based z-test.

Quite clearly the number of observations required is always less with the sequential comparison method. This reduction is characteristic of sequential comparison schemes (Ghosh, 1970). Most important is the fact that the sequential comparison method produces quick responses but must also produce errors. In this theory errors are an inherent part of the function of the comparison process. These errors are not due to Gaussian-based error per the 19th-century view of the sensory system. Rather the errors are inherent in the system itself. Faster responses occur at the cost of more errors. Slower responses are more accurate, but there exists a small error probability that is measurable when a sufficient number of observations are taken for the errors to occur.

In some remarkable studies of response time (e.g., Steinman, 1944) any series of trials containing an error was run again in an effort to produce error-free performance. The belief, common at that time, was that errorless performance provides a better view of performance. In contrast, Wave Theory suggests that errors are a necessary outcome of the comparison system. Errors must occur. Therefore, a better view of performance requires that errors be treated as data, as a consequence of the comparison process. To do otherwise is to capitalize on the chance that no errors will occur, thereby generating a biased view of the comparison process. On a more positive note, even this biased view provides data for the rigorous analyses of error-free performance.

SUMMARY

Part I shows how experimental results came to focus on response proportions and response times. This chapter introduces concepts fundamental to the analysis of discrimination experiments. The stimulus, always an abstraction in psychological theory, is no longer treated as a psychological variable producing a single value for a stimulus presentation. Rather the stimulus is an electrical wave that changes over time during an experimental trial. The probability distribution of signal amplitudes plays an analytical role in expressing response times and response probabilities as a function of the theoretical parameters θ^* and μ. By and large, the analysis of the theory shows that a number of predictions are not dependent on the form of the probability distribution of amplitudes. Because these rather general predictions are distribution-free they apply to a wide range of stimuli used in typical psychological experiments.

These new theoretical ideas clarify previous experimental work, generate new predictions and, as the remaining chapters reveal, provide the basis for the empirical observations of Weber, Fechner, and Stevens. These extensions follow from imposing on the general theory additional assumptions regarding the exact nature of the waves undergoing comparison. Also a new mechanism for the generation and transmission of these signals provides an alternative to the 19th-century view that stimulation is cloaked by Gaussian error.

12 A New Theory of Variability

POISSON WAVES AND THE GENERALIZED WEBER LAW

As Laming (1986) shows, Weber's Law is of even greater use beyond the confines of experimental psychophysics. His rewarding summary of its many applications (Laming 1986, p. 70) also reveals that some 30 models of Weber's Law occurred after the 19th-century. Although Laming distilled the various models into five explanatory principles that are correlates of the empirical facts defining Weber's Law, the source of Weber's Law remains a mystery.

A fundamentally new interpretation of Weber's, Fechner's, and Stevens' Laws, designed to dispel much of the mystery, results from the application of Wave Discrimination Theory to stimuli characterized as Poisson waves. These waves represent the joint output of many sensory transceivers that maintain a relatively low probability of responding, a feature necessary for a sensory system to remain sensitive to external stimulation over a broad range of stimulus intensities and yet adapt to changes in the level of stimulus intensity. The process of comparing Poisson signals generates Weber's, and simultaneously, Fechner's Law. The parameter θ^* proves to be Weber's Constant, and a new psychophysical law, including Weber's, is derived and applied to experimental results presented in Part I.

POISSON SIGNALS

The surface of the body contains a layered mosaic of sensory transceivers that forms a dense shield against certain bands of energy and a sensitive, quantized,

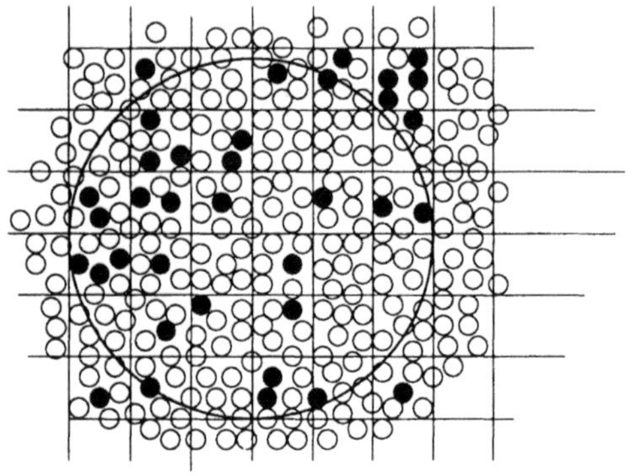

FIG. 12.1. The surface of the body and interior tissues contain transceivers that propagate signals through the nervous system. Filled circles represent transceivers that, having recently sent a signal, are in a refractory period.

receptive interface for other stimulation. This surface acts as a filter, transmitting only certain energies to and from the body. The interactive element of this filter of peripheral stimulation is a discrete transceiver unit idealized in Fig. 12.1. These units are sensitive transceivers that respond selectively to energy applied to the body's surface.

The existence of such transceivers suggests that a suitable force applied to the sensory surface will cause at least some transceivers to respond. To create a simple but accurate model of this energy transfer let the probability that a transceiver responds as a consequence of the energy transferred to it be greater than zero. The probability that κ of M independent transceivers respond is binomially distributed, $b(\kappa; \mathrm{M}, p_{\mathrm{M}})$ where p_{M} is the non-zero probability that a transceiver responds. The probability that no transceiver responds is $(1 - p_{\mathrm{M}})^{\mathrm{M}}$. As M increases, the probability of no transceiver responding approaches a limiting value greater than zero. For, if this limiting value were zero then p_{M} must also be zero, but this conclusion is inconsistent with the original premise that there is a non-zero probability that a transceiver responds.

These are the classic assumptions[1] that give rise to the Poisson probability distribution, defined in this instance for a spatial region containing a number of sensory transceivers (Feller, 1957, 1966; McGill, 1964; Pirenne & Marriott, 1959). For large M, beyond about 20, the Poisson probability of exactly κ responding transceivers is

$$Pr\{\kappa; \alpha\} = \frac{e^{-\alpha}\,\alpha^{\kappa}}{\kappa!} \qquad \kappa = 0, 1, 2, \ldots \qquad (12.1)$$

where α is the expected value of K.

[1]The same conclusions also follow from considerations that drop the assumption of independence of units and include such refinements as refractory periods during which the transceiver remains dormant.

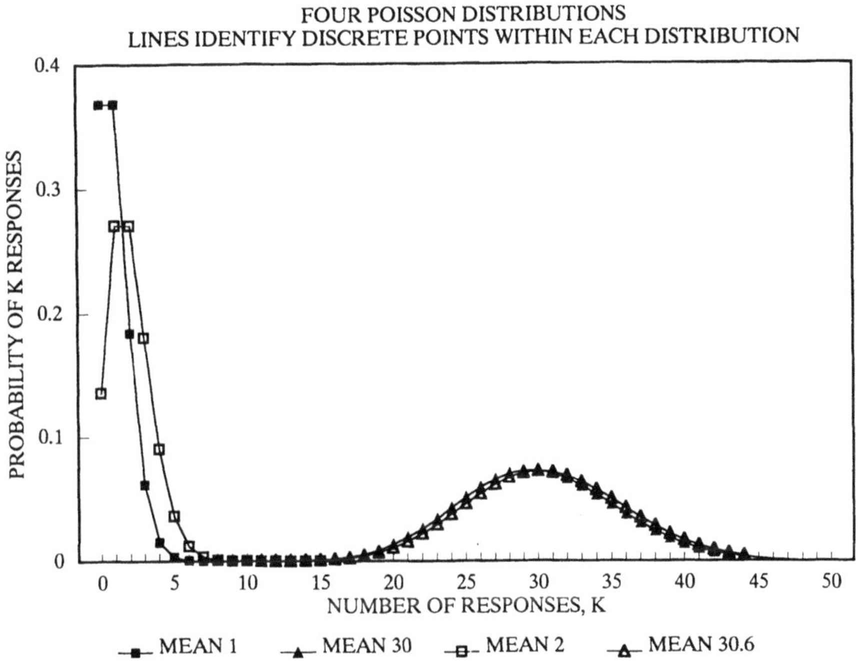

FIG. 12.2. Four Poisson distributions with intensity parameters of 1, 2, 30, and 30.6. The distributions are discrete. Therefore the ordinate shows probability, not probability density.

Examples of the distribution of $Pr\{\kappa; \alpha\}$ for $\alpha = 1$ and $\alpha = 2$ are shown in Fig. 12.2. The distribution is discrete. The ordinate gives the probability of the number of transceivers indicated on the abscissa responding. The abscissa shows the number of transceivers that are able to respond to stimulation. Although the distributions are discrete, the intensity, α, is continuous. Because the average number of transceivers emitting a quantum pulse equals α, the means of these distributions slide along the abscissa continuously. Thus the four discrete distributions shown in Fig. 12.2 vary along the abscissa depending upon their mean value.

When $\kappa = 0$ no transceivers respond. This event has non-zero probability equal to $e^{-\alpha}$, a value that decreases exponentially as the Poisson intensity parameter α increases. When $\alpha = 1$ the value of $e^{-\alpha}$ is 0.381, but when $\alpha = 30$ the value of $e^{-\alpha}$ is essentially zero. The probability that at least one transceiver responds equals $1 - e^{-\alpha}$. Because $1 - e^{-\alpha}$ is the probability distribution function for an exponential variable, an increase in α will appear to generate an exponential increase in the probability of at least one response. Thus, at a level of analysis focussed on whether any response occurs, the probability of at least one response will appear to increase exponentially as a function of stimulus intensity.

These Poisson assumptions capture the on/off quantal behavior of many discrete, spatially separate, transceiving units. As a mathematical model it characterizes important features of a sensory surface containing many transceivers (McGill, 1964). The spatial arrangement of these units is functional, not geometrical. Thus the probability distribution of the number of neurons firing in a volume of brain tissue or the number of raisins in a piece of bread (Parzen, 1962), or the number of bomb hits on a city block in London during World War II (Feller, 1957) may be modelled by application of these Poisson variables.

The Poisson intensity parameter defines the sensitivity of the transceiver. For a very sensitive transceiver the value of α is high, whereas a transceiver is less sensitive when the probability that the transceiver responds is small. Sensitivity measures a local characteristic of response to the sensory stimulus. An easy computation shows that α equals both the mean and variance of the number of responding transceivers. Thus as α increases the mean and variance of the number of responding units increases and the probability that no transceivers respond decreases, as is easily seen in Fig. 12.2. Interestingly, the probability that no transceiver responds is a global characteristic of responding that depends on the sensitivity of the local transceiver.

In more familiar theoretical surroundings Rashevsky (1938, 1960, p. 67) introduced the Poisson sensory model of mathematical biophysics to psychophysical discrimination. Like many theorists, Rashevsky assumed that when a large number of Poisson variables are added together the result is approximately a Gaussian distribution. To permit an easy determination of the probability distribution of a difference between two signals, subsequent calculations assumed that the process being investigated is also Gaussian, because the difference between two Gaussian signals is also Gaussian. As will be seen, this path from the approximation to the belief that sensory processes are based on Gaussian error is a fundamental pitfall into which many psychophysical theories tumble.

The Poisson model of sensory responses due to surface stimulation applies equally well to the number of electrical pulses emitted during a unit of time. If δt represents a small unit of time, and the probability of a single electrical event occurring within δt approaches a non-zero limit as δt approaches zero then the number of events occurring within a unit of time, t, is also Poisson distributed. This time-based Poisson process arises in a vast array of practical applications of stochastic processes (Parzen, 1962).

Spatial and temporal Poisson processes may transmit information by creating a chain, or sequence, of transceivers that convey a peripheral sensory signal to a new physical location. The transformation of peripheral stimulation into a Poisson signal necessarily occurs in time. Considered by an external observer measuring the transmitted signal somewhere in the transmission path, the signal will appear to be a sequence of Poisson amplitudes as shown in Fig. 9.1. The temporal signal transmits over time a sequence of electrical events that codes the original sensory signal. At the end of the chain the temporal signal may activate a region of spatially arranged receivers of the temporal signal and, thereby, transform the temporal signal into its original spatial information at a new physical location. In this way the Poisson model offers a unique method of digitally recording external events and digitally transmitting their characteristics through a chain of Poisson processes that maintain the integrity of the original event. Although such a transmission system exhibits variability, the variability is not due to error added to the transmitted signal. Instead the variability is an inherent part of the signal itself.

Peering more deeply into characteristics of such processes is greatly facilitated by use of the moment-generating function (mgf) to simplify the analysis of sums and differences of random variables. In this regard the mgf for a random variable X is defined as the expected value of $e^{-\theta X}$, that is

$$G_X(\theta) = E\{e^{-\theta X}\}. \tag{12.2}$$

Expanding $e^{-\theta X}$ gives,

$$= E\{1 - (\theta X)^1/1! + (\theta X)^2/2! - (\theta X)^3/3! + \ldots\}$$

$$= 1 - \theta E\{X\} + \theta^2 E\{X^2\}/2 - \theta^3 E\{X^3\}/6 + \ldots.$$

Successive differentiation with respect to θ followed by setting θ to 0 produces the various raw moments of the probability distribution for X. Thus the name *moment-generating function*. This tool appears often in Appendix A as a means of simplifying otherwise formidable computations. In this chapter mgfs are used to derive properties of sums and differences of random variables. Under conditions that generally characterize the physical world, the form of a moment-generating function identifies its associated probability distribution. Often the form of a probability distribution can be determined by examining the form of a mgf.

Poisson random variables play an important role in the quantum transfer of energy, and the mgf of the Poisson random variable assists in the development of properties of sums and differences of Poisson variables. Let $Pr\{X_A; \alpha\}$ be the Poisson probability function for a random variable X_A. Then the moment-generating function for X_A is

$$G_A(\theta) = E(e^{-\theta X_A}) \tag{12.3}$$

$$= \sum_{X_A=0}^{\infty} Pr\{X_A; \alpha\}(e^{-\theta X_A})$$

$$= e^{[\alpha(e^{-\theta}-1)]}. \tag{12.4}$$

This mgf depends upon two parameters, α and θ. The parameter α equals the mean and variance of the number of transceivers responding to stimulation. The value θ is a real variable. When $G_A(\theta)$ is differentiated and θ set to zero the result equals minus the expected value of X_A, that is, $-E(X_A) = -\alpha$. Successive differentiation yields higher moments.

SUMS OF POISSON SIGNALS

The moment-generating function is also a theoretical tool used to show that when the outputs of a number of Poisson processes are combined, the resulting sum is still a Poisson process. An important property of the new Poisson process is that it preserves values of the Poisson parameters that are summed to create the final output. This property is the basis for showing that the overall Poisson intensity equals the number of transceivers stimulated multiplied by the average intensity of response. This important relation establishes that the output of the Poisson process is a direct transformation of the physical stimulus energy.

Starting with two independent random variables X_A and X_B, and applying the convolution theorem for the sum of two independent random variables (cf. Townsend & Ashby, 1983), shows that the sum, $S = X_A + X_B$, has moment-generating function,

$$G_S(\theta) = G_A(\theta)G_B(\theta).$$

When X_A and X_B are Poisson random variables with intensity parameters α and β, respectively, the mgf becomes,

$$\begin{aligned} G_S(\theta) &= e^{[\alpha(e^{-\theta}-1)]}\, e^{[\beta(e^{-\theta}-1)]} \\ &= e^{[(\alpha+\beta)(e^{-\theta}-1)]}. \end{aligned} \tag{12.5}$$

Because this moment-generating function is of the same form as Equation 12.4 the sum of two Poisson variables with different parameters is still Poisson distributed, but with an intensity parameter equal to the sum of the individual intensity parameters, in this case, $\alpha + \beta$.

This feature is commutative. When there are L Poisson variables, that is, L independent regions or spaces with Poisson outputs, then the sum of their outputs is also Poisson distributed. Let these processes have parameters $\alpha_1, \alpha_2, \ldots, \alpha_L$. Then the moment-generating function for their sum is,

$$\begin{aligned} G(\theta) &= \prod_{L=1}^{L} G_L(\theta) \\ &= e^{[(\alpha_1+\alpha_2+\ldots+\alpha_L)(e^{-\theta}-1)]} \end{aligned} \tag{12.6}$$

which is also of the same Poisson form as in Equation 12.4 and is therefore the mgf of a Poisson variable with intensity $(\alpha_1 + \alpha_2 + \ldots + \alpha_L)$. Apparently some information is lost through this addition because the individual Poisson components can no longer be distinguished once they form a common signal.

When these L areas, or transceivers, are similar in the sense that their parameters are equal, the total output is Poisson distributed with parameter $L\alpha$. That is,

$$G(\theta) = e^{[(L\alpha)(e^{-\theta}-1)]}. \tag{12.7}$$

The form in Equation 12.7 is that of a Poisson distribution with intensity parameter $L\alpha$. Thus, adding together receptive areas produces an apparent *multiplication* of the intensity of the output.

When the parameters are unequal their sum still equals the average of all the parameters multiplied by the number of transceivers. For example, let the average of L Poisson intensities be

$$\alpha = (\alpha_1 + \alpha_2 + \ldots + \alpha_L)/L.$$

Then,

$$L\alpha = (\alpha_1 + \alpha_2 + \ldots + \alpha_L) \tag{12.8}$$

is the sum of Poisson parameters for L transceivers or receptive areas. This Poisson intensity parameter also equals the average of all intensity parameters times the number of transceivers, that is, $L\alpha$ is proportional to the number of transceivers or receptive areas. In this way the Poisson intensity is a similarity transformation[2] of the number of transceivers, even when the transceivers respond with different intensities. The broad conditions under which a sum of random variables converges in mean (Shiryayev, 1984) assures that the sum of transceiver outputs equals the number of transceivers firing multiplied by the average intensity. This suggests that as stimulus intensity increases, the resulting increase in firing may appear to be a similarity transformation of the intensity of the stimulus.

Representing the Poisson process parameter as a constant times the number of transceivers is particularly useful in characterizing the effect of stimulus intensity. In many sensory experiments the physical application of a stimulus to a larger area creates an increase in stimulus intensity. The probability of a Poisson event remains constant although the number of sensory areas and the potential number of stimulated transceivers increases. The physical structure of the periphery helps bleed off energy to the surrounding area. In this way the Poisson intensity parameter can be viewed as a similarity transformation of the stimulus intensity. The greater the stimulus magnitude, the larger the number of areas it affects. The larger the number of areas, the greater the value of the Poisson intensity parameter.

In summary, let $S_{A,L}$ be the sum of L Poisson variables with *average* intensity parameter α. Then the mgf of the sum is, by Equation 12.5,

$$M_{A,L}(\theta) = e^{[L\alpha(e^{-\theta}-1)]}.$$

---

[2]A similarity transformation of a variable X is defined as $Y = cX$ where c is a constant. Multiplying kilometers by c = 1000 to create meters is an example. A critical feature of this transformation is that $X = 0$ implies $Y = 0$.

Because the sum is Poisson distributed it follows immediately that the average output is

$$E\{\boldsymbol{S}_{A,L}\} = L\alpha$$

and,

$$Var\{\boldsymbol{S}_{A,L}\} = L\alpha.$$

These values are similarity transformations of the number of stimulated transceivers.

SENSITIVITY VERSUS DISCRIMINABILITY

Sensitivity refers to the ability of a transceiver to respond to stimulation. Low sensitivity is reflected in a small value for α. Increases in sensitivity are increases in the ability of the transceiver to respond, that is, increases in α. Discriminability is a more global measure of the sensory system's ability to respond to a stimulus. The average output, $L\alpha$, measures discriminability of the stimulus as distinct from sensitivity of the transceiver. Although sensitivity may be low, the recruitment of a large number of transceivers can result in high discriminability. Or, for example, the same pressure applied to a larger surface area results in greater discriminability but no change in sensitivity.

DIFFERENCES BETWEEN POISSON SIGNALS

Sums of Poisson impulses result from stimulation, but differences between Poisson signals give rise to Weber's Law. Determining how a difference between two Poisson signals gives rise to Weber's Law requires a brief discussion of the distribution of the difference between Poisson variables, such as the difference between the signals shown in Fig. 9.2. The analysis is greatly facilitated by using properties of moment-generating functions.

Let a sum of M Poisson variables with average intensity parameter β be $\boldsymbol{S}_{B,M}$ with mgf,

$$G_{B,M}(\theta) = e^{[M\beta(e^{-\theta}-1)]}.$$

The difference between the two Poisson variables $\boldsymbol{S}_{A,L}$ and $\boldsymbol{S}_{B,M}$ is

$$\boldsymbol{D} = \boldsymbol{S}_{A,L} + (-\boldsymbol{S}_{B,M}).$$

The familiar properties of a difference are easily derived. The expected value is

$$E\{\boldsymbol{D}\} = L\alpha - M\beta,$$

and, from assumptions of independence, the variance of a difference equals the variance of the sum, that is,

$$Var\,\{\boldsymbol{D}\} = \mathrm{L}\alpha + \mathrm{M}\beta.$$

Thus, the mean and the variance of the difference between these two Poisson variables contain quite different information. The variance is a measure of the total intensity of stimulation, the strength of the Poisson wave. The mean difference is a more subtle measurement of stimulus difference that may remain constant across changes in intensity characteristics of the standard and comparison stimuli.

The mgf continues to play an important theoretical role. For this difference the mgf equals the product of mgfs for $\boldsymbol{S}_{\mathrm{A,L}}$ and $-\boldsymbol{S}_{\mathrm{B,M}}$. Noting that when $\boldsymbol{S}$ is a random variable with mgf $G(\theta)$ the mgf of $-\boldsymbol{S}$ is $G(-\theta)$, the mgf for $\boldsymbol{D} = \boldsymbol{S}_{\mathrm{A,L}} - \boldsymbol{S}_{\mathrm{B,M}}$ is

$$G_{\boldsymbol{D}}(\theta) = G_{\mathrm{A,L}}(\theta)G_{\mathrm{B,M}}(-\theta)$$

gives,

$$= e^{[\mathrm{L}\alpha(e^{-\theta}-1) + \mathrm{M}\beta(e^{+\theta}-1)]} \tag{12.9}$$

for the mgf of the difference random variable $\boldsymbol{D}$.

WEBER'S LAW

Weber's Law is easily confirmed by experimental techniques applied to all sense modalities. Indeed, Fullerton and Cattell commented as early as 1892 that

> Since Weber published this result, an amount of experiment and theorizing has been devoted to this subject, which has probably never been surpassed in the history of any science. It is commonly claimed that experiments confirm Weber's generalization. At all events, the difference which can just be noticed is usually found to become greater as the stimulus is taken greater. Weber's law can be expressed by the equation
>
> $$N = C\,\frac{\Delta S}{S}$$
>
> in which N is the least difference which can be noticed, and ΔS the increase in the stimulus S which causes this difference. No objection can be made to this statement of Weber's law so long as the meaning of N be left an open question. We ourselves consider it a physical quantity, dependent on the error of observation, and this indeed, if Weber's law hold, would increase in direct proportion to the stimulus. (pp. 21–22)
>
> Fechner, followed by many others, regards N as an equal increase in sensation, always the same whatever value may be given to S. If N be such a unit fit for measuring the intensity of sensation, it follows directly from Weber's law that the sensation increases in arithmetic ratio as the stimulus increases in geometric ratio. Fechner further assumes that the least sensation which can be noticed is a mental quantity equal to the other N's, whence follows the equation:[FECHNER'S LAW]

$$E = C \log S$$

that is, [AMOUNT OF] the sensation is equal to the logarithm of the stimulus multiplied by a constant. This equation, which leads to negative sensations and other difficulties, has been discussed at great length by many writers. (pp. 22)

To derive Weber's Law's frequent occurrence in discrimination experiments let $\boldsymbol{X}_{\mathrm{A}}$, the response to stimulus $\mathbf{S}_{\mathrm{A}}$, be Poisson distributed with parameter $\mathrm{L}\alpha$. Thus L units with average intensity parameter α generate the values of $\boldsymbol{X}_{\mathrm{A}}$. As a referent against which $\boldsymbol{X}_{\mathrm{A}}$ is compared, define $\boldsymbol{X}_{\mathrm{B}}$ to be the output of M units generating outputs with average intensity β. Such random variables exhibit the characteristics illustrated in Fig. 9.2. The mgf of the difference $\boldsymbol{D}_{\mathrm{A}} = \boldsymbol{X}_{\mathrm{A}} - \boldsymbol{X}_{\mathrm{B}}$ is identical to Equation 12.9. This is the mgf of the amplitudes shown in Fig. 9.3 which have mean $\mathrm{L}\alpha - \mathrm{M}\beta$ and variance $\mathrm{L}\alpha + \mathrm{M}\beta$. The mean measures the difference between intensities, whereas the variance measures the total intensity.

The probability that response R_{A} occurs depends on the non-zero value of θ for which $G_D(\theta^*) = 1$, for this value of θ^* enters into the response-probability Equations 10.1 and 11.2. For the difference between Poisson signals the value θ^* is easily determined by solving for θ^* in the equation,

$$0 = [\mathrm{L}\alpha(e^{-\theta^*} - 1) + \mathrm{M}\beta(e^{+\theta^*} - 1)]$$

which has as a non-zero solution,

$$\theta^* = ln\left[\frac{\mathrm{L}\alpha}{\mathrm{M}\beta}\right]. \tag{12.10}$$

For a fixed standard, increases in the magnitude of the comparison produces increases in L which in turn produce increases in θ^*. Thus θ^* is a discriminability parameter that depends on the ratio of stimulus intensities. The larger the value of θ^* the greater the difference between the standard and comparison stimuli and the more accurate the discrimination between them.

Increases in stimulus magnitude that change the intensity of comparison stimuli are basic to the study of discrimination. Of the many possible increases in stimulation that might be used, multiplicative increases to $\mathbf{S}_{\mathrm{A}}$ and $\mathbf{S}_{\mathrm{B}}$ are the basis for Weber's Law. Such increases in the standard $\mathbf{S}_{\mathrm{B}}$ and comparison $\mathbf{S}_{\mathrm{A}}$ by a constant w yields a multiplicative increase in the number of receptive areas recruited for stimulation. The multiplicative increase in the parameters $\mathrm{L}\alpha$ and $\mathrm{M}\beta$ changes their value to $\mathrm{wL}\alpha$ and $\mathrm{wM}\beta$.

Yet the value of θ^* remains constant when multiplicative increases in stimulus magnitude occur. For constant response thresholds with values A and $-A$, the probability of judging the comparison to be larger than the standard is,

$$P_{\mathrm{A}} = \frac{e^{\theta^* A} - 1}{e^{\theta^* A} - e^{-\theta^* A}}.$$

Because neither θ^* nor A change under multiplicative increases in stimulus magnitude, P_A must also remain constant. Consequently, Fechner's conjecture that the response probability should remain constant under these experimental conditions receives support from the Poisson representation of discrete transceiver units and the accumulation of their differences over time.

The typical study of Weber's Law uses the Method of Multiplied Stimuli to obtain response probabilities for comparisons between stimuli at different physical magnitudes. Regardless of the physical magnitudes of the stimuli presented for comparison, the Gaussian Theory of Error is used to determine the comparison stimulus value that yields 75% correct responding; what is often called the "threshold" stimulus. The estimated threshold stimuli are rarely actually used as comparison stimuli against a fixed standard stimulus. To demonstrate the occurrence of Weber's Law, the estimated (rarely the actual) increment in the comparison stimulus magnitude needed to produce 75% correct responding is plotted as a function of the magnitude of the standard. The result so often reported is that for 75% correct responding the increment in stimulus magnitude versus the magnitude of the standard remains constant. That is,

$$(\mathbf{S}_A - \mathbf{S}_B)/\mathbf{S}_B = \Delta s/\mathbf{S}_B.$$

$$= \text{Weber's Constant.}$$

This constant changes from one sense modality to another and from one site of stimulation to another within a sense modality. The Weber Constant for discriminating linear extent is 1/60 (Kiesow 1925–1926 cited in Laming 1986, p. 20) but for lifted weights 0.029 (Oberlin, 1936). The numerical values are generally small, less than 0.10, often less than 0.05, and often showing that about a 3% difference between standard and comparison stimuli is needed for 75% correct discrimination. These measurements are also confirmed, as Fechner (1860/1966) showed, by using experimental methods that do not depend upon the laborious Method of Right and Wrong Cases.

In spite of its frequent measurement in different sense modalities, Weber's Constant remained a mystery for so long that it is often forgotten that it must have a source, that it must be a derivable consequence of more fundamental theoretical considerations. Thus it comes as something of a surprise to discover that in those experiments that use a 75% threshold, θ^* is Weber's Constant.

To derive this famous constant from the quantum transfer of energy to the sensory transceivers let $L\alpha$ and $M\beta$ be the intensity parameters of the two stimuli $\mathbf{S}_A$ and $\mathbf{S}_B$, respectively. The value of θ^* for the difference between $\mathbf{S}_A$ and $\mathbf{S}_B$ is from Equation 12.10,

$$\theta^* = ln(L\alpha/M\beta).$$

The difference between these parameters, $L\alpha - M\beta$, equals $\Delta M\beta$, the mean difference between the two stimuli, and also the amount of stimulation necessary to increment $M\beta$ to $L\alpha$. Re-writing θ^* gives

$$\theta^* = ln[(M\beta + \Delta M\beta)/M\beta]$$
$$= ln[1 + \Delta\beta/\beta].$$

Expanding as a Maclaurin series yields,

$$\theta^* = [\Delta\beta/\beta] - (1/2)[\Delta\beta/\beta]^2 + (1/6)[\Delta\beta/\beta]^3 - \ldots$$

Empirical observation shows that $\Delta\beta/\beta$ is very much less than one ($\Delta\beta \ll 1$), often on the order of 0.05, so that the higher order terms [such as $0.5(0.05)^2 = 0.00125$] of this expansion contribute little to the numerical value and can be ignored. Therefore,

$$\theta^* \cong \Delta\beta/\beta. \qquad (12.11)$$

This is Weber's Constant with respect to the Poisson intensity parameters that create the signals representing the two stimuli. Under the broad conditions discussed earlier the Poisson intensity can be characterized as a similarity transformation of stimulus magnitude. Thus the Poisson intensity parameters may be set equal to the stimulus magnitudes multiplied by a constant, say w. Therefore,

$$\theta^* = w\Delta\beta/w\beta$$
$$= \Delta s/\mathbf{S}_B \qquad (12.12)$$

That is, θ^* equals Weber's Constant.

If β is small, near 0, this relation breaks down, and additional terms of the Maclaurin series must be used to compute θ^*. Examining the previous Maclaurin series shows that for small β,

$$\theta^* \cong \Delta\beta/\beta - (1/2)(\Delta\beta/\beta)^2$$
$$= \Delta s/\mathbf{S}_B - (1/2)(\Delta s/\mathbf{S}_B)^2.$$

In this case a graph of Δs versus $\mathbf{S}_B$ is decidedly non-linear, containing a quadratic component. The size of Δs needed for "Threshold" discrimination must be larger than that predicted by Weber's Law. In fact,

$$\Delta s = \mathbf{S}_B[1-(1 - 2\theta^*)^{1/2}]. \qquad (12.13)$$

Note that zero is the physical measurement of zero and not the physical measure that corresponds to a stimulus at absolute threshold.

A NEW PSYCHOPHYSICAL LAW

The derivation that $\Delta s/\mathbf{S}_B = \theta^*$ did not depend upon actually observing 75% threshold responding, only that $\Delta s/\mathbf{S}$ is typically small. Thus the demonstration that θ^* equals Weber's Constant produces a more general law because the ratio $\Delta s/\mathbf{S}$ applies more widely than to just 75% threshold responding. More generally, the probability of judging a comparison to be larger than a standard is:

$$P_A = \frac{e^{(\Delta s/S_B)A} - 1}{e^{(\Delta s/S_B)A} - e^{-(\Delta s/S_B)A}}.$$

$$= \frac{1}{1 + e^{-(\Delta s/S_B)A}}. \qquad (12.14)$$

where $\Delta s/\mathbf{S}_B$ is called the **Generalized Weber Constant**.

The response probability P_A is now written in terms of stimulus magnitude and the unknown value of the response threshold, A. Weber's Law applies only to threshold discrimination, that is, to either 75% correct responses for difference thresholds and/or 50% correct responses for studies of absolute thresholds. This law, Equation 12.14, applies across the entire range of response probabilities.

An excellent testing ground for this new psychophysical law is the paired comparisons of lifted weights conducted by Guilford (1931; chap. 4, Fig. 4.6). The weights ranged to 215 g from 185 g in steps of 5 g. Thus values of $\Delta s/\mathbf{S}$ range from the smallest value of $(185 - 215)/215 = -0.14$ to the largest, $(215 - 185)/185 = 0.16$. The comparisons, presented 100 times in each order, generate the results in Table 4.2. If Weber's Law applies to lifted weights, as all previous experiments show, then the results from Guilford's study should also conform to Weber's Law. If so, only a single parameter, A, accounts for the entire set of results (Link, in press).

An estimate of A is easily obtained from the logarithms of the reciprocals of the values of P_A. From Equation 10.4,

$$ln[P_A/(1 - P_A)] = \theta^* A,$$

and from Equation 12.12,

$$\theta^* A = (\Delta s/\mathbf{S}_B)A.$$

Therefore,

$$ln[P_A/(1 - P_A)] = (\Delta s/\mathbf{S}_B)A. \qquad (12.15)$$

Moreover, because the values of $\Delta s/\mathbf{S}_B$ are fixed, rather than random, variables, the method of least squares can be used to estimate A, the unknown value of the Slope in Equation 12.15.

Noting that

$$\theta^* A = (\Delta s/\mathbf{S}_B)A$$
$$= ln[P_A/(1 - P_A)]$$

gives the relation,

$$A = \frac{\mathbf{S}_B\{ln[P_A/(1 - P_A)]\}}{\Delta s}.$$

This dimensionless quantity measures the subjects resistance to respond in units of stimulus difference, Δs. Thus for experiments using constant stimuli, and thus

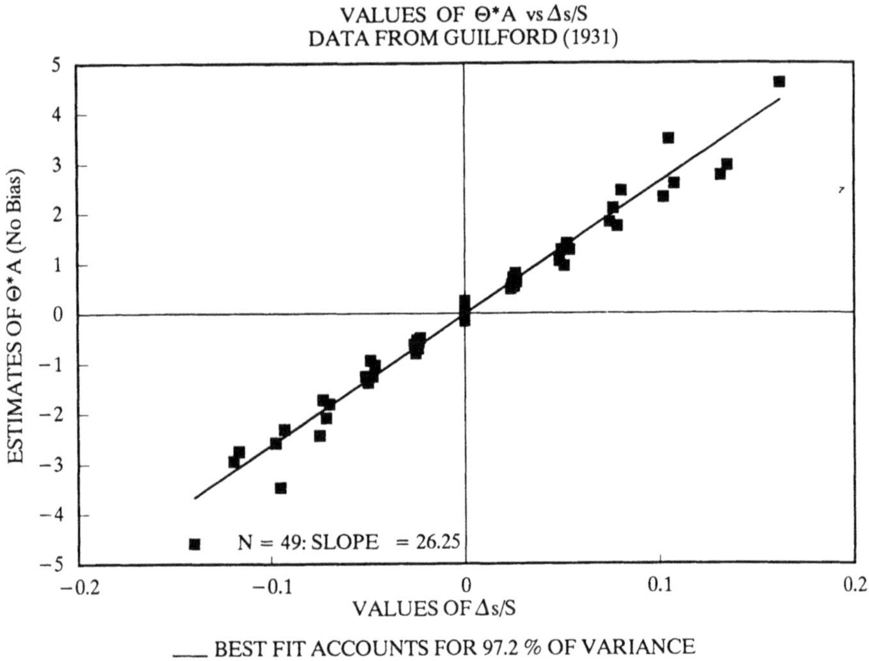

FIG. 12.3. Values of θ^*A estimated from response probabilities are nearly a linear function of $\Delta s/\mathbf{S}$.

known values of Δs, this measure provides a method of determining whether the experimenter's admonitions to control the value of A actually influence the subject's behavior. In experiments such as Guilford's (1931) study of lifted weights the small range of weight, 185 g to 215 g, suggests that the assumption of a constant value for A is reasonable.

Assuming A to be constant during the experiment reported by Guilford (1931) means that all response probabilities depend upon the known values of $\Delta s/\mathbf{S}_B$. A least-squares analysis, depicted in Fig. 12.3, determined the percentage of variance accounted for by the linear relation between θ^*A and $\Delta s/\mathbf{S}$ to be 97.2. For theoretical consistency this analysis assumed the Intercept to be zero but estimation of the Intercept yielded a value of only -0.06, a value that is quite close to zero given that the range of values of θ^*A is nearly 10. Finally the slope, which is 26.25, is an estimate of the common value for A. Thus, knowing that θ^* is a similarity transformation of $\Delta s/\mathbf{S}$ means that all of Guilford's results are accounted for by a single parameter.

The predicted and observed response proportions are plotted in Fig. 12.4 as a function of the comparison stimulus values. The goodness of this single-parameter fit to Guilford's data is measured by a chi-squared analysis of the observed and predicted response proportions. Guilford stated that the probabilities in opposite cells of the paired-comparison matrix were obtained indepen-

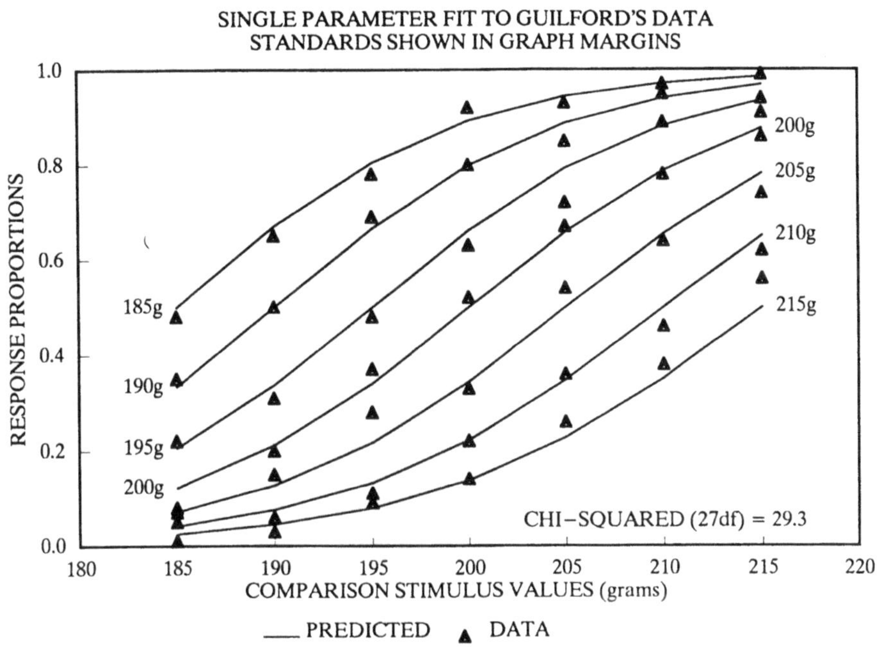

FIG. 12.4. Guilford's (1931) paired comparisons of lifted weights generated response proportions ranging from 0.01 to 0.99. A new psychophysical law using only one parameter predicts the results.

dently. This seems unlikely because they always sum to unity. Depending upon what is to be believed, the number of parameters for the chi-square statistic is either $49 - 1$ or $[(49 - 7)/2 + 7 - 1] = 27$. In any case the goodness of fit is quite remarkable. The chi-squared value is only 29.3. A fit based on estimating no parameters at all is described by Link (in press).

This theory makes no appeal to error as an explanation of discrimination. Rather, discrete receptive units combine their responses to sensory stimulation to produce electrical signals. Differences between these signals drive the decision process. Mathematical analysis of the decision process shows that the response probabilities are Logistic Response Functions. The relation between θ^* and $\Delta s/S$ yields a general psychophysical Law that includes Weber's Law. As well as illustrating that Gaussian error is unnecessary, there is no longer a need to imagine stimuli as ordered on a Psychological Continuum. Instead, the value of θ^* captures properties of the difference between waves representing standard and comparison stimuli. Analyzing how θ^* changes, given a fixed-standard and variable-comparison stimulus, produces a rewarding view of changes in discrimination.

The question of how large $\Delta s/S$ can become before the approximated values for response probabilities become too inaccurate is answered by Fig. 12.5. To illustrate how the probabilities vary a range of Δs values somewhat greater than

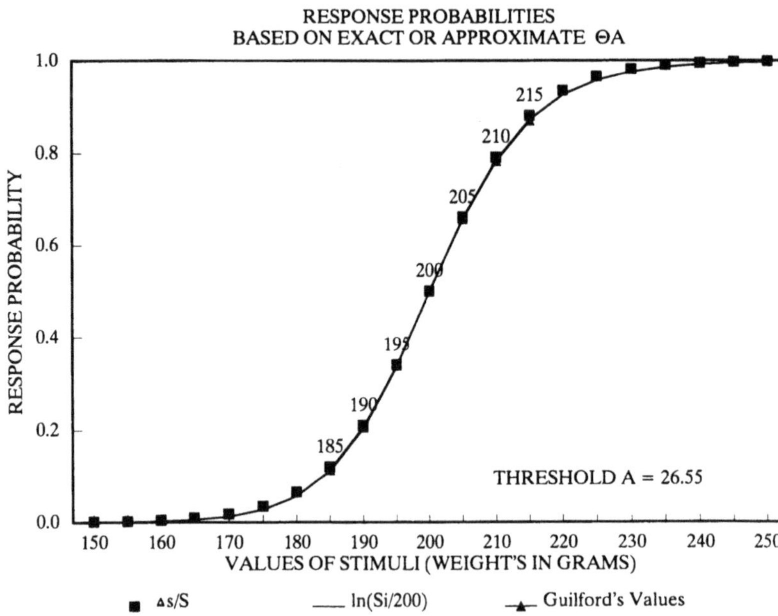

FIG. 12.5. The predicted probabilities and observed response proportions for a new psychophysical law.

those used by Guilford spans 100 g with a fixed standard at 200 g. Actual response proportions from Guilford's study are shown by closed triangles which are mostly obscured due to the closeness of the theoretical fit. The exact and approximate values of P_A are plotted as a function of the comparison stimulus, $\mathbf{S} + \Delta s$. The smooth curve shows the exact value for predicted probabilities. The form of the curve is similar to that for a Gaussian distribution function. Approximations of these values using the approximation $\theta^* \cong \Delta s/\mathbf{S}$ show how very close the resulting response probabilities come to their exact value.

There is no question of the closeness of fit in the range of values used by Guilford and used commonly in psychophysical studies. Inaccuracies might be expected to occur at the extremes where the Maclaurin series expansion of θ^* may require additional terms. But at the extremes the probability of responding R_A is either near zero or close to one. Thus, for extreme values of $\Delta s/\mathbf{S}$, where inaccurate predictions might occur, both the predictions and performance have only a small range within which to vary. A vast number of trials may be necessary to distinguish between the exact result and this accurate approximation.

13 A Theory of Sensation

"We call the psychological elements of ideas Sensations," Wundt announced in the Second *Lecture on Human and Animal Psychology* (1896, p. 15). Later Titchener (1929) summarized Wundt's thoughts:

> Analysis of the processes of inner life brings us, in the last resort, to pure sensations, constituted originally of intensity and quality. Sensations carry no reference; they look neither before nor after; they tell us nothing of their stimuli, whether external or organic, and nothing of their point of excitation, whether peripheral or central, nor do they forecast the ideas in which we find them synthesized. They simply run their course, qualitatively and intensively, . . . Ideas, in their turn, are originally constituted of these sensations . . . (p. 14)

Wundt's proclamation met with little resistance because the sensory basis of mental life was already a cornerstone for theories of mind. In this view, a view consistent with the mental chemistry of Mill, the *petites perceptions* of Leibnitz, the sense psychology of Herbart, and many other mental theorists, sensations are such fundamental elements of psychology that they constitute the raw atoms of mind. Even William James (1893), a principal antagonist of Fechner's Psychophysics, observed that, "Every sensation presents itself as an indivisible unit" (p. 23).

This 19-century belief held that these unitary elements of mind arise through the action of sensory receptors: Sensation results from the transfer of energy to internal receptive sites from the peripheral receptors. The view gained support from the observation that both the quality and intensity of sensation are easily changed by application of analgesics, intoxicants, and medications that appear to diminish, mask, or cloud internal reception.

Yet, the emergence of a sensation appears to require more structure than just the transfer of external energy to an internal receptive site. Studies of Yogic practices, for example, reveal that the physiological-receptor processes can be supplemented by a form of mental control that alters the intensity and quality of sensation. This ability to enhance mental awareness is not only possible but is also a highly sought after goal. A more common but often overlooked form of the mental control of sensation is the ability to diminish awareness. The ability is easily demonstrated by such personal experiences as becoming "absorbed" in reading a book or other activities. A stimulus presented during the time of this absorption goes unnoticed when at other times the same stimulus may evoke a quick response. The fact that the awareness of a stimulus depends on a mental state suggests that sensations are not simply naive, inner, representations of the external world. In some way sensation, the awareness of the stimulus, depends upon the mental state of the mind creating the sensation.

To extend the analysis of sensations, and peer more deeply into the processes by which they arise, requires a theory to integrate both the physical and mental components of sensation. Defining these separate components of sensation splits the atom of the mind into two parts. One part depends upon the physical stimulus, the other upon the subject's resistance to respond. The theory of sensation shows how these components work together to cause awareness.

The first concern is to produce a model of the phenomenon and from the model develop a measure of the object of study. Fechner, realizing that a tool for measuring the outcome must be developed before experiments can begin, approached the problem of measuring the quantity of sensation through a brilliant extension of Weber's Law. Fechner sought to determine the sensation magnitude of a comparison stimulus that produced a just noticeable difference from a standard. Experimental results showed that not just one, but two different comparison stimuli produced a just noticeable difference (*jnd*). One comparison stimulus appeared as just noticeably different when the magnitude of the comparison increased from the value of the standard. The other just noticeably different comparison appeared when a large comparison stimulus decreased in value toward the standard. The existence of two different stimuli that evoked the same sensation, the same awareness of difference, troubled early psychophysicists.

In spite of this anomaly, one result remained consistent across a multitude of experiments: As the magnitude of the standard increased, the magnitude of the physical difference required for a just noticeable difference also increased. Weber discovered the necessary increases to be related to the magnitude of the standard. When the increment necessary to produce a *jnd*, Δs, was divided by the magnitude of the standard, $\mathbf{S}_B$, the ratio remained constant across a wide range of $\mathbf{S}_B$ values. This ratio, $\Delta s/\mathbf{S}_B$, is Weber's Constant. In an extraordinary insight Fechner saw that Weber's Constant expressed a constant increment of psychological experience, that is, a fixed sense of awareness, in terms of a relative increment in physical stimulation.

To follow Fechner's now famous argument, let $\mathcal{S}$ be the sensation produced by a standard stimulus $\mathbf{S}_B$ and let a just noticeable difference be an increment in sensation of size Δs. The *jnd* is produced by a comparison stimulus $\mathbf{S}_A = \mathbf{S}_B + \Delta\mathrm{s}$ which produces a sensation equal to the original sensation plus the new increment, that is, $\mathcal{S} + \Delta s$. Whatever the Psychophysical Function, say $f(\ )$, that relates sensation to physical magnitude it must conform to Weber's Constant. That is, a *jnd* occurs when the increment in stimulation to $\mathbf{S}_B + \Delta\mathrm{s}$ results in an increment in sensation to $\mathcal{S} + \Delta s$. Stated in terms of the unknown Psychophysical Function $f(\ )$, the increment in sensation is

$$\Delta s = f(\mathbf{S}_A) - f(\mathbf{S}_B).$$

Fechner saw in Weber's empirical discovery the relation,

$$\Delta s = \frac{\Delta \mathrm{s}}{\mathbf{S}_B}.$$

Therefore, on dividing both sides by $\Delta\mathrm{s}$,

$$\frac{\Delta s}{\Delta \mathrm{s}} = \frac{1}{\mathbf{S}_B}.$$

Substituting for the value of Δs gives,

$$\frac{f(\mathbf{S}_A) - f(\mathbf{S}_B)}{\Delta \mathrm{s}} = \frac{1}{\mathbf{S}_B}. \tag{13.1}$$

As $\Delta\mathrm{s}$ approaches zero the limiting value of the left-hand side defines the derivative of the unknown Psychophysical Function (cf. Luce & Edwards, 1958, for an extensive critique of this approach). The right-hand side is the value of the derivative. Because functions can be identified by their derivatives the unknown function $f(\ )$ can be identified. The function that has $1/\mathbf{S}_B$ as a derivative is the (natural) logarithmic function, $ln(\mathbf{S}_B)$. Fechner proposed a boundary condition that zero sensation occurs when the absolute-threshold stimulus is presented. This condition imposes a solution that yields a measure of the quantity of sensation called

Fechner's Law:

$$\mathcal{S} = \mathrm{k}\ ln(\mathbf{S}_B/\mathbf{S}_0) \tag{13.2}$$

where k is a constant

and $\mathbf{S}_0$ is the absolute threshold stimulus.

The amount of sensation is a logarithmic function of stimulus magnitude. According to this Law the Psychophysical Function is a logarithmic transformation of the stimulus magnitude scaled in units of the absolute threshold stimulus, $\mathbf{S}_0$. Fechner graciously named this law after Weber, but later theorists properly renamed it after Fechner.

THE POISSON MODEL FOR DISCRIMINABILITY

The Poisson model of the quantum response from a sensory surface or volume, and the associated accumulation of wave differences, produces Fechner's Law from new considerations. Let $\mathbf{S}_B$ be the standard stimulus and let $\mathbf{S}_A$ be a just noticeably different comparison. Let the presentation of $\mathbf{S}_B$ generate a Poisson wave with mean amplitude Mβ, based upon M transceivers, and let the just noticeably different comparison, $\mathbf{S}_B + \Delta\mathbf{S}_B = \mathbf{S}_A$, also have a Poisson wave with mean amplitude Lα, based upon L transceivers. The Poisson intensities for the two transceivers may be identical, that is, $\alpha = \beta$, but this is not necessary.

When $\mathbf{S}_A$ is compared against $\mathbf{S}_B$, differences between the waves for the two stimuli, Informative Values, accumulate until one or the other response threshold is first exceeded, as in Fig. 9.4. When the upper threshold is exceeded the subject responds that $\mathbf{S}_A$ occurred. If the lower threshold is exceeded first the subject decides in favor of $\mathbf{S}_B$.

The comparison of $\mathbf{S}_A$ to $\mathbf{S}_B$ generates a wave difference with expected value $L\alpha - M\beta$, and a value of

$$\theta^* = ln(L\alpha/M\beta)$$

which is a measure of the mean amplitude, and/or variance, of the comparison stimulus scaled in units of the mean amplitude, and/or variance, of stimulus $\mathbf{S}_B$. From the analysis of Poisson transceivers presented in chap. 12 these Poisson parameters are similarity transformations of the physical stimuli. That is, $L\alpha = w\mathbf{S}_A$ and $M\beta = w\mathbf{S}_B$. Therefore,

$$\theta^* = ln(\mathbf{S}_A/\mathbf{S}_B).$$

Because $\mathbf{S}_B$ is an arbitrary standard it may be taken to be $\mathbf{S}_0$, the absolute-threshold stimulus. Then,

$$\theta^* = ln(\mathbf{S}_A/\mathbf{S}_0).$$

This relation between θ^* and stimulus magnitude is shown in Fig. 13.1. This logarithmic law is the same as Fechner's Law except for a multiplicative constant. Thus the logarithmic relation is a necessary consequence of the Poisson foundations of Wave Discrimination Theory.

COMPONENTS OF SENSATION

Sensation stirs our daily lives, provides feedback about success and failure, and provides awareness of the many different feelings that distinguish animate creatures from their inanimate surroundings. Measuring sensation is fundamental, as all psychophysicists agree, but sensation is more than just an object of wonder and curiosity. Sensation is the psychological energy that creates mental sub-

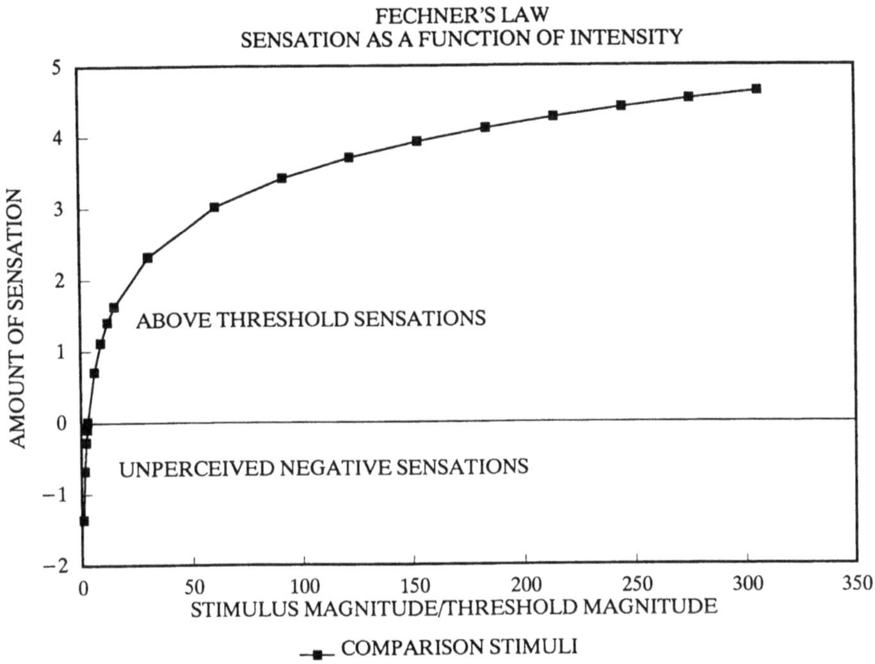

FIG. 13.1. Fechner's Law originally derived from the assumption that Weber's Constant represented a derivative of the Psychophysical Function shown here. In Wave Discrimination Theory, Fechner's Law is a consequence of the theory of discrimination.

stance. It is as fundamental to psychology as energy is to physics, for sensation is the energy that drives psychological processes. Representing sensation in a manner that fulfills this fundamental role leads to a major unification of psychophysical theory.

Although the mechanism for generating sensation is unknown, much is known about sensation through personal experience. The absolute-threshold stimulus, for example, is the stimulus that creates a psychological state at the edge of awareness and is the stimulus that is compared against an internal referent in order for a judgment about its presence to occur. Formally the absolute threshold is the physical amount of stimulation needed to evoke an awareness of the stimulus with probability 0.50. These attempts to "feel" a stimulus depend upon an accumulation of Informative Values that lead either to success or to failure with probability 0.50. Thus the absolute threshold stimulus is not a stimulus without a sensation. Rather it is a stimulus that gives rise to sensation with a probability of 0.50.

For an unbiased comparison process the value of θ^* consistent with a response probability of 0.50 equals zero. Yet it is not θ^*, but the process of accumulation that leads to the awareness of a stimulus. The accumulation ceases when the value A is first reached. Then, the subject senses the stimulus. The awareness, or sensation, appears to depend upon both θ^* and A. To impose Fechner's boundary condition, the sensation generated by the absolute-threshold stimulus is zero. In order for the amount of sensation to be zero, and yet depend on A, sensation is postulated to equal discriminiability multiplied by the resistance to respond,

$$\mathcal{S} = \theta^* A.$$

That is,

$$\text{Sensation} = \text{Discriminability} \times \text{Resistance}. \tag{13.3}$$

Substituting for the value of θ^* yields

$$\mathcal{S} = ln[(\mathbf{S}_A/\mathbf{S}_0)]A. \tag{13.4}$$

which maintains the property that when the comparison stimulus $\mathbf{S}_A$ equals the absolute-threshold stimulus $\mathbf{S}_0$ the sensation equals zero.

Formally this is Fechner's Law with the multiplicative constant. The relation between θ^* and A maintains the logarithmic increase in sensation as a function of stimulus magnitude. Furthermore, the relation $\mathcal{S} = \theta^* A$ derives directly from the accumulation of wave differences rather than the differential equation so soundly criticized by Luce and Edwards (1958).

Discriminability depends upon the two parameters, $L\alpha$ and $M\beta$, that characterize the absolute threshold and comparison stimulus waves. The amount of sensation depends upon discriminability and the subject's resistance to respond. This measure of sensation is consistent with previous meanings (e.g., Wundt, 1896) but is enriched by the expression of subjective control over the amount of sensation. The inclusion of this subjective component of sensation changes dramatically the interpretation of many experimental results.

There is no better place to examine the application of this relation than the experiments first conducted by Fechner. A major achievement of Fechnerian psychophysical theory was the discovery that three different experimental methods produced similar measures of the just noticeable difference. The stimulus difference producing a unit of sensation, the *jnd*, found by using the Method of Limits generated about 75% correct responses using the Method of Right and Wrong Cases. It was this result that gained the Gaussian-error theory favor and the general acceptance of 75% correct responding as an indicator of a difference threshold.

In Wave Discrimination Theory, when $\theta^* A$ equals 1, the probability of correctly discriminating the comparison from a standard equals 0.731. The closeness of 0.731 to 0.750, the classical value used to separate stimuli by a *jnd*, suggests that when there is a unit value for $\theta^* A$ there is a unit value of sensation, one *jnd*. This value occurs when the discriminability of the stimulus difference is balanced exactly by the resistance to respond. The value of θ^* derives from the experimental stimuli and the physical features of the subject's sensory system, whereas the value of A is controlled by the subject. Thus, in the Method of Right and Wrong Cases the subject may control A in order to discriminate correctly in about 75% of the cases. Moreover, as Johnson's data suggest, the subject may adjust A to provide performances compatible with the experimenter's instructions, explicit or otherwise. Increasing A generates a greater resistance to respond and promotes greater accuracy at the expense of longer response times.

A subject controlling the resistance to respond is a subject who also controls the amount of sensation, θ^*A, by varying the value of A. As resistance decreases small stimulus differences produce swift awareness of a stimulus because their accumulation quickly exceeds the threshold for awareness, A. Clinically, the subject is alert. At the other extreme a large value for A allows much unwanted stimulation to pass by unnoticed, as if the subject were absorbed by a different task. Such mental control of awareness is an inherent part of adapting to a changing sensory environment.

A theoretical difficulty that dogged the heels of Fechnerian theory is evident in Fig. 13.1. The ordinate represents amount of sensation and the abscissa, the physical magnitude of the stimulus. The abscissa is scaled in units of the absolute-threshold stimulus; the stimulus that is perceived just 50% of the time. Fechner proposed that the absolute-threshold stimulus be taken to define the point of zero sensation. In fact 50% of the time, the so-called absolute-threshold stimulus generates awareness of its presence, not zero awareness. Furthermore, stimuli much smaller than the absolute-threshold stimulus can also generate an awareness of their presence although the probability of this awareness occurring is small.

The placement of this zero point for sensation creates the problem of "negative" sensations. According to Fechner's Law any stimulus less in physical value than the absolute-threshold stimulus produces negative sensation. These are the "sub-threshold," or as they are sometimes called, "subliminal sensations," which are usually not perceived. Paradoxically, they arise from stimuli that are most often not felt. Yet, the stimuli generating these sensations are no different from other stimuli, their only distinctive property is that when compared against the absolute-threshold stimulus they are perceived to be greater than the absolute-threshold stimulus on less than 50% of the trials (cf. Wundt, 1896).

The Wave Discrimination Theory of discriminating the presence of a stimulus from no stimulus at all requires the subject to compare the experimenter's stimulus against an internal referent. If anything is absolute, it is this internal base against which a sample of the sensory surround is compared. Treating the internal referent as a standard, $\mathbf{S}_B$, and the stimulus used to determine an absolute-threshold as $\mathbf{S}_A$ provides an average difference between the two Poisson waves equal to, $L\alpha - M\beta$. Thus, a comparison stimulus which on the average is smaller than the absolute-threshold stimulus tends to produce negative values. The accumulation of these values, as in Fig. 9.4, tends to drift in a negative direction. Furthermore, the value of $\theta^* = ln[L\alpha/M\beta]$ is negative because the average, $L\alpha$, is smaller than the average, $M\beta$. Thus, the sensation $\mathcal{S} = \theta^*A$ is also negative.

When presenting stimuli for comparison against this absolute referent the process of accumulation may have no lower boundary.[1] Indeed, the experimenter must explicitly inform the subject of termination of the stimulus. Otherwise the subject continues to sample the sensory display, as in experiments on vigilance. Of course, there is some probability that a so called subliminal stimulus will still evoke response R_A because Informative Values sometimes exceed A, even when θ^* is negative, thereby causing the subject to detect a sub-threshold stimulus. By presenting stimuli sufficiently smaller than the absolute-threshold stimulus, $\mathbf{S}_0$, the probability of generating response R_A is made arbitrarily small. An occasional R_A response may appear to be the intrusion of unwanted experimental error. On a more disturbing note, repeated stimulation with such remote stimuli must eventually generate response R_A. In this way even stimuli that are rarely felt may still prompt explicit behavior. The production of a response may be simply inexplicable to the subject. To a Freudian analyst the response may be taken as a conscious manifestation of a subliminal process.

THE EFFECTS OF RESPONSIVENESS ON SENSATION

As a unit for measuring sensation none is better than the value of θ^*A which co-occurs with a *jnd*. In this case the value of θ^*A actually equals 1.0 and corresponds to a single unit of psychological experience representing a balance between the force of discriminability and resistance to respond. How this balancing act occurs is discussed in chap. 17. For now, to show how the value of A affects performance, let the standard be $\mathbf{S}_B$ and the comparison that is just noticeably different be $\mathbf{S}_A$. The Poisson waves corresponding to these stimuli have mean amplitudes $M\beta$ and $L\alpha$, respectively. The similarity transformation between Poisson means and physical-stimulus intensity provides that $\mathbf{S}_B = wM\beta$ and $\mathbf{S}_A = wL\alpha$. The just noticeable difference is a unit of sensation because

$$\begin{aligned}\mathcal{S} &= \theta^*A \\ &= 1.\end{aligned} \tag{13.5}$$

Substituting for the value of θ^*A yields

$$1 = ln[(L\alpha/M\beta)]A.$$

Multiplying the numerator and denominator by w does not change the value of

[1]See Emerson (1970) for an example of the application of Wald's sequential model to simple reaction time.

the expression. Therefore, $wL\alpha = \mathbf{S}_A$ may be placed in the numerator and $wM\beta = \mathbf{S}_B$ may be put in the denominator to give,

$$1 = ln[(\mathbf{S}_A/\mathbf{S}_B)]A.$$

Now, the unknown value for responsiveness equals Weber's Constant, as the following calculation shows:

$$\begin{aligned} 1/A &= ln[(\mathbf{S}_A/\mathbf{S}_B)] \\ &= \Delta s/\mathbf{S}_B \\ &= \text{Weber's Constant.} \end{aligned} \qquad (13.6)$$

The subject's responsiveness contributes as much to the determination of response probability as stimulus intensity. The influence of these various factors is evident in Fig. 10.8 where three Psychometric Functions show how performance improves as the value of A increases. These increases in A correspond to less responsive but more accurate subjects.

Noting that responsiveness equals $1/A$ gives another relation,

$$\text{Responsiveness} = \frac{\text{Discriminability}}{\text{Sensation}}. \qquad (13.7)$$

From which it follows that,

$$\text{Discriminability} = \text{Responsiveness} \times \text{Sensation}. \qquad (13.8)$$

At 73% threshold these various relations simplify because the value of sensation equals 1. Therefore, at (73%) threshold, discriminability equals responsiveness, that is, $\theta^* = 1/A$. This result produces important interpretations in the analysis of response times, where responsiveness is a major contributor to performance.

Summarizing the relations among many theoretical concepts proves simple when the basic variables of Sensation, Discriminability and Resistance are considered. Figure 13.2 is a schematic of their relationships, taken from a similar expression of Ohm's Law. The variable at the top of the triangle is Sensation. Dividing Sensation by Discriminability produces the value of Resistance. Dividing Sensation by Resistance yields the value for Discriminability. Finally, multiplying Discriminability by Resistance gives Sensation. Thus, knowing any two of these variables allows for calculation of the unknown third.

Figure 13.3 illustrates how joint changes in Discriminability and Responsiveness modify performance. The ordinate represents values of A, the response threshold, and the abscissa values of θ^*. Along any equal probability contour for P_A the subject's probability performance remains constant, but the contributions from the two different sources, θ^* and A, change dramatically. For what is called "threshold responding" there is a single curve, demonstrating that so-called "threshold responses" are available across a range of Discriminability.

Because the amount of sensation equals Discriminability multiplied by Re-

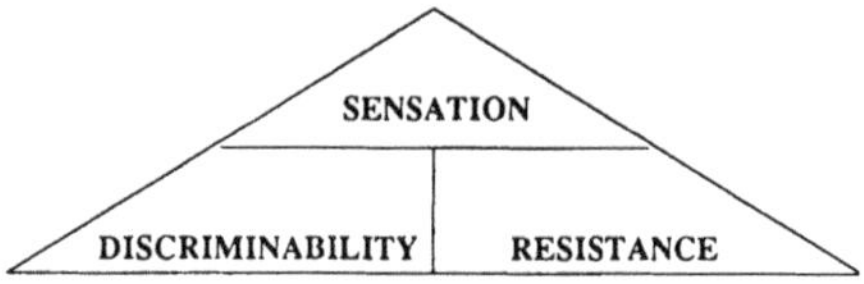

SENSATION = DISCRIMINABILITY x RESISTANCE

RESISTANCE = SENSATION / DISCRIMINABILITY

DISCRIMINABILITY = SENSATION / RESISTANCE

FIG. 13.2. The schematic representation of relations among the major theoretical variables of Sensation (S), Discriminability (D) and Resistance (R) shows that multiplying R by D gives Sensation. Dividing S by D gives R and dividing S by R gives D.

sistance each of these contours also represents a contour of equal sensation. The total amount of sensation is greater for contours farther from the origin. A unit of sensation occurs along a contour (not shown) for 73.1% correct responding that lies slightly nearer the origin than the contour for 75% correct responding. The figure emphasizes the relationship between response probability and the sense of the stimulus that we call "sensation." The two measures, one empirical, the other personal, are indicators of a common experience.

When the resistance to respond, A, is fixed, the subject's responsiveness is controlled. Then response probability changes due to θ^* occur along lines paral-

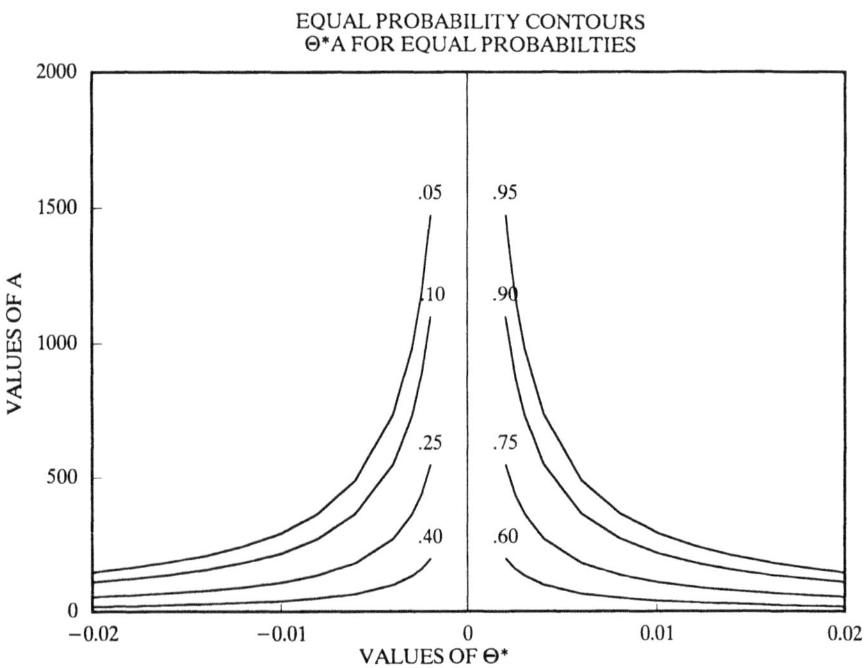

FIG. 13.3. Equal probability contours are shown in the plane relating values of θ^* to A. Because the subject can control A different performances can result from a fixed value of θ^*.

lel to the abscissa. The lines of equal responsiveness cut across the contours of equal probability. For a fixed value of A increasing the value of θ^* increases the value of P_A. If the probabilities are plotted against θ^* the result is a Logistic Response Function, that is, a Psychometric Function.

A line perpendicular to the abscissa produces increases in P_A due to increases in Responsiveness for a fixed value of θ^*, that is, a fixed Discriminability. If these probabilities are plotted as a function of A then, again, a logistic response curve results. Thus the Logistic Response Function has two sources, either changes in Responsiveness for a fixed stimulus which leads to undesirable variability in determining the value of a threshold stimulus, or changes in Discriminability for a fixed Responsiveness—the goal of many psychological studies.

Figure 13.3 also illustrates the tradeoff between sensitivity and responsiveness that maintains constant P_A. A very sensitive subject is sensitive because the amount of stimulus needed to evoke a transceiver response is small. As the receptive surface is degraded through injury or age, compensation for losses in sensitivity occur by increasing the response threshold. As the value of A increases, the subject becomes less responsive and longer periods of time are

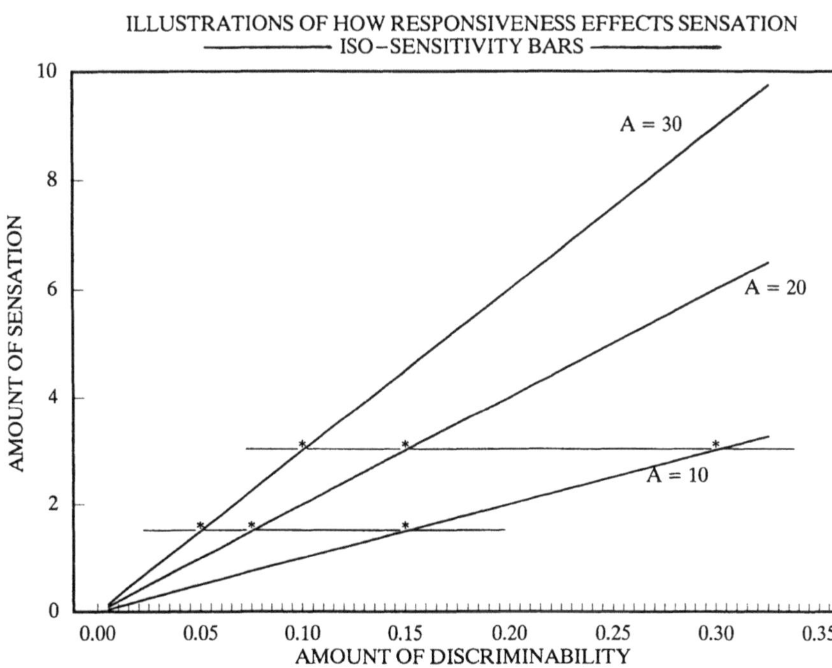

FIG. 13.4. Sensation is a direct function of Discriminability but the amount of Sensation increases as A increases.

required before a response occurs. Responses occur more slowly but performance remains high.

MEASURING THE RESISTANCE TO RESPOND

Responsiveness captures a behavioral characteristic of performance. The cause of this behavior is the subject's control of the resistance to respond. Measuring the resistance to respond is facilitated by inverting Equation 13.6 to obtain

$$A = -ln[(\mathbf{S}_B/\mathbf{S}_A)].$$

This measure assumes that the stimuli under comparison are those that generate a *jnd*. Thus $\mathbf{S}_B$ is the standard and $\mathbf{S}_A$ is the just noticeably different comparison. Note that the logarithm of $\mathbf{S}_B$ divided by $\mathbf{S}_A$ is approximately the reciprocal of Weber's Constant.

Figure 13.4 shows that the amount of sensation increases linearly with increases in discriminability. The slopes of the three examples are for values of A equal to 10, 20, and 30, respectively. The two horizontal lines in Fig. 13.4 are iso-sensation contours, equal sensations evoked along each line by stimuli of different magnitudes presented to a subject using response threshold values of either 10, 20, or 30. Along each line, different stimuli feel the same, but there are dramatic changes in performance. For a fixed value of θ^*, which defines discriminability as $ln(\mathrm{L}\alpha/\mathrm{M}\beta)$ or by approximation as $\Delta s/\mathbf{S}$, the three different values of A indicate larger response thresholds. For larger response thresholds more Informative Values must accumulate before a response threshold is met. Thus, for fixed discriminability, larger values of A produce longer times for responses to occur.

14 Discriminability and Responsiveness

SPEED VERSUS ACCURACY

The Wave Theory of Sensation suggests that performance variables such as response proportion and response time depend on two separate components; discriminability as measured by θ^*, and resistance as measured by A. Although many theorists suspected that subjective factors such as the "attitude of the subject" (Fernberger, 1914) influence performance, subjective factors were difficult to estimate even in those few quantitative theories (Cartwright & Festinger, 1943; Thomson, 1920) that might produce such measures. Signal Detection Theory did offer a measure of response bias but this is not the only subjective component of performance and is not related to the resistance to respond at all. The subject's attitude, molded by pressure to complete the experiment within certain time constraints, may reduce the resistance to respond so that trials take less time, thereby reducing accuracy. Previous measurements of thresholds, notoriously variable, could profit from a reanalysis of this subjective component. But such a theory did not exist at the time the measurements were made and the experiments were not conducted in a way that allows the joint influence of discriminability and resistance to respond to be revealed.

However, the experiments of Link and Tindall (1971), Link (1971), and Link (1978) permit a distinction to be drawn between changes in response time and proportion due to θ^* and changes due to A. These experiments extend the results of D. M. Johnson (1939) and Henmon (1911) who in turn advanced the ideas of Cattell (1902) and Peirce and Jastrow (1884). Link and Tindall (1971) also consolidated within a single design various features of previous experiments by Ollman (1966), Yellott (1967, 1971), and Link and Tindall (1970) that showed subjects capable of controlling their response times. The experiment sought more information about the mechanism governing this control because at the time a common belief posited that "reaction time" represented a measurement of the time required to activate a chain of synaptic elements within a stimulus-response path. Such a model of the stimulus-response framework quickly rejected the notion of voluntarily controlled response times. The experimental method is of major importance because the mathematical model of discrimination and the experimental method must complement each other (cf. Link, 1990b).

Experimental Details

A novel feature of the experimental method was the continuous visibility of the comparison stimulus until the end of the trial. In some classical studies the presentation duration of the comparison was fixed, sometimes very short by use of a tachistoscope (Titchener, 1901a). This procedure changes the physical nature of the comparison stimulus as soon as the visual display terminates. When the physical stimulus is removed from the visual field the visual field may still be sampled by the receptor system. Some experimental procedures introduced visual noise after a comparison stimulus presentation in the belief that the introduction of this additional stimulus (noise) facilitated the study of the comparison process. However, the introduction of external noise is simply the introduction of another exceptionally complex stimulus whose effect must also be modelled. To simplify the modelling of the comparison stimulus Link and Tindall (1971) displayed a fixed, visible, comparison until the subject responded.

The stimuli were white horizontal line lengths presented on a computer controlled X-Y display screen (Tektronix 602 with television phosphor). The standard, a 20 mm length viewed at a distance of 1 m preceded each presentation of the comparison stimulus. The comparison differed from the standard by 0, 1, 2, 3, or 4 mm. To control response bias either no difference or a fixed-size fixed-direction difference occurred on 50% of the 60 trials defining a single experimental block. Two subjects always saw a positive increment from $\Delta s = 0$, the remaining two subjects always saw a decrement. Subjects were instructed to respond *different* if the comparison stimulus appeared different in length from the standard, and to respond *same* when the two successively presented line lengths appeared equal in length. Although the subjects were to respond *same* or *different* the direction of any difference remained fixed for a particular subject.[1] Thus,

[1]At the time, Link believed the judgments to be based on the sameness or difference of the stimuli. The unidirectional design of stimulus differences suggested to Dalbir Bindra that the judgments did not involve sameness at all. As chap. 17 shows judgments of *sameness* produce results quite different from Link and Tindall. Therefore, I believe Bindra. These judgments are most likely based on an internal referent that permits the subject to judge a comparison stimulus to be either more or less than the internal referent.

from the subject's point of view the comparison stimulus either did not differ from the standard or differed in a fixed direction. This design feature maintained consistency with previous psychophysical research. At the end of each trial subjects were informed on the X-Y display device whether the response was *Correct* or in *Error*.

The second feature of this experiment was the method used to demonstrate the subject's control over their decision response times—that is, the subjective control of response time. To examine the influence of instructions to control response time subjects were instructed to be as accurate as possible, or to respond before a 460 ms response-time deadline (RTD), or to respond within a 260 ms RTD. The instruction remained in effect during a block of 60 trials. With the aid of feedback following each trial, (TOO SLOW or OK) subjects mastered this task without difficulty. Response times were well within the deadline time. Only about 5% of response times exceeded a deadline. Further, as Link (1971) showed, subjects easily respond to random, trial by trial, variation in the RT deadline.

Each trial began when the subject depressed with the right forefinger a trial initiation key placed on top of a response panel. This caused the presentation of a standard horizontal line segment 2.0 cm wide to appear at the center of the X-Y display screen for one second. Following the termination of the standard the X-Y display remained blank for 0.5 sec. Then a horizontal comparison stimulus presented at the center on the X-Y display remained visible until the subject released the forefinger from a trial initiation key (TIK) and then depressed one of two choice keys [*same* or *different*] about 10 cm distant from TIK. The response keys were counterbalanced across subjects, but for any one subject the keys remained fixed for the duration of the experiment.

For a particular subject, the comparison stimuli differed from the standard in only one direction. For this analysis the instances of a standard followed by itself were identical for all subjects. An exact analysis requires that a distinction be drawn between those subjects where the comparison stimuli are greater than the standard and those where the comparisons are less than the same standard. Analysis of the data revealed that if any differences due to direction were to appear, many more trials than those obtained here would be necessary to reveal them. To illustrate certain features of discrimination that do not depend upon this distinction the cases of difference are treated as absolute differences of 1Δs to 4Δs.

Subjects were very well practiced at the task by completing nearly 1000 trials before the data contributing to this analysis were collected. Each subject contributed a total of 4,800 trials making for a total of 19,200 for the four subjects. The first 10 trials of each 60-trial block were treated as practice. They do not enter into this analysis. The details of an experimental trial are shown in Fig. 14.1, which illustrates the sequence of events that comprised these self-paced trials.

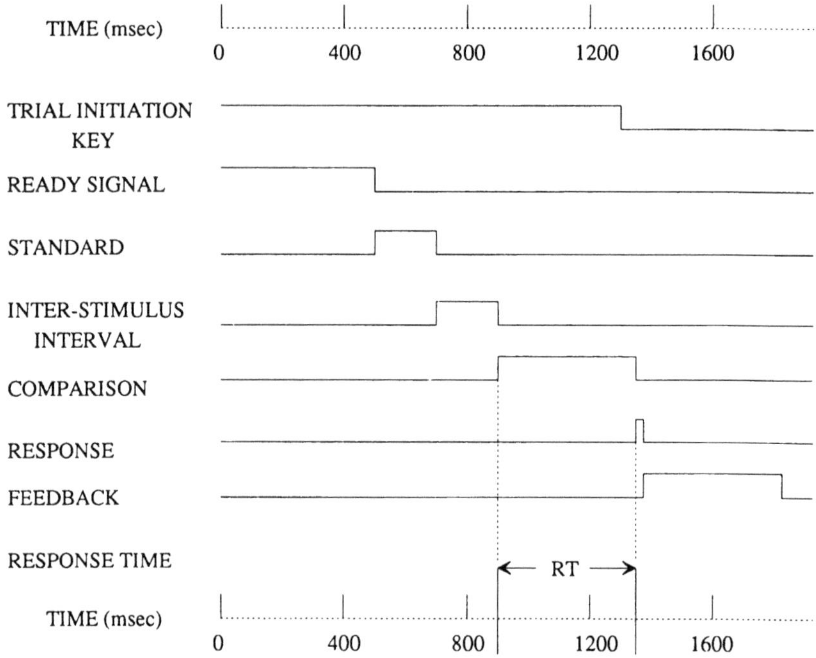

FIG. 14.1. A typical timing sequence for self-paced psychophysical experiments.

EXPERIMENTAL RESULTS

Figure 14.2 presents the mean response times for each response to each stimulus. Mean correct response time is longer than mean error response time in all response-time deadline conditions and for all sizes of stimulus-difference. Both the stimulus difference, which ranged from 1 mm to 4 mm and the response-time deadline (RTD) dramatically affected response times. The RTDs especially pushed response time down to remarkably low levels of only 180 msec.

In spite of these fast responses there is no evidence for guessing. The ROC space in Fig. 14.3 presents the proportion of *different* responses when a difference occurred, $Pr\{D \mid D\}$, as a function of the proportion of *different* responses when no difference occurred, $Pr\{D \mid S\}$. The three RT deadlines and four stimulus-difference conditions yield 12 data points. All data points lie above the main diagonal which represents guessing. Also all data points lie near the anti-diagonal suggesting that there is no response bias.

Within each RT deadline condition the data points in Fig. 14.3 are ordered. A smaller stimulus difference produces a point nearer the main diagonal. The nearer a point is to the main diagonal the closer the performance is to chance

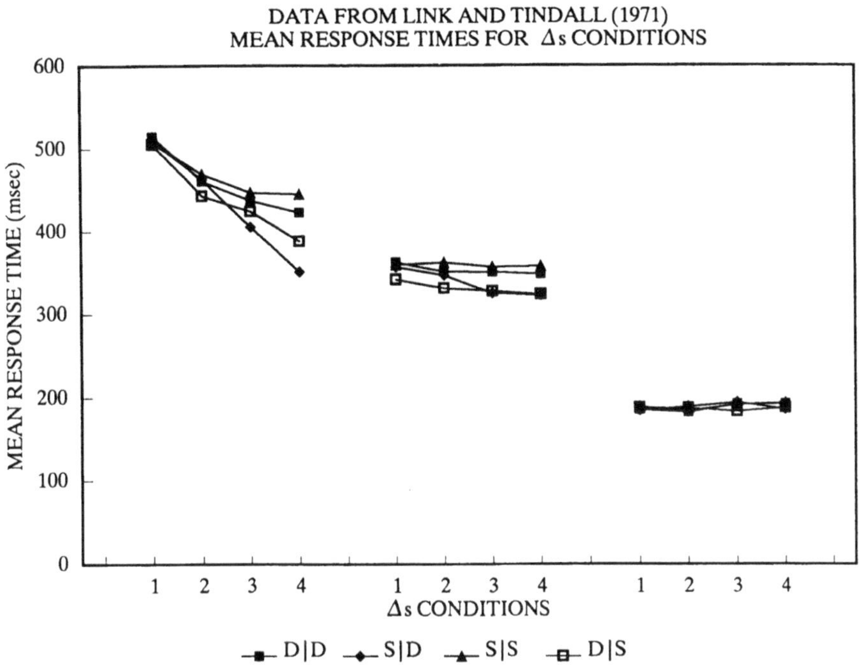

FIG. 14.2. Mean response times across subjects for each combination of response-time deadline and stimulus-difference condition.

guessing (chap. 8). Thus, reducing the stimulus difference to 1 mm from 4 mm promotes a decline in performance under each RT instruction. However, the four data points that are so nicely ordered within an RT deadline are not so easily ordered across deadline conditions. For example, the presentation of a 1-mm difference in the accuracy condition generates worse performance than the presentation of a 4-mm difference in the 460-ms deadline condition. This somewhat confusing aspect of the ordering of performance is straightened out by analyzing the data in terms of the theory.

These experimental operations affect the values of both θ^* and A. The value of θ^* should increase as a function of stimulus-difference. Values of A control the resistance to respond. For a fixed RT deadline, a fixed value of A should apply to all comparisons. According to the Poisson representation the value of θ^* derives from the equation,

$$\begin{aligned}\theta^* &= ln(\mathbf{S}_i/\mathbf{S}_0)\\ &\cong \Delta s/\mathbf{S}_0.\end{aligned}$$

Because $\mathbf{S}_0$ is fixed, estimates of $\theta^* A$ should increase as a linear function of Δs. When plotted as a function of Δs the values of $\theta^* A$ should increase with constant slope that depends upon A. Presumably the more strict the response-

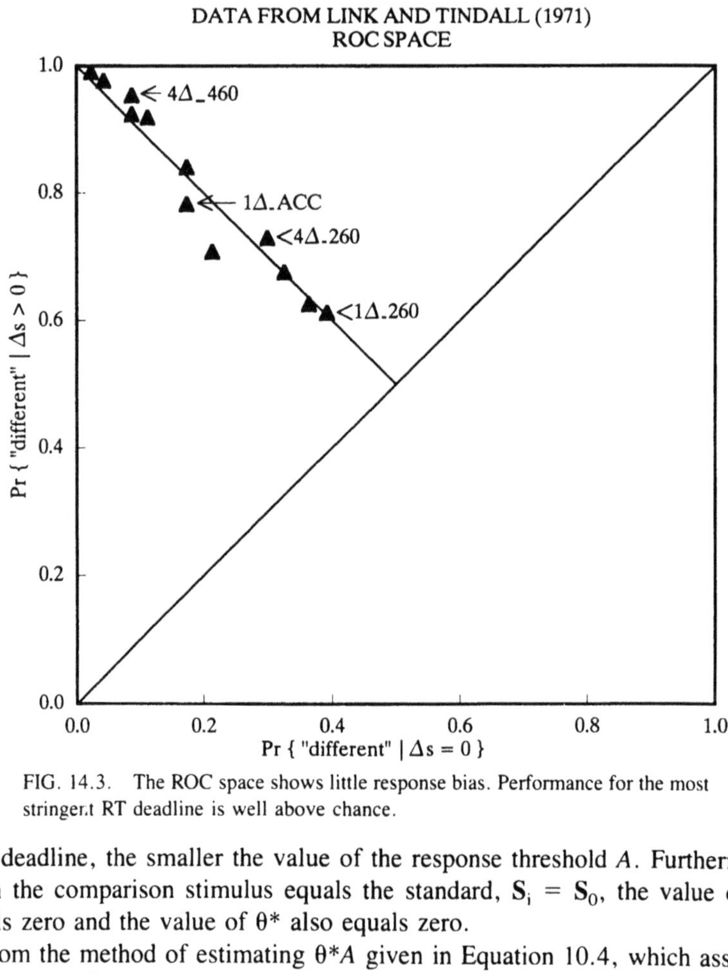

FIG. 14.3. The ROC space shows little response bias. Performance for the most stringent RT deadline is well above chance.

time deadline, the smaller the value of the response threshold A. Furthermore, when the comparison stimulus equals the standard, $\mathbf{S}_i = \mathbf{S}_0$, the value of Δs equals zero and the value of θ^* also equals zero.

From the method of estimating θ^*A given in Equation 10.4, which assumes no response bias,

$$\theta_i^*A = ln\left[\frac{Pr\{different|S_i\}}{Pr\{same|S_i\}}\right] \tag{14.1}$$

or,

$$\theta_0^*A = ln\left[\frac{Pr\{different|S_0\}}{Pr\{same|S_0\}}\right] \tag{14.2}$$

The estimates of θ^*A obtained from each experimental condition appear in Fig. 14.4. The upper three straight lines provide estimates of θ_i^*A obtained from the data when a difference between the standard and comparison did exist, Equation 14.1. The larger the value of θ_iA the larger the probability of a correct

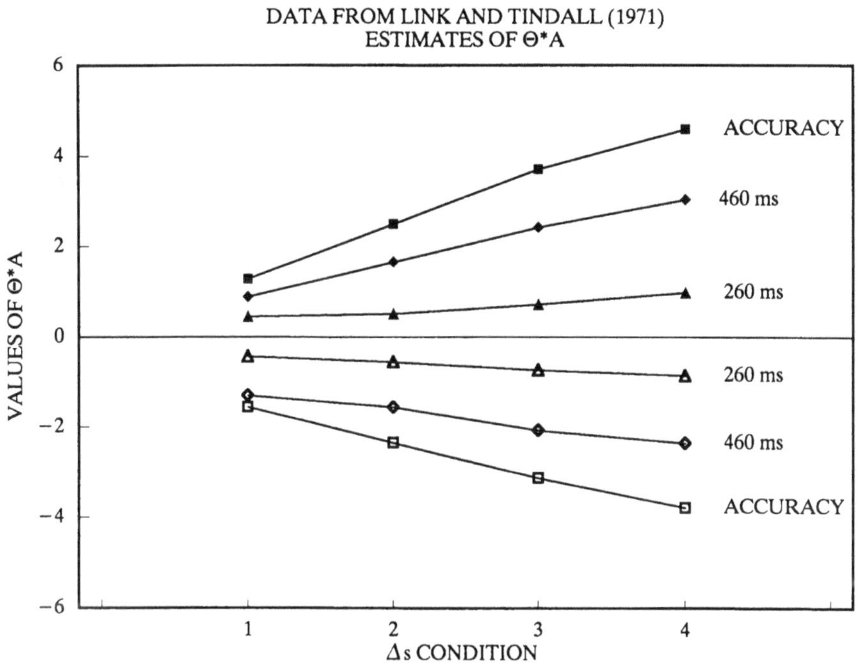

FIG. 14.4. Estimates of θ^*A for each RT deadline and each value of Δs show the predicted linear increase as a function of Δs and a change in slope as a function of the RT deadline.

response. The uppermost line shows estimates of θ_i^*A for the accuracy condition. The intermediate line shows estimates of θ_i^*A for the 460-ms deadline condition. Again the predicted linearity is evident. Even the lower line showing θ_i^*A estimates for the 260-ms condition where discrimination appears to be the worst shows a slight increase as a function of Δs.

Estimates of θ_0^*A derive from comparisons when there is no difference between the standard and comparison stimulus (Equation 14.2). The data hardly demonstrate chance performance, as one might expect when there is no difference presented to the subject. Instead the estimates of θ_0^*A are nearly mirror images of those obtained using the data for differences. Again the slope of the lines depends upon response-threshold values that increase as the time deadline increases. Values of θ_0^*A increase linearly across Δs conditions even though for the data used to obtain these estimates no difference is presented to the subject. This result disturbs the ancient belief that the subject judges a comparison stimulus against the standard presented by the experimenter, but is consistent with many results suggesting the existence of a psychophysical referent used for the purpose of comparison (Wever & Zener, 1928).

The values of θ_0^*A obtained when there is no physical difference between the standard and comparison are not zero at all, but systematically change in the same way as when a stimulus difference does exist. This result is contrary to the predictions of either Fechner or Thurstone. Signal Detection Theory may assume that the comparison is not judged against the standard but is judged against a criterion equal to the midpoint between the standard and comparison stimulus. Several other possibilities suggest themselves. First, these well-practiced subjects may simply use the standard as a warning signal. Then the comparison stimulus is judged against an internal referent that differs from the standard. In any case, the average value of the referent adapted to this discrimination experiment must lie between the standard and comparison in order to generate the mirror-image values of θ^*A shown in Fig. 14.4. Second, the difference between the standard and comparison may itself be compared against a referent. In this case a non-zero referent divides the range of stimulus-difference. In either case the stimuli are compared against an internal referent unequal to the standard.

The data conform to the general concept of how subjects can control response time by controlling the resistance to respond, that is, the response threshold value, A. If so, the data should conform to the ***RT***-versus-***Z*** relation which relates mean response time and response probability. From Equation 10.6, when two stimuli are compared, the mean number of samples before a decision is,

$$E\{\boldsymbol{N}_{ij}\} = \frac{A_j(2PC_{ij} - 1)}{\mu_i} \quad (14.3)$$

where PC_{ij} is the probability of a correct response given RT deadline j (accuracy, 460, 260) and stimulus-difference condition i (0, 1, 2, 3, 4).

How may this relation be tested when the values of A_j, which change across RT deadlines, and the values of μ_i, which change across stimulus differences, are unknown? A solution to this problem is to choose from among a huge set of parameters those values that meet a criterion of goodness of fit. However, the function generating the goodness of fit may be quite shallow in the range of the optimum parameter values, making the procedure for selecting parameters insensitive and thus unsatisfactory.

A different approach is to modify the relation to conform to the specific circumstances in which it is to be employed. In the case of the Link and Tindall experiment the most critical feature is the control of the response-threshold values. These are unknown but the values of θ^*A are known for each RT deadline and stimulus difference. Multiplying the numerator and denominator of Equation 14.3 by $\theta_j^*$and adding $E\{\boldsymbol{K}\}$, the mean nondecision time, gives,

$$E\{\boldsymbol{RT}_{ij}\} = \frac{\theta_i^* A_j(2PC_{ij} - 1)}{\theta_i^* \mu_i} + E\{\boldsymbol{K}\}. \quad (14.4)$$

The values $\theta_i^*A_j$, which now appear in the numerator, are estimated from response proportions. The only unknown is the value of the denominator. Even this may be simplified by reducing the data to only two cases—those where a stimulus difference did exist and those where it did not. This reduction leads to an equation where the numerator is estimable from the response proportions, the denominator is constant, but unknown, and the intercept value, $E\{\boldsymbol{K}\}$ is also unknown. Nevertheless the theory predicts that by plotting the mean response time for each deadline against the numerator, estimated for each deadline, the result should be a linear function. That is,

$$E\{\boldsymbol{RT}_{\mathrm{j}}\} = \frac{[\theta^* A_{\mathrm{j}}(2PC_{\mathrm{j}} - 1)]}{\theta^*\mu} + E\{\boldsymbol{K}\} \tag{14.5}$$

where the numerator can be estimated from marginal response proportions for each RTD.

The result of this analysis appears in Fig. 14.5. The intercept value proves to be 160 ms, which is quite reasonable given that the fastest average response time in the 260-ms deadline condition is 182 ms. Although some straight line fits any set of three points, in this case the intercept is sensible and the points do not deviate from the predicted linearity by less than the measurement accuracy of the clock recording the response times. Thus, the data satisfy the form of the ***RT***-versus-***Z*** relation previously shown to apply to the brightness-discrimination data of Kellogg but untestable in the previous experiment of D. M. Johnson (1939).

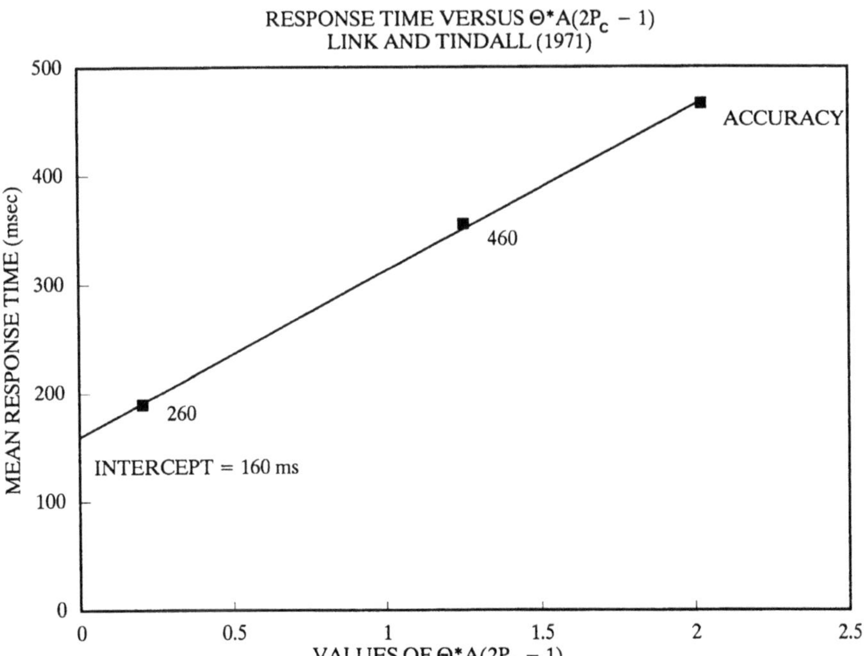

FIG. 14.5. The predicted linearity between mean response time and ***Z*** comes from the data of Link and Tindall (1971).

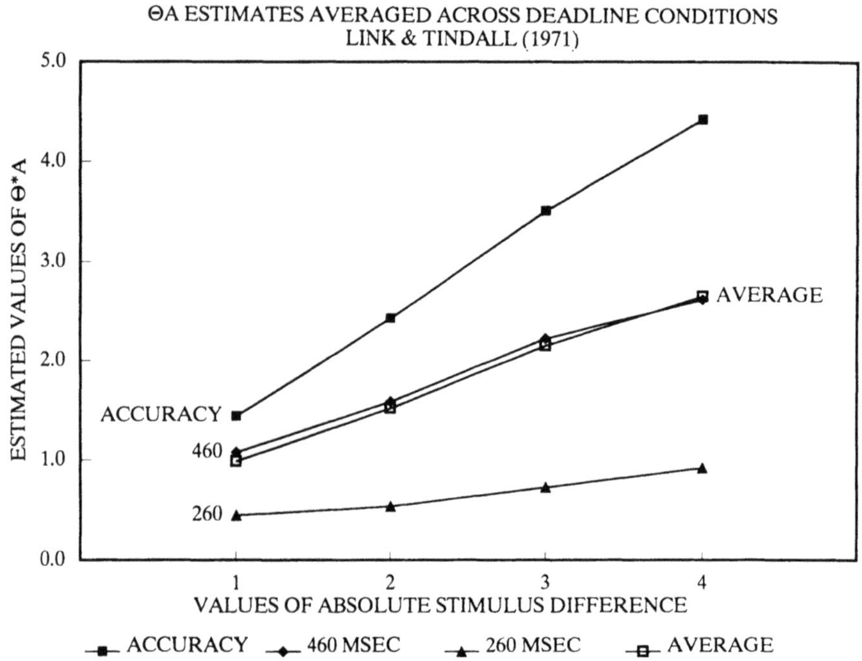

FIG. 14.6. θ^*A estimates for each stimulus difference and deadline were combined to produce this representation of the sensations evoked in Link and Tindall's (1971) experiment.

How do the subjects manage to control response times and also make judgments that show sensitivity to the stimulus differences. Theoretically the subject controls the resistance to respond, A. But how can this response threshold be manipulated on the basis of the stimuli presented to the subjects? There is no internal clock postulated by this theory, and indeed, no internal clock is necessary for the subtle control of response time. What is required is sensation, because by setting a value for A, the subject simultaneously senses the stimulus difference and also controls response time.

Estimates of θ^*A measure the experience of the subject. Averaging the individual-subject estimates of θ^*A within a stimulus-difference and RT deadline yields the surprising result shown in Fig. 14.6. These estimates of θ^*A vary linearly with stimulus-difference, yielding increases in sensation like those pictured in Fig. 13.4. The average values of θ^*A versus stimulus-difference are sufficiently similar to the estimates of θ^*A in the 460-ms deadline condition to suggest that the value of A used in the 460-ms condition is simply midway between the two values used in the accuracy and 260-ms deadline conditions. The measure of sensation, θ^*A, straightens out the difficult interpretation of the probability results in the ROC space by making evident differences in sensation that the ROC analysis cannot reveal. Quite clearly, as the RT deadline is relaxed, increases in response proportion come from systematic changes in the resistance to respond.

RESPONSE CONFIDENCE

Peirce and Jastrow

C. S. Peirce, the father of American philosophical pragmatism, discovered the first empirical relation between correct response probability and response confidence (chap. 2). Using a four point linear scale to measure confidence, Peirce and Jastrow (1884) found that the average confidence, m, was predicted closely by

Peirce's Law:

$$m = c\, ln\left(\frac{p}{1 - p}\right) \tag{14.6}$$

where p is the proportion of correct responses and c is a constant called by Peirce the "index of confidence."

If average confidence is a function of probability, then response probability must be a function of average confidence. Letting the scaling constant $c = 1/c'$, exponentiating Equation 14.6, and solving for p gives,

$$p = \frac{1}{1 + e^{-mc'}}. \tag{14.7}$$

This is the familiar logistic relation, Equation 10.1, that relates response probability and the amount of sensation, θ^*A. Here the relation appears between response probability and average confidence. Judging by similarity, the value mc' must equal θ^*A, or $m = c\theta^*A$. The demonstration that average confidence equals the amount of sensation, θ^*A, follows easily by noting that when the proportion of correct responses estimating P_A is substituted into Equation 14.6, the average confidence is

$$m = c\left[ln\left(\frac{P_A}{1 - P_A}\right)\right] \tag{14.8}$$

$$= c[\theta^*A]$$

$$= c[\text{Sensation}]. \tag{14.9}$$

Thus, sensation may be taken as the basis for Peirce's Law.

Average confidence is also proportional to the Generalized Weber Constant. From previous calculations the value of θ^* closely approximates $\Delta s/\mathbf{S}$, the Gen-

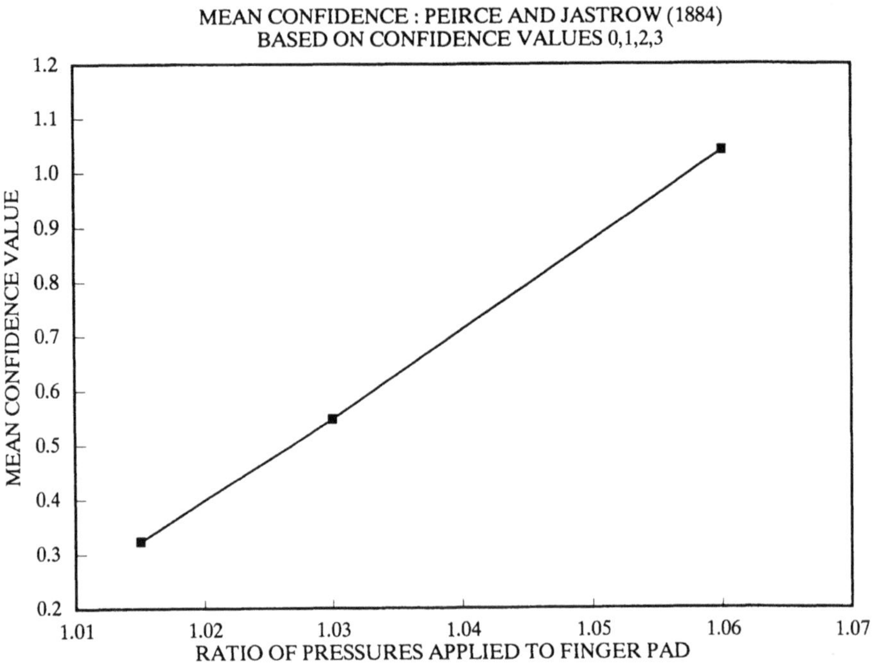

FIG. 14.7. The mean response confidences obtained by Peirce, with subject Jastrow, provide a fairly linear function of the ratio of stimulus values. These are mean confidences, the correct and error responses included together.

eralized Weber Ratio. Thus response confidence depends upon Sensation, which depends upon $\Delta s/\mathbf{S}$. To a close approximation, average confidence equals,

$$m = c \times A \times \Delta s/\mathbf{S} \ . \qquad (14.10)$$

Surely the classic data of Peirce and Jastrow should provide the first test for this relation. The confidence scale ranged from 0 to 3 and the ratios of pressures, comparison divided by standard, were 1.015, 1.030, and 1.060. For these three conditions the average confidence equalled 0.32, 0.55, and 1.04, respectively. The analysis of these historical data, shown in Fig. 14.7, supports a linear relation between average confidence and $\Delta s/\mathbf{S}$. Critics may complain that there are only three data points that must yield something of a straight line, and that there are no data to test the Intercept value, that is $\theta^* = 0$ when $\Delta s = 0$. Yet the success of this analysis provides a starting point for the developing story of how the psychophysical analysis of response confidence reached its present state.

JOHNSON'S EXPERIMENT

D. M. Johnson (1939) approached the measurement of confidence in a different manner. Johnson instructed the subjects:

> In this experiment the left side of the line will remain constant while the right side will vary. As before you are to make your judgment by pressing either the right or left key. Press the right key if you think the right side of the line is longer, the left key if you think it is shorter. (p. 30)

To obtain a measure of confidence Johnson measured certainty by instructing the subjects to,

> Let the right end of the scale (a 10 cm long line on a piece of paper with vertical cross bars placed near the ends) represent complete certainty that the right side of the line is longer than the left. Let the left end of the scale represent complete certainty that the right side is shorter, i.e., that the left side is longer. The center of the scale represents entire lack of certainty in the judgment. Intermediate points on the scale represent intermediate degrees of certainty. Instead of reporting your confidence in words you can represent the precise degree of your confidence by marking the appropriate point on the scale. (p. 30)

These instructions encouraged subjects to treat the 10-cm line as a scale of certainty. Although certainty may be related to confidence, certainty appears to be a measure of the probability that a judgment is correct. For example, taking the relative distance from the left edge of the scale as a measure of certainty, the measure acts very much like the probability of a *longer* response. In the middle of the scale there is a probability of 0.50 that the response is correct. This corresponds to the middle of the stimulus scale, $\Delta s = 0$, as well. The larger the deviation from the left side the greater the probability of a *longer* response.

This measure is quite different from the assignment of numbers to confidence that prompted Peirce to form his law. Here there are no numbers at all, only relative certainty defined on a linear scale. If Johnson's measures are more like probabilities than Peirce's confidences, then Johnson's relations between confidence and response time shown in Fig. 5.11 require reevaluation.

To determine whether Johnson's certainty judgments can be interpreted in the same way as response probabilities, the data from all three subjects for the accuracy condition were combined to yield an average measure of certainty, C. This measure ranged from -100, the value assigned to the left end of the scale by Johnson, to $+100$ at the right edge. Thus a relative measure of certainty is,

$$RC = [(C+100)/200].$$

This measure varies from zero to one as C, the average subject's mark, varies from the scale's left to right.

If relative certainty, RC, is another measure of response probability then RC should behave like a response probability. From previous results

$$\begin{aligned} ln[P_A/(1 - P_A)] &= \theta^* A \\ &= (\Delta s/\mathbf{S})A. \end{aligned} \tag{14.11}$$

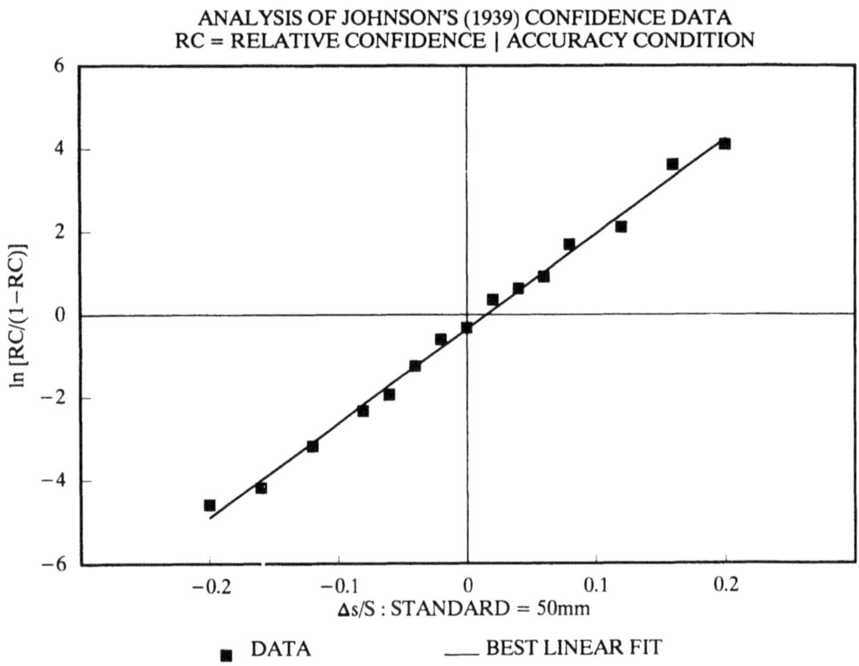

FIG. 14.8. Values of relative certainty (*RC*) from Johnson's confidence experiment provide a means of estimating θ^*. The obtained values are a linear function of $\Delta s/S$ suggesting that *RC* is actually an indicator of response probability.

Treating *RC* as if it measures response probability suggests that

$$ln[RC/(1 - RC)] = (\Delta s/\mathbf{S})A. \qquad (14.12)$$

Thus a graph of the left side of Equation 14.12 versus the $\Delta s/\mathbf{S}$ should yield a similarity transformation of $\Delta s/\mathbf{S}$.

The result shown in Fig. 14.8 demonstrates the remarkable linearity of this function. All 15 data points are quite near the best fitting linear function, which accounts for 99% of the variance. The function should equal 0 when the standard is presented (and $\Delta s/\mathbf{S} = 0$). In fact, the predicted line crosses the abscissa at $\Delta s/\mathbf{S} = 0.015$, a value close to 0 given that the range spans −0.20 to 0.20.

Because *RC* acts as an index of response probability, a measure of the unknown response threshold, *A*, follows directly. This unknown response-threshold value, the amount of accumulated comparative difference that generates a response, equals the slope of the best fitting linear function. The value proves to be 22.8. The reciprocal of this value 0.044 is quite near the Weber Constant for length of 0.043 suggested by Laming's (1986) reanalysis of Oberlin's (1936) data, although exact values depend, of course, on certain experimental conditions, lack of response bias, and constant response thresholds. For all this, however, the agreement is surprising.

There is little doubt about the interpretation of Johnson's measure of confidence. It is, instead, a predictor of response probability. This analysis suggests that an experimenter can estimate a response probability by using only a single observation—an indicator on a linear scale. The procedure permits substantial reductions in the number of observations needed in the study of psychological issues, especially in clinical settings.

FECHNER'S EXPERIMENTS (1855–1856)

The joint influence of θ^* and A exists even in Fechner's own work. From an extensive series of observations on his own ability to discriminate differences between weights, Fechner obtained the discrimination data reported in Table 1.1 (chap. 1). The standard weights $\mathbf{S}_{Bj}$ ranged from 300 g to 3000 g with comparisons equal to 1.04 or 1.08 times the value of the standard. Some 24,576 liftings led to the discovery that Fechner's Conjecture did not hold. In fact, as shown in Fig. 1.3, the proportion of correct responses rose steadily as the magnitude of the standard increased, rather than remaining constant.

Fechner was a careful experimenter, in fact, an inventor of rigorous psychological experimentation. This was not a novice practicing at improving experimental technique. Thus, some importance must be attached to the fact that these response proportions failed to remain constant across multiplicative increases in stimulus magnitude, as is required by Weber's Law. There must be cause behind this effect.

Because the trials were run in blocks using a fixed standard within each block, ample opportunity existed for the subject to adopt a different response-threshold value, A, for each standard. Perhaps changes in the subject's response threshold from standard to standard caused the deviation from Weber's Law. There is no way of intervening in the experiment now to test this hypothesis, but there is a method of analyzing the response proportions that sheds new light on this classic experiment.

Assume that the value of A depends only upon the magnitude of the standard so that for each standard there is a fixed value of A, say A_j ($j = 1, \ldots, 6$) for the six different standards. The strength of each standard corresponds to a Poisson value β_j, whereas the Poisson values for the two comparison stimuli equal α_i ($i = 1, 2$) for the 1.04 and 1.08 comparisons, respectively. If Weber's Law holds then, the ratio

$$\theta^*_{ij} = ln(\alpha_i/\beta_j) \tag{14.13}$$

remains constant for a fixed comparison ($i = 1, 2$) across all magnitudes of the standard (β_j, $j = 1, \ldots, 6$). Thus, $\theta^*_{11} = \theta^*_{12} = \ldots = \theta^*_{16}$ and $\theta^*_{21} = \theta^*_{22} = \ldots = \theta^*_{26}$.

The unknown value of A_j may be removed from consideration by computing the ratio of the two values of $\theta^*_i A_j$ for a fixed standard. Then,

$$\frac{\theta^*_{1j}A_j}{\theta^*_{2j}A_j} = \frac{ln(\alpha_1/\beta_j)A_j}{ln(\alpha_2/\beta_j)A_j}$$

$$= \frac{ln(\alpha_1/\beta_j)}{ln(\alpha_2/\beta_j)}. \quad (14.14)$$

Written in terms of the response proportion used to estimate values of $\theta^*_{ij}A_j$, the ratio to be calculated is,

$$\frac{\theta^*_{1j}A_j}{\theta^*_{2j}A_j} = \frac{ln[P_{1j}/(1 - P_{1j})]}{ln[P_{2j}/(1 - P_{2j})]} \quad (14.15)$$

where P_{1j} is the probability of a correct response when using standard $\mathbf{S}_j$ and a comparison equal to $(1.04)\mathbf{S}_j$. P_{2j} is the corresponding correct response probability for the $(1.08)\mathbf{S}_j$ comparison.

Thus, if Fechner's conjecture is true then multiplication of the weights by a constant results in multiplying the values of α_i by a constant, leaving the ratio unchanged. When the value of A changes across the magnitudes of the standard and the values of θ^*_i remain constant, as they should according to Fechner's conjecture, the ratio θ^*_1/θ^*_2 should remain constant.

The individual estimates of θ^*A given in Table 14.1 are each an average of four estimates obtained from analysis of the response frequencies in Table 1.1. The obtained values are given in the rightmost columns of Table 14.1. The ratios of the values computed for each standard are shown in Fig. 14.9. There is little doubt that the ratio remains constant as the magnitude of the standard increases.

Thus, after many years Fechner's suggestion of 1860 (see Fechner, 1860/1966, p. 155) that the demonstration of Weber's Law failed because the additional weight of the arm influenced judgments more at lower-stimulus magnitudes is easily displaced by an explanation that leads to a testable consequence. Although the individual discriminability values are unknown, the fact that their ratio remains constant is convincing evidence that when responsiveness is taken into account, even Fechner's own, seemingly unsatisfactory, results satisfy

TABLE 14.1
Averages of Estimates of $\theta^*_i A_j$
Based on Fechner's Results (1855–1856)

j	*Standard (g)*	*i = 1* *1.04 × S_{Bj}*	*i = 2* *1.08 × S_{Bj}*
1	300	.664	.981
2	500	.669	1.183
3	1000	.870	1.300
4	1500	.924	1.549
5	2000	.965	1.473
6	3000	.914	1.681

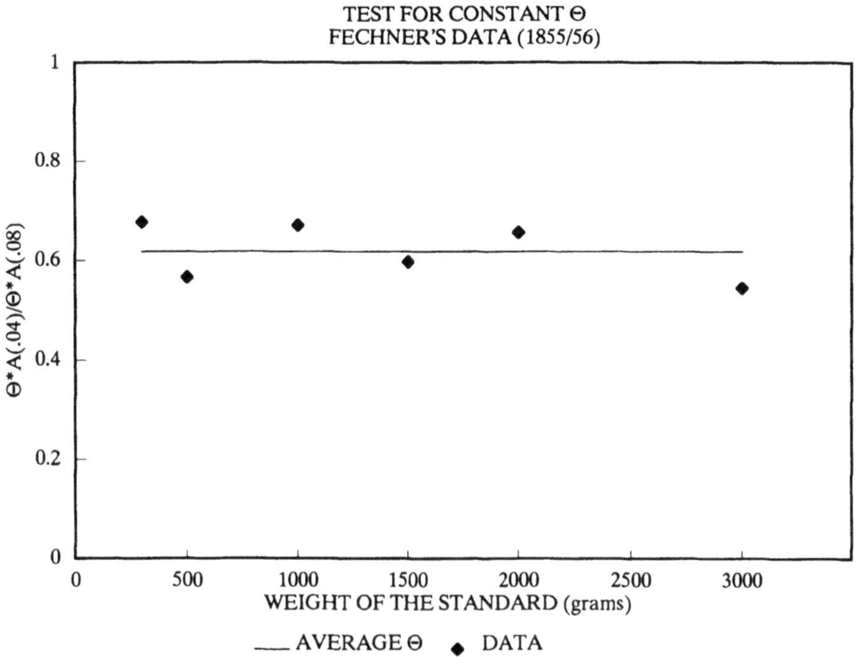

FIG. 14.9. Values of response thresholds may change across standards. However, if Weber's Law is true, then ratios of θ^*A values for a fixed standard should remain constant across increases in the magnitude of the standard. Fechner's data confirm this prediction.

Fechner's conjecture that multiplication of stimulus values by a constant leaves discriminability unchanged.

BEYOND THE PSYCHOMETRIC FUNCTION

Stimulus $\mathbf{S}_j$ with discriminability $\theta_j^*$and response threshold A, produces a sensation relative to a stimulus, $\mathbf{S}_1$, with sensation θ_1^*A, that is measured by the ratio,

$$\frac{S_j}{S_1} = \frac{\theta_j^*A}{\theta_1^*A}.$$

The cancellation from the numerator and denominator of the value of the response threshold, A, shows that the ratio of sensations is preserved under changes in the values of A. Thus in experiments where the response threshold is constant, the influence of A is easily overlooked.

When the resistance to respond balances the discriminability of the stimulus, the result is a unit value of sensation, $\theta_1^*A = 1$. The stimulus that gives rise to the feeling of a unit increment in sensation is by tradition called the "difference-threshold stimulus." Because the "difference-threshold stimulus" generates a unit of sensation, other stimuli may be scaled with respect to this unit.

Response probabilities also depend upon sensation. In fact, for unbiased discrimination only the values of θ_i^*A are needed to determine response probability. For the threshold stimulus the sensation value θ_i^*A equals 1.0 and the response probability equals 0.731. Thus response probabilities may be plotted as a function of values of θ^*A, which have a natural unit of measurement related to a unit of sensation. In this way response probability and the magnitude of sensation are related. In this representation the absolute-threshold stimulus, the stimulus that produces 50% correct discriminations, is the same as the standard. This observation suggests that in absolute-threshold experiments, where the experimenter defines the absolute threshold by the stimulus that is perceived 50% of the time (the traditional definition), the subject uses the absolute-threshold stimulus as the standard against which other stimuli are compared.

Figure 14.10 shows the relation between response probability and sensation. Because there is an isomorphic function relating sensation and probability there is an inverse function relating probability to sensation. Thus, regardless of the physical magnitudes of the stimulus generating the response probabilities, the amount of sensation is determined by

$$\theta_i^*A = ln[P_L/(1 - P_L)],$$

where P_L is the probability of responding *larger*. If these probabilities were plotted as a function of the physical stimuli Fig. 14.10 would define a Psychometric Function.

As the response probabilities approach 1.0, little room is left to show the influence of increases in sensation. The function reaches a practical ceiling, as in Guilford's experiment (Fig. 12.5). Subsequent increases in the magnitude of the comparison stimulus barely effect response proportions. The subject can perceive yet greater increases in the magnitude of the comparison stimulus, but response probability is no longer a useful tool for the study of feeling.

Thus, beyond the range of stimulation created by studies of the Psychometric Function is a larger range of stimulation, often of more practical interest than the rather small changes that dominated early psychophysical studies. Within this larger range are stimuli that differ by many orders of magnitude and yet are easily compared against each other to determine the relative amount of sensation evoked by one stimulus versus another.

SENSATION MATCHING

S. S. Stevens (1975) argued forcefully that the sensations of practical interest are related by their ratios. As an illustration of this relation Stevens cleverly asked subjects to match sensations produced by a physical stimulus to the sensations

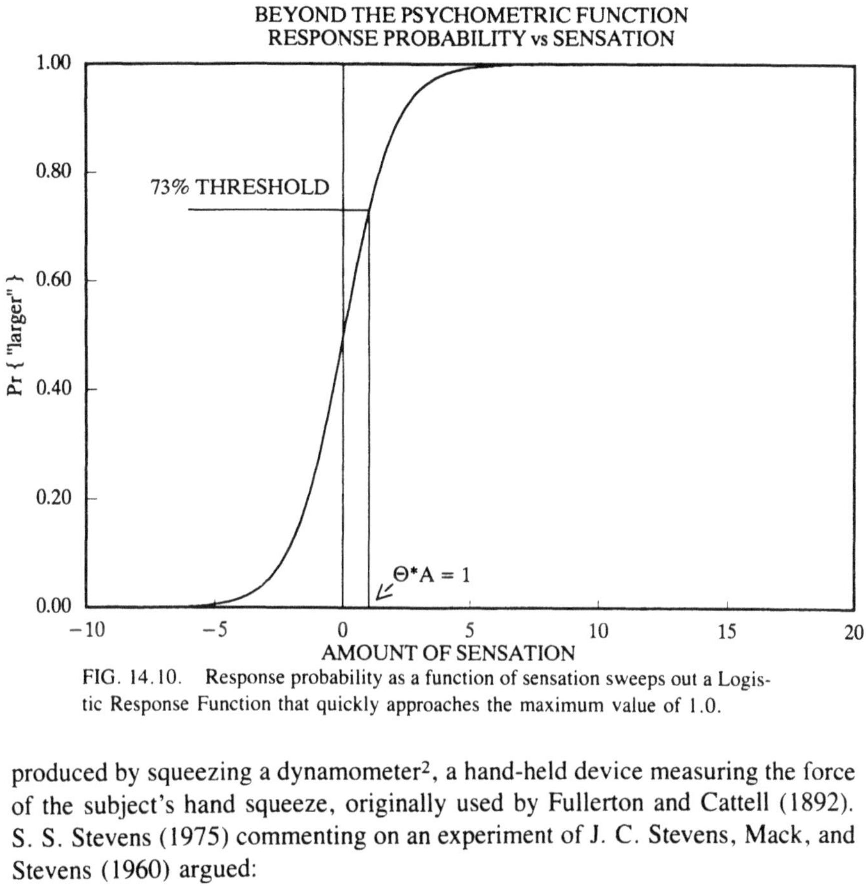

FIG. 14.10. Response probability as a function of sensation sweeps out a Logistic Response Function that quickly approaches the maximum value of 1.0.

produced by squeezing a dynamometer[2], a hand-held device measuring the force of the subject's hand squeeze, originally used by Fullerton and Cattell (1892). S. S. Stevens (1975) commenting on an experiment of J. C. Stevens, Mack, and Stevens (1960) argued:

> When you squeeze a hand dynamometer you feel a sense of strain that can vary in degree from weak to very intense. You yourself can regulate the intensity of the sensation. Suppose then that you also hear a noise of a certain loudness, can you squeeze the dynamometer with such a force that the sensation of strain appears as intense as the sensation of loudness? (p. 111)
>
> The results of matching force of handgrip to nine different continua are shown in Figure [14.11]. What we see there is quite a splendid family of power functions. On each of nine continua, as the intensity of the criterion stimulus was changed by the experimenter, the observer squeezed a precision dynamometer with a force that he judged appropriate to indicate the perceived magnitude. The remarkable fact is that the force of handgrip increased along a straight line when data were plotted in log-log coordinates. (pp. 111–112)

[2]Matching implies that the ratio of sensations equals 1.

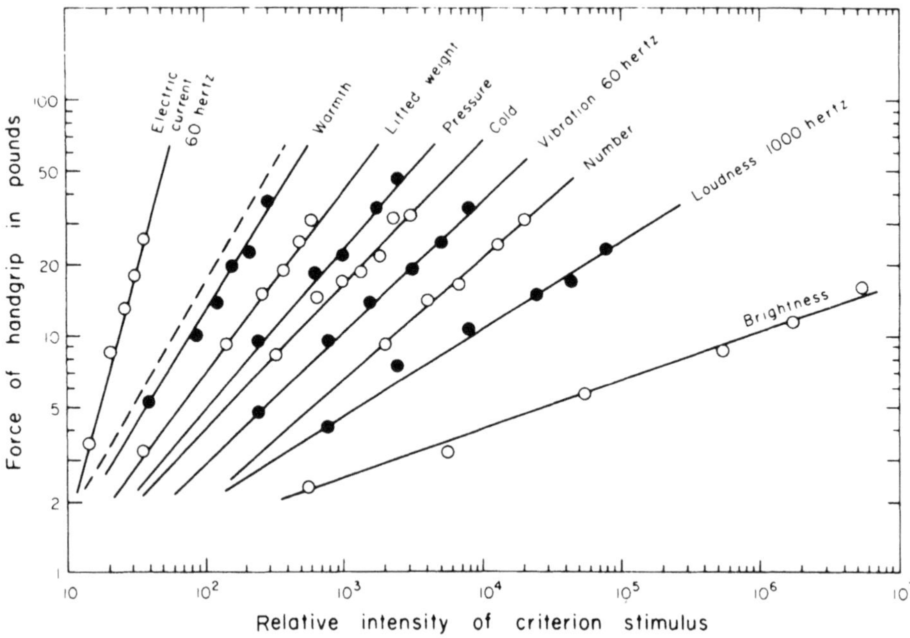

FIG. 14.11. Equal sensation functions obtained by matching force of handgrip to various criterion stimuli. Each point stands for the median force exerted by 10 or more observers to match the apparent intensity of a criterion stimulus. The relative position along the abscissa is arbitrary (J. C. Stevens, Mack, & Stevens (1960) and from Stevens (1975)).

The empirical results are indisputable. When matching sensations across sense modalities the data generally conform to the rule that the logarithm of the dynamometer reading is a linear function of the logarithm of the physical stimulus intensity. The two physical measures, dynes and intensity, are the result of equal sensations produced by different sense modalities.

What does this mean? Let the sensation produced by squeezing a dynamometer, D, be measured in units of a standard squeeze D_0. The amount of sensation produced by the squeeze is

$$S_D = \theta_D^* A_D, \tag{14.16}$$

where the value of A_D is the response threshold when squeezing the dynamometer. Let the sensation produced by a tone T, again determined with respect to a standard tone T_0, be

$$S_T = \theta_T^* A_T. \tag{14.17}$$

The subject equates these two sensations by applying greater or lesser force to the dynamometer. Therefore,

$$S_D = S_T$$

or,

$$\theta^*_D A_D = \theta^*_T A_T. \tag{14.18}$$

Replacing values of θ^* by their logarithmic equivalents gives,

$$[ln(D/D_0)]A_D = A_T[ln(T/T_0)]. \tag{14.19}$$

Finally, dividing both sides by the response threshold needed to feel the force applied to the dynamometer gives the result S. S. Stevens often observed: The logarithm of the dynamometer reading is a linear function of the logarithm of the physical measure of the comparison stimulus.

$$ln(D) = \frac{A_T}{A_D}[ln(T) - ln(T_0)] + ln(D_0).$$

Isolating the constants, everything except D and T, gives the linear relation,

$$ln(D) = \frac{A_T}{A_D}[ln(T)] - \frac{A_T}{A_D}[ln(T_0)] + ln(D_0)$$

$$ln(D) = \frac{A_T}{A_D}[ln(T)] + \text{Constant}. \tag{14.20}$$

Thus, according to the Wave Theory of Sensation, the linear relation between logarithms of physical-stimulus measures for equated sensations must occur. S. S. Stevens many results confirm this prediction.

Stevens considered the slope (A_T/A_D) to be important but failed to attach to it any theoretical significance. J. C. Stevens, Mack, and Stevens (1960) showed remarkably linear functions from cross-modality sensation matches for nine different continua. In describing these functions S. S. Stevens (1975) commented that

> The results of matching force of handgrip to nine different continua are shown in Figure [14.11]. What we see there is a quite splendid family of power functions.[3] On each of nine continua, as the intensity of the criterion stimulus was changed by the experimenter, the observer squeezed a precision dynamometer with a force that he judged appropriate to indicate the perceived magnitude. (p. 111)

The previously unknown meaning of these slope measures is now clear. The slope proves to be a measure of the ratio of response thresholds used to sense the stimuli. When the resistance used to sense the force applied to the dynamometer remains constant, the value of A_D is constant. Then the difference between slopes

[3]These are not, of course, power functions but linear functions. Stevens believed them to arise from power functions.

is a measure of the different sized resistances, response threshold values, for different sense modalities. It is a small matter that the value for A_D is not known, because this value simply stretches or shrinks all the resistances for the sense modalities.

Undiscovered relations exist in the data of J. C. Stevens, Mack, and Stevens (1960). Consider, for example, the slopes of the various linear functions for sensory matching shown in Fig. 14.11. The slopes are measures of the response thresholds used to form a sensation and are unique to each sense modality. From Stevens' many experiments comes his observation that the response-threshold value (what he called the "exponent of the Power Law") "serves as a kind of signature that differs from one sensory continuum to another" (S. S. Stevens, 1975, p. 13). If these response thresholds, values of A, are unique to sense modalities do they bear the same relation to each other regardless of the magnitude of the stimuli?

Previously reciprocals of Weber's Constants were shown to provide estimates of the unknown values of the response threshold, A. Is it possible that the reciprocals of Weber's Constants might be substituted for S. S. Stevens' "Exponents" in order to account for the linear relations shown in Fig. 14.11? Are Weber's Constants and Stevens' Exponents so unique to a sensory modality that they remain relatively unchanged as stimulus magnitude increases?

Weber's Constants are well known although the values are sometimes the subject of disputes that arise over the different ways in which the constants are estimated. Nevertheless Teightsoonian (1971) in a study of the "dynamic range" of sensation provided a list of Weber's Constants that may be compared to the slopes of the sensory-matching functions shown in Fig. 14.11. Of course the functions in Fig. 14.11 are based on dynamometer settings for which we have no corresponding Weber Constant. Therefore, to provide a common basis for comparison of results from discrimination and sensory-matching experiments the Weber Constants and Stevens' Exponents must be referred to some common sensory experience. The stimulus that most closely generates psychological values close to physical values is length. The Stevens' Exponent for length is 1.04 ($\log_{10}$) and the Weber Constant for length is 0.029.[4]

Each of Stevens' Exponents obtained by matches to the dynamometer may be divided by 1.04 to obtain a value relative to length rather than the dynamometer. Each Weber ratio may be divided by 0.029 to obtain a Weber ratio relative to the judgment of length. Then the reciprocals of the relative Weber ratios can be compared to the relative Stevens' Exponents. The results are quite extraordinary. Figure 14.12 shows the values of the response thresholds (relative to length) versus the relative reciprocal Weber Constants for each modality. There is really

[4]This estimate is reasonable. It comes from estimation using only the higher stimulus magnitudes in Oberlin's study, whereas Laming's estimate comes from the entire range which appears to be two tiered.

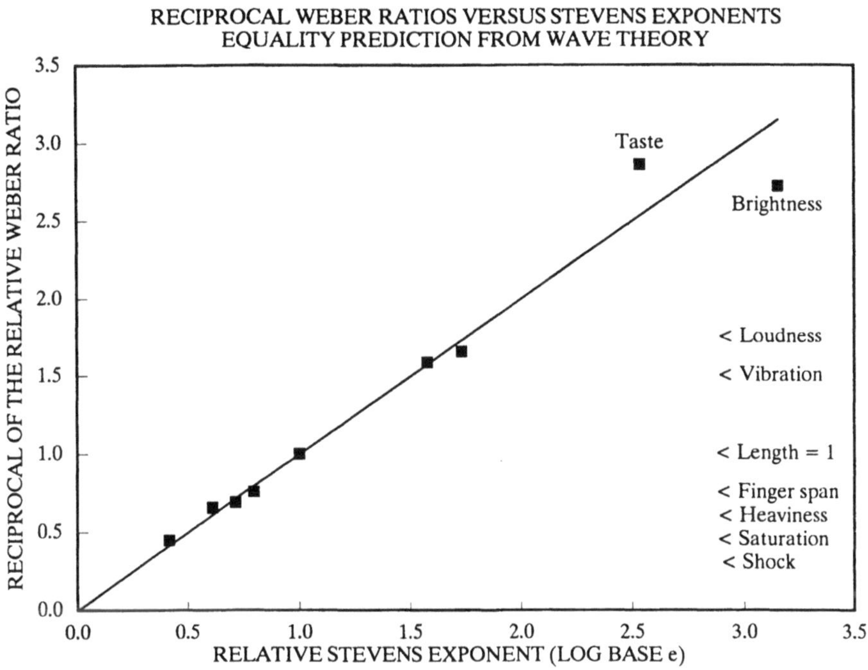

FIG. 14.12. The equality of resistances to respond in Weber-class and Stevens-class experiments is supported by data drawn from discrimination and sensory-matching experiments.

no question about the relation, the values are equal, as shown by the regression line fitted to the results from Weber-class experiments and Stevens-class experiments for nine sensory continua. Stevens' Exponents are resistances to respond. The relative size of the resistance is unique to each sense modality and independent of the magnitude of the stimuli presented for judgment.

EQUAL STIMULUS RATIOS

S. S. Stevens (1975) contended as did Hering before him (cf. Hurvich & Jameson, 1989) that equal stimulus ratios produced equal sensation ratios. This result, argued Stevens (1961), implies that the Psychophysical Function relating stimulus magnitude to sensation magnitude is a power function. A power function is not a logarithmic function. Therefore, Fechner's Law should be repealed (Stevens, 1961).

A closer analysis based on Wave Theory considerations shows that Stevens and Fechner simply felt different parts of the same elephant. Imagine two stimuli $\mathbf{S}_1$ and $\mathbf{S}_2$ with Poisson intensities of α_1 and α_2 and a ratio $\alpha_1/\alpha_2 = \alpha_r$. Because the values of α_1 and α_2 are similarity transformations of the physical stimulus values, α_1/α_2 equals $\mathbf{S}_1/\mathbf{S}_2$. The sensation evoked by the comparison of these stimuli depends upon the response threshold value A and equals $\theta^*_{12}A = ln(\alpha_1/\alpha_2)A = ln(\alpha_r)A$.

Imagine a second pair of stimuli $\mathbf{S}_3$ and $\mathbf{S}_4$ that produce Poisson intensities of α_3 and α_4 with a ratio, $\alpha_3/\alpha_4 = \alpha_r$ which equals $\mathbf{S}_3/\mathbf{S}_4$. The sensations produced by these two equal stimulus ratios are then equal. That is,

$$\begin{aligned} \theta^*_{34}A &= ln(\alpha_3/\alpha_4)A \\ &= ln(\mathbf{S}_3/\mathbf{S}_4)A \\ &= ln(\alpha_r)A. \end{aligned}$$

But,

$$\begin{aligned} ln(\alpha_r)A &= ln(\alpha_1/\alpha_2)A \\ &= ln(\mathbf{S}_1/\mathbf{S}_2)A \\ &= \theta^*_{12}A\ . \end{aligned}$$

Within the domain of the Psychometric Function, when values of $\mathbf{S}_1$ and $\mathbf{S}_3$ are constant multiples of $\mathbf{S}_2$ and $\mathbf{S}_4$, S. S. Stevens' observation simplifies to Weber's Law because the comparison of $\mathbf{S}_1$ against $\mathbf{S}_2$ produces the same probability of a correct response as comparing $\mathbf{S}_3$ against $\mathbf{S}_4$. To pursue S. S. Stevens' (1961) admonition to honor Fechner and repeal his Law requires that Weber's Law be repealed as well. The results of the preceding section showing that equal sensation ratios are consistent with the Wave Theory derivation of Weber's and Fechner's Laws suggest that Stevens' results are a product of the same sensory processes that generate Weber's Law. If Stevens were to repeal Weber's and Fechner's Laws he would also repeal his own law that equal stimulus ratios produce equal sensation ratios!

SUMMARY

Even the most disparate experimental results are often reconciled by a theory that predicts empirical phenomena rather than follows from them. The Wave Theory of Sensation proposes that Poisson signals characterize external stimuli. The mathematical fact that under very broad conditions a number of receptors with potentially different sensitivities converge to a stable mean output shows that the discriminability of a stimulus may be taken as a similarity transformation of the physical stimulus intensity. Ratios of mean amplitudes equal the ratios of stimulus magnitudes. The fundamental relation derived from Wave Theory, that the discriminability between two stimuli with Poisson intensities of α_i and α_j, is

$$\theta^*_{ij} = ln(\alpha_i/\alpha_j).$$

This is Fechner's Law with respect to mean signal amplitudes. When the actual stimulus values replace the mean signal amplitudes, because the values of θ^*

may be considered as similarity transformations of the physical stimuli, the relation becomes,

$$\theta^*_{ij} = ln(\mathbf{S}_i/\mathbf{S}_j).$$

This is Fechner's Law.

Letting the difference between the stimuli be $\Delta s = \mathbf{S}_i - \mathbf{S}_j$ leads to the relation

$$\theta^*_{ij} = \Delta s/\mathbf{S}_j.$$

For threshold responding this is Weber's Law. The idea generalizes to the full range of response probability to produce a new Psychophysical Law that generalizes Weber's Law. Thus the three major features of psychophysical theory, Weber's Law, Fechner's Law, and Stevens' observation that equal stimulus ratios give equal sensation ratios are now reconciled.

These sensory processes are neither seen nor felt. Their absence deludes us into believing that sensation is a raw atom of experience, the basis for all mental processes. Yet, sensation depends upon quantum energy transfers that bring the proximal world into our inner experience. The transformation operates seamlessly—without our knowledge—because the awareness of the stimulation does not occur until the response threshold is exceeded. The actual accumulation of stimulation occurs below threshold. The generation of sensation is invisible. The transparency of the process makes exposing it all the more difficult. Yet mathematical properties of the process reveal its transparent functioning and suggest that the same mechanism operates for discriminations between all stimuli.

15 A Theory of Feeling

FEELING AND THE PSYCHOLOGICAL FUNCTION

During the years 1865–1876, following publication of *Elemente der Psychophysik* in 1860, Fechner used the newly crafted tools of psychophysical analysis to open the field of experimental aesthetics. The tools soon struggled with an insurmountable problem. How could stimuli with no physical measure be forced to yield a measure of their aesthetic value? In 1876 *Vorschule der Aesthetik* presented Fechner's many findings, but the problem of measuring aesthetic value remained unsolved.

While the measurement of aesthetic value awaited future developments, Wundt (1896) took an important step by carefully analyzing the difference between sensation, the object of psychophysical theory, and feeling.

> The ultimate fact is, that we sense and feel. The logical separation of feeling from sensation can only come about after we have distinguished subject from object. Then, and then only, is the elementary process of sensation analyzed into a subjective factor, the feeling, and the objective factor, the sensation. The sense-feeling may in this way be considered as an integral element of sensation itself; and for that reason it is also termed the *affective tone of sensation*. (p. 212)

Sense impressions may give rise to feelings but other higher-order processes give rise to other feelings such as pleasantness and unpleasantness. Often these feelings are evoked by complex stimuli, a Bach fugue, a Mozart concerto, a painting, or poem. Furthermore these feelings are not without intensity but vary sufficiently to allow judgments to be made between two stimuli that each yield an affective tone value. Fechner's 19th-century methods of psychophysical study require that the stimuli have physical measures in order to determine the sensitivity to the objects under study. Failing this requirement, judgments between such stimuli as poems, paintings, concertos, and fugues lacked a theoretical basis.

But, Thurstone (1927b) once again thrust psychophysical measurement onto the field of experimental aesthetics. The Law of Comparative Judgment pushed beyond the obstacles hindering Fechner's progress by creating psychological values for stimuli that had no physical measure. Later Thurstone (1954) remarked,

> Some time ago I attended a series of seminars on aesthetics at the home of one of my colleagues. . . . In some of those discussions it occurred to me that the question at issue could be treated as a question of experimental fact, and I ventured to suggest how the psychophysical methods could be adapted to obtain an empirical answer to the question at issue. It was an illuminating experience to discover that some of my friends in the humanities were hostile to the very idea of subjecting questions of aesthetic theory to empirical inquiry. . . . Unfortunately, I have not been able to induce many students of psychology to study experimental aesthetics. (p. 58)

Thurstone's breakthrough also encountered strong opposition from outside the humanities because, like Fechner, he impressed error into the service of measurement. As S. S. Stevens (1975) argued, "Thurstone improved the mathematical machinery but left intact the assumption that out of variability and confusion we can forge the units of measurement" (p. 10). Whereas the Law of Comparative Judgment brought new meaning to the results of many paired-comparison experiments (cf. Beebe-Center, 1932), the measurement of the feelings produced by stimuli were, frankly, nonexistent. What measures there are measure the differences between stimuli with respect to the standard deviation of a hypothesized Gaussian distribution, rather than the feelings aroused by the stimuli themselves.

In contrast to the use of additive error or Gaussian distributed dispersions, Wave Discrimination Theory proposes that amplitude differences between stimulus-generated waves are the basis for aesthetic judgment. The amplitudes are not the result of error but the consequence of the inherent variability in the transmitted signal. Like comparisons studied in previous chapters aesthetic judgments occur through the accumulation of wave differences, but the stimuli are not the sense stimuli that dominate psychophysical research. Rather they are the broader class of stimuli of general interest to all psychologists, stimuli that please, disturb and sometimes soothe the human spirit.

The essential problem is to create an intensive measure of feeling that applies to all stimuli, and then develop methods for its estimation. Feeling differs from sensation. Thus, in spite of the previous synthesis of the conflicting views of Weber, Fechner, and Stevens, the elusive measure of feeling is still at large. The view put forth in Wave Theory is that sensation is the awareness of a stimulus but feeling is the measure of awareness. The feeling created by a stimulus is the perceived magnitude of the stimulus as seen by the subject who is aware of it.

To pursue judgments of these more general stimuli requires an experimental method independent of physical stimulus measures. This method proves to be a new theory of the Method of Paired Comparison. No more clear distinction between Wave Theory and Gaussian error based theories occurs than in this analysis of paired comparisons. Here Wave Discrimination Theory advances toward the goal of Fechner and Thurstone by providing a measure of feeling described by the *Psychological Function.*

THE PSYCHOLOGICAL FUNCTION

Physical stimuli generate a sensation equal to $\theta^*_{ij}A$ but they are measured, learned about, and manipulated with respect to physical units, not the logarithmic scale value that results from the process of accumulating wave differences. To see the stimulus from the vantage point of the subject the logarithmic function must be removed by converting the sensation back into units of the physical world. To see through the sensation to the original stimulus by removing the logarithmic function requires that the value of $\theta_{ij}A$ be exponentiated. The result is the feeling, the psychological magnitude, of the stimulus generating the sensation. That is,

$$\begin{aligned} \Psi(\theta^*_{ij}A) &= e^{\theta^*_{ij}A} \\ &= e^{ln(S_i/S_j)A} \\ &= \left[\frac{\mathbf{S}_i}{\mathbf{S}_j}\right]^A . \end{aligned} \tag{15.1}$$

Feeling, psychological magnitude, is a power function of stimulus ratios, a result of which both Hering and S. S. Stevens would no doubt approve.

There are, then, three dimensions to the modelling of sensation and feeling. There is, first of all, the dimension of the stimuli. Second, the stimuli give rise to sensation, which is another dimension. Third, the dimension of feeling characterizes the apparent magnitude of the stimulus given the sensation. As shown in Fig. 15.1, feelings are a power function of physical magnitudes, but feelings are also an exponential function of sensation. The Psychological Function that creates feeling is, in the parlance of Part I, a Response Transfer Function. If, in some way the values of $\theta^*_{ij}A$ may be obtained for stimuli lacking physical measures, then the Psychological Function will reveal the magnitudes of the otherwise unmeasurable feelings associated with these stimuli.

At the absolute threshold, the point where a sensation is experienced 50% of

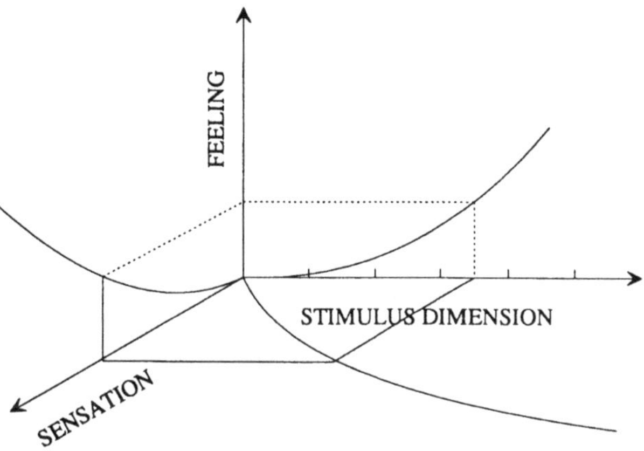

FIG. 15.1. The three dimensional representation of stimulus magnitude, sensation, and feeling. The values of sensation can be negative but values of feeling are greater than zero.

the time, the magnitude of the sensation, θ^*A, equals zero. At absolute threshold the Psychological Function yields a value of feeling,

$$\begin{aligned}\Psi(\theta^*A) &= e^{\theta^*A}\\ &= e^0\\ &= 1.\end{aligned}$$

That is, the stimulus that is just at absolute threshold, the stimulus that just generates an awareness of its presence with probability 0.50, creates a unit of feeling. This unit may be taken as a measuring increment for feelings generally. One unit of feeling is created by the stimulus that is just felt 50% of the time.

Stimuli below absolute threshold are those that produce an awareness of themselves with probability less than 0.50. Such sub-threshold stimuli produce a detection response with low probability. They do so because the accumulation of wave differences drifts in a negative direction. Because the value of θ^* must match in sign the average wave difference, the value of θ^* must be negative. Therefore θ^*A must be negative, indicating the presence of negative sensations.

Feelings are never negative because an exponential value for such negative sensations must lie between zero, no feeling at all, and one, the feeling generated by the absolute-threshold stimulus. Thus, feelings resulting from instances where a negatively drifting accumulation of wave differences actually reaches the response threshold at A generate an awareness of the stimulus and a faint feeling. The less likely the stimulus is to generate such an event, the less feeling is evoked by the sensation. There are no negative feelings, only feelings that can vary from nearly zero to those of overwhelming magnitude.

The foundation for the preceding psychophysical studies is the Poisson wave and the relation of its parameter to the intensity of the physical stimulus. Because the intensities of the physical stimuli were known some strong tests of Wave Theory predictions followed. Now the set of stimuli is to be expanded to include those with no physical measure. The Poisson waves still characterize stimuli but the development of a new theory of paired comparisons obviates the need for the physical measures. The resulting estimates of sensation feed into the Psychological Function to yield measures of the intensity of feeling.

First the Wave Theory of Paired Comparisons is developed. The Adaptation Level of Helson (1959) proves to be compatible with this theory. Then the theory is applied to the lifted weights from Guilford's (1931) experiment to show how sensation values and feelings are estimated. The theory is applied to Thurstone's (1927a) data for the seriousness of crimes to reveal the striking increases in feeling associated with serious crimes.

WAVE THEORY OF PAIRED COMPARISONS

The full paired comparison design (David, 1969; Thurstone, 1927a) requires judgments between every pair of stimuli ($\mathbf{S}_i$, $\mathbf{S}_j$) with $\mathbf{S}_i$ and $\mathbf{S}_j$ the standard on different occasions. These two stimuli are said to have symmetric effects in the sense that a stochastic path of wave differences generated when $\mathbf{S}_j$ is compared against $\mathbf{S}_i$ is exactly opposite to that generated when $\mathbf{S}_i$ is compared against $\mathbf{S}_j$. That is, the accumulation shown in Fig. 9.3 has a mirror image, with respect to the starting value of $\mathbf{V}_0$, obtained by reversing the roles of the two stimuli. Figure 15.2 shows that the two comparisons are reflections, mirror images, of each other. Appendix C describes mirror symmetry in detail.

The paired-comparison experimental design and attendant features are illustrated in Table 15.1. At the left of each row is the value of the standard, $\mathbf{S}_i$, whereas the comparison, $\mathbf{S}_j$, is indicated at the top of each column. Entries in the matrix are response proportions, P_{Aij}, obtained through repeated, randomized stimulus presentations. Also shown are the unknown values of θ^*_{ij}. When $\mathbf{S}_i$ is the standard and $\mathbf{S}_j$ the comparison the Poisson intensity for the difference between $\mathbf{S}_i$ and $\mathbf{S}_j$ is,

$$\theta^*_{ij} = ln(\alpha_j/\alpha_i).$$

Each cell in Table 15.1 contains values of θ^*_{ij} that together with the response threshold value, A, generate the response probability within the cell.

When the roles of the stimuli are reversed, and $\mathbf{S}_j$ becomes the standard against which $\mathbf{S}_i$ is compared, the value

$$\begin{aligned}\theta^*_{ji} &= ln(\alpha_i/\alpha_j)\\ &= -ln(\alpha_j/\alpha_i)\\ &= -\theta^*_{ij}.\end{aligned}$$

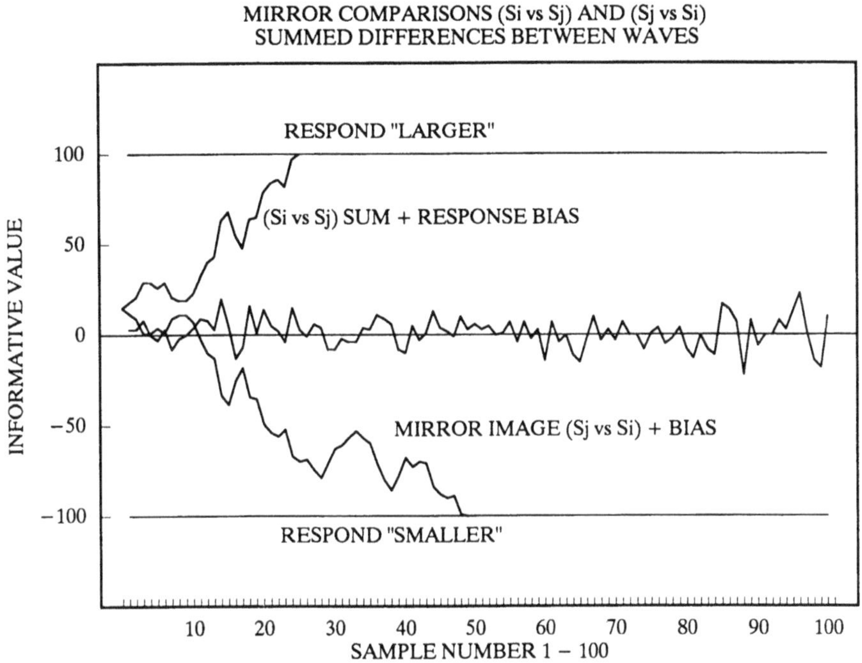

FIG. 15.2. When stimulus $\mathbf{S}_i$ is compared against $\mathbf{S}_j$ the accumulation of Informative Values has a mirror image corresponding to the comparison of $\mathbf{S}_j$ against $\mathbf{S}_i$.

TABLE 15.1
The Full Paired Comparison Design Matrix.
Entries Are Probabilities, P_{Aij}, and θ^*_{ij} Values.
Comparisons

Standards	S_1	S_2	. . .	S_N
S_1	$P_{11} = 0.50$ $\theta^*_{11} = 0$	P_{12} $\theta^*_{12} = -\theta^*_{21}$	. . . . . .	P_{1N} $\theta^*_{1N} = -\theta^*_{N1}$
S_2	P_{21} $\theta^*_{21} = -\theta^*_{12}$		. . . . . .	P_{2N} $\theta^*_{2N} = -\theta^*_{N2}$
S_3	P_{31} $\theta^*_{31} = -\theta^*_{13}$		. . . . . .	
. . .	. . .		. . . . .	. . .
S_N	P_{N1} $\theta^*_{N1} = -\theta^*_{1N}$		. . . . . .	$P_{NN} = 0.50$ $\theta^*_{NN} = 0$
Averages	θ^*_1	θ^*_2		θ^*_N

TABLE 15.2
Guilford's Paired-Comparison Design Matrix. Entries Are Proportions, P_{Aij}, and $\theta^*_{ij}A$ Values

		Comprisons (g)							
Standards		185	190	195	200	205	210	215	Mean
185	P_{1J}	0.48	0.65	0.78	0.92	0.93	0.95	0.99	
	$\theta^*_{1J}A$	−0.08	0.62	1.27	2.44	2.59	2.94	4.60	2.05
190	P_{2J}	0.35	0.50	0.69	0.80	0.85	0.97	0.94	
	$\theta^*_{2J}A$	−0.62	0.00	0.80	1.39	1.74	3.48	2.75	1.36
195	P_{3J}	0.22	0.31	0.48	0.63	0.72	0.89	0.91	
	$\theta^*_{3J}A$	−1.27	−0.80	−0.08	0.53	0.94	2.09	2.31	0.53
200	P_{4J}	0.08	0.20	0.37	0.52	0.67	0.78	0.86	
	$\theta^*_{4J}A$	−2.44	−1.39	−0.53	0.08	0.78	1.27	1.82	−0.07
205	P_{5J}	0.07	0.15	0.28	0.33	0.54	0.64	0.74	
	$\theta^*_{5J}A$	−2.59	−1.74	−0.94	−0.71	0.16	0.58	1.05	−0.60
210	P_{6J}	0.05	0.03	0.11	0.22	0.36	0.46	0.62	
	$\theta^*_{6J}A$	−2.94	−3.48	−2.09	−1.27	−0.58	−0.16	0.49	−1.43
215	P_{7J}	0.01	0.06	0.09	0.14	0.26	0.38	0.56	
	$\theta^*_{7J}A$	−4.60	−2.75	−2.31	−1.82	−1.05	−0.49	0.24	−1.82
θ^*A Means		−2.08	−1.36	−0.56	0.09	0.65	1.39	1.89	0

The trivial comparison results from comparing a stimulus against itself. Then the value

$$\theta^*_{ii} = ln(\alpha_i/\alpha_i)$$
$$= 0.$$

All non-zero values have symmetric partners of opposite sign. Thus, the sum of all θ^*_{ij} values must equal zero.

The classic psychophysical data are from Guilford's paired comparisons of lifted weights. Table 15.2 gives the response proportions, P_{ij} for choosing the *heavier* weight. These proportions also generate the response surface of Fig. 4.6. The θ^*_{ij} are not directly observable, but from the response proportions the values of $\theta^*_{ij}A$ can be derived from the logistic relation (Equation 10.1). Thus for each cell an estimate of $\theta^*_{ij}A$ exists and is proportional to θ^*_{ij}.

The average value of a column of $\theta^*_{ij}A$ values is generally not zero. For column j it equals,

$$\theta_j^* A = \frac{1}{n} \sum_{i=1}^{n} \theta_{ij}^* A$$

$$= \frac{1}{n} A \sum_{i=1}^{n} [ln(\alpha_j) - ln(\alpha_i)]$$

$$= A\left[ln(\alpha_j) - \frac{1}{n} \sum_{i=1}^{n} ln(\alpha_i) \right]$$

$$= A[ln(\alpha_j) - ln(\alpha)]$$

where $\alpha = [\alpha_1\, \alpha_2\, \alpha_3 \ldots \alpha_n]^{(1/n)}$ is the geometric mean of the Poisson intensities. Therefore,

$$= A\; ln\left[\frac{\alpha_j}{\alpha}\right]$$

$$= ln\left[\frac{\alpha_j}{\alpha}\right]^A. \qquad (15.2)$$

The value α plays a role similar to Helson's famous Adaptation Level (Helson, 1964; Restle & Greeno, 1970), as an apparent standard against which stimulus $\mathbf{S}_j$ is judged.

Within a column the average of θ_{ij}^* values can be considered an estimate of the Poisson intensity for the stimulus at the head of the column, divided by the Adaptation Level. Thus, the column averages produce estimates of the Poisson intensity for a particular stimulus relative to a measure of the average stimulus intensity. The Poisson intensity determined for stimulus j is,

$$\theta_j^* = ln(\alpha_j/\alpha).$$

Letting $\alpha_j = \alpha + \Delta\alpha_j$ gives,

$$\theta_j^* = ln[1 + (\Delta\alpha_j/\alpha)]$$

which, expanding as a Maclaurin series gives,

$$\theta_j^* \cong \Delta\alpha_j/\alpha.$$

In terms of the physical stimuli this is the Generalized Weber Constant, $\Delta s/\mathbf{S}$.

The value of **S** is a physical magnitude that corresponds to the geometric mean of all the wave-amplitude averages. The comparison-stimulus wave amplitude that equals the Adaptation Level cannot be the arithmetic average of the comparison-stimulus wave averages, because **S** equals the geometric mean. However, when a comparison-stimulus value produces an average amplitude equal to **S** the resulting value of θ_j^* equals zero and the (unbiased) response probability equals 0.50.

The psychophysical standard that produces 0.50 response proportions appears to deviate from the arithmetic average at the midpoint of the stimulus range. A Psychometric Function based on the values of θ_j^* appears to be off center, moved away from the arithmetical mean of the stimuli. This is another source of the so-called Constant Error. The Constant Error is hardly an error, rather its occurrence is an expected outcome of the theory.

The value of θ_j^* can be measured indirectly by estimating values of $\theta_j^* A$, the sensation generated by stimulus $\mathbf{S}_j$. To determine the unknown values for A and **S**, note that the sensation

$$\theta_j^* A = ln(\alpha_j/\alpha)A.$$

Substituting actual stimulus values for the unknown values of α_j estimates the unknown value of A, assuming that the Poisson parameters actually are a similarity transformation of the stimulus magnitudes. Then, the sensation is

$$\begin{aligned} \theta_j^* A &= ln(\mathbf{S}_j/\mathbf{S})A \\ &= A[ln(\mathbf{S}_j) - (1/7)\Sigma\, ln(\mathbf{S}_i)] \end{aligned} \tag{15.3}$$

Plotting the values of $\theta_j^* A$ versus the value in brackets provides a slope estimate of the unknown threshold value A. This estimate proves to be 26.55.

Using Equation 15.3 to estimate the comparison stimulus that would reduce θ^* to zero, yields a value of 199.72 g. A theoretical value hardly distinguishable from the predicted Adaptation Level of 199.75 g obtained from computing the geometric mean of the stimuli.

Figure 15.3 shows the values of $\theta_j^* A$ as a function of the comparison weights. These estimates of sensation appear to be a linear function of the actual stimulus differences. Although the zero-crossing point, which occurs at the position of the hypothetical standard, or Adaptation Level, is not equal to the midpoint of the stimulus range, the linearity of the function is still impressive, indicating that the average sensation corresponds closely to the physical weights.

In the context of paired comparisons the concepts of Adaptation Level Theory are characteristics associated with a particular analysis of the experimental design. In the present case, for example, the subject did no more than compare one weight against another. Thus, the Adaptation Level appears in the equations but not in the subject's mind.

The values of $\theta_j^* A$ represent estimated sensations using the geometric mean as the standard. The Derived Psychometric Function, defined by the probabilities that $\mathbf{S}_j$ is heavier than **S**, is computed from the relation

$$P_{Aj} = \frac{1}{1 + e^{-\theta_j^* A}}, \tag{15.4}$$

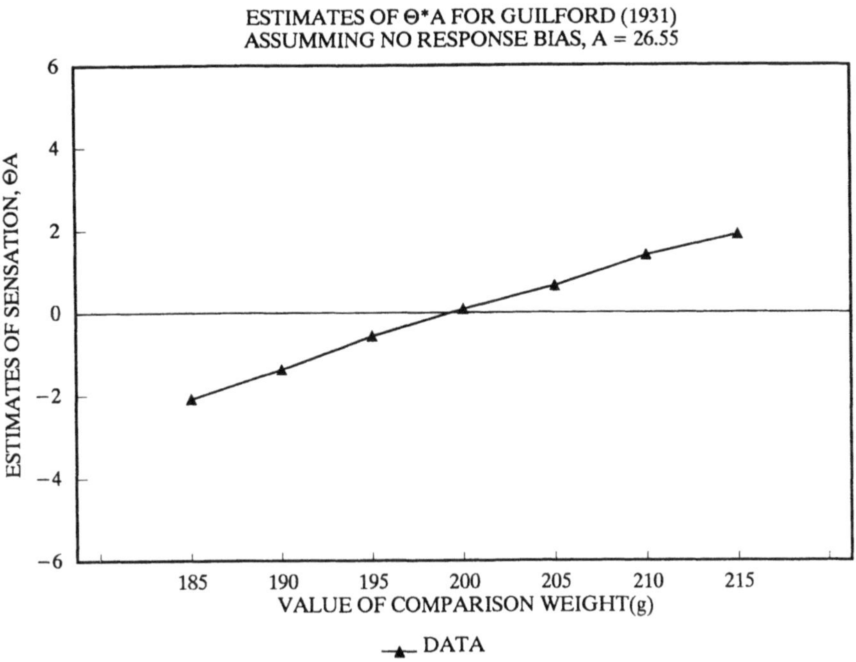

FIG. 15.3. Values of θ^*A are plotted as a function of the comparison weight values.

and are shown in Fig. 15.4. The function must be logistic, and must be shifted from the average standard of 200 g giving the impression of a small Constant Error.

The column averages provide θ_j^*A values that estimate the sensation evoked by stimulus $\mathbf{S}_j$. The differences between column averages, say for columns j and k, estimate the difference between sensations for these two stimuli, that is the value of $\theta_{jk}^* A$. For example, the difference between the averages in Columns 1 and 2 is

$$\begin{aligned}\theta_1^*A - \theta_2^*A &= ln\left[\frac{\alpha_1}{\alpha}\right]A - ln\left[\frac{\alpha_2}{\alpha}\right]A \\ &= ln\left[\frac{\alpha_1}{\alpha_2}\right]A \\ &= \theta_{12}^*A.\end{aligned}$$

Thus differences between column averages determine a predicted value of $\theta_{ij}^* A$ for each paired-comparison cell. This difference depends only on the two stimuli and not on the Adaptation Level.

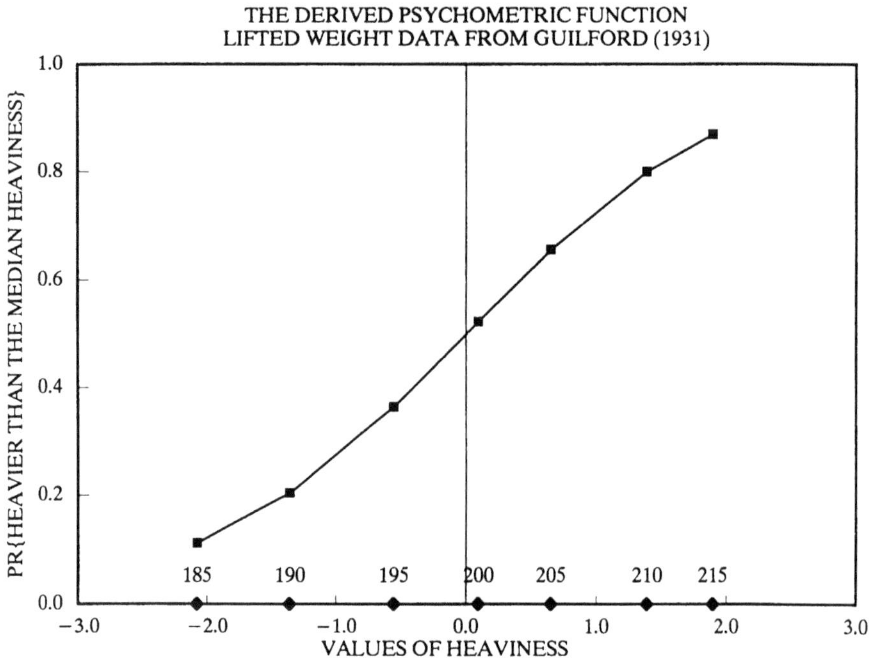

FIG. 15.4. Estimates of θ_i^*A predict the probability that a comparison stimulus will be judged heavier than the adaptation level.

These estimates are used to determine how well the theory predicts observed response proportions. The response proportions are those for choosing between two stimuli, presumably on the basis of their weight difference. The estimates of θ_j^*A at the bottom of Table 15.2 were used to compute values of θ_{ij}^*A. For computational convenience the predicted value of θ_{ij}^*A for each cell is obtained by adding the cell's column average to the cell's row average. Using these estimates, the predicted response probability for cell ij is,

$$P_{Aij} = \frac{1}{1 + e^{-(\theta_i^* - \theta_j^*)A}}. \tag{15.5}$$

Because $\theta_i^* - \theta_j^* = -(\theta_j^* - \theta_i^*)$, and there is no response bias, $P_{Aji} = 1 - P_{Aij}$. Substituting estimates of $(\theta_i^* - \theta_j^*)A$ into Equation 15.5 provides a complete reconstruction of the entire paired-comparison response-probability matrix.

A chi-squared statistic measured the nearness of these predicted response probabilities to the observed proportions. Deviations between the theoretical probabilities and the observed proportions yielded a chi-squared value of 25.0 based on 21 degrees of freedom. Thus the estimates of sensation correspond closely to the physical weights and also generate very close predictions for the paired comparison response probabilities. The corresponding chi-squared statistic for the fit of the Law of Comparative Judgment is 36.3 (21 *df*).

The predicted response probabilities also permit the construction of a Psychometric Function for lifted weights based upon stimulus differences ranging to 30 g from −30 g. All response proportions predicted for a fixed stimulus difference appear in Fig. 15.5. Slight deviations of the predicted response proportions for a fixed stimulus-difference occur because certain pairs of stimulus magnitudes that may produce equal stimulus-differences also produce different values of θ^*. Thus a single value of Δs may show several different response proportions. The probabilities sweep out a logistic-looking response function. In Fig. 15.5 the lowest response proportion is the value that the theoretical line connects with first. The theoretical line leaves a fixed Δs from the highest response proportion. There is no asymmetry, as would be expected from Adaptation Level Theory, because these probabilities depend only upon stimuli compared against each other. The Adaptation Level does not enter into the individual comparisons only into the estimates of mean values. The subject making the judgments never used the Adaptation Level. Its influence simply disappears.

The Adaptation Level is a consequence of the type of experimental method that is applied to the subject and the method of analysis applied to the resulting

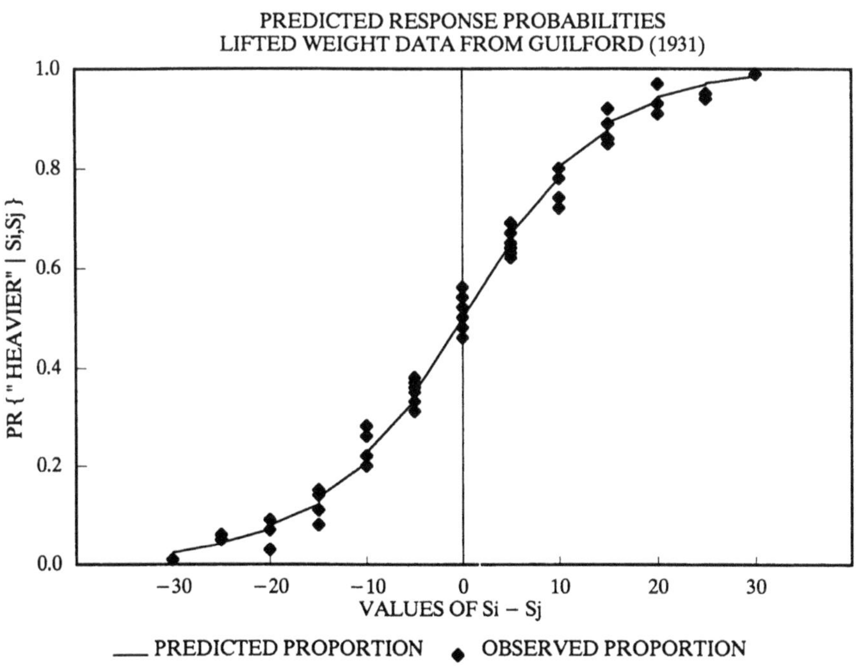

FIG. 15.5. Predicted response probabilities for lifted weights based on estimates of $\theta_i^* - \theta_j^* = \theta_{ij}^*$ from Guilford's (1931) experiment.

data. In the analysis of the Method of Paired Comparisons the column average involves a fixed-comparison stimulus but many standards, against which the average amplitude of the wave for stimulus $\mathbf{S}_j$ is compared. If the experimental design changes then so does this Adaptation Level, the **S**, against which $\mathbf{S}_j$ is measured. By changing the design, as Parducci (1991) often shows, the value of **S** may be seen to vary. There is nothing mystical about the appearance of an Adaptation Level; it is a mathematical consequence of the experimental design and analysis of data, just as the mean is a mathematical consequence of the analysis of data. The use of the Method of Paired Comparisons makes its appearance and source obvious.

Some pleasant surprises reward a deeper analysis of the mechanism that transforms physical weight to sensation. This mechanism provides a sensitive digital transmission of external signals to internal sites. The signal is transmitted and then compared against the signal for a standard. The difference between these two signals accumulates over time until a threshold is exceeded. Then the subject is aware that one stimulus appears heavier than the other.

This transformation from physical weight to sensation is transparent yet quite accurate. The accuracy is evidenced by the direct relation between the physical measure of weight and its expression as heaviness, $\theta^*A = [ln(\alpha_i/\alpha_j)]A$. This is Fechner's Law with respect to the average amplitude of the comparison and

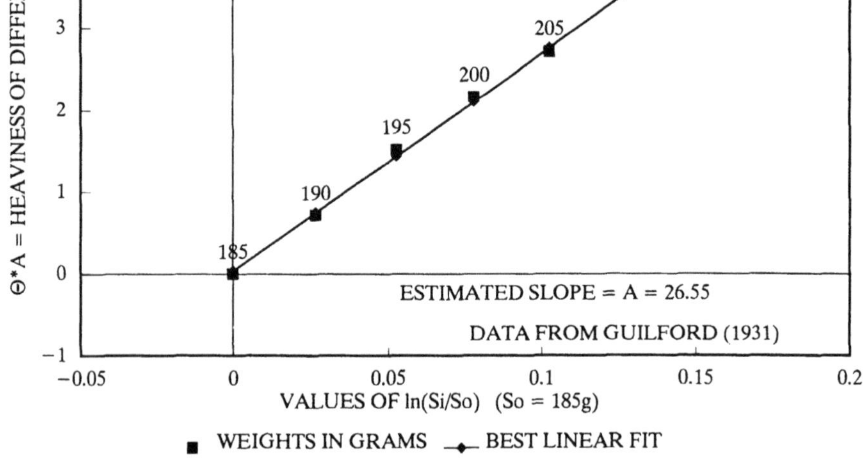

FIG. 15.6. The relation between sensation as measured by θ^*A and the logarithm of stimulus ratios is nearly perfect.

standard waves. One step farther, substituting the physical magnitudes of the stimuli for the average wave values gives the logarithmic relation between feeling and stimulus magnitude, $\theta^*A = [ln(\mathbf{S}_i/\mathbf{S}_j)]A$, Fechner's Law.

In this regard an important assumption, justified in chap. 12, is that the values of α_i and α_j are similarity transformations of $\mathbf{S}_i$ and $\mathbf{S}_j$. Then the ratio of the Poisson parameters, α_i/α_j, equals the ratio of the physical stimulus magnitudes, $\mathbf{S}_i/\mathbf{S}_j$. A test of this assumption is shown in Fig. 15.6 where the estimates of θ_i^*A obtained from the columns of Table 15.2 are compared to the logarithms of the stimulus ratios. Taking the smallest weight 185 g as a standard, $\mathbf{S}_0$, for all comparisons provides an estimate of θ_i^*/θ_0^*. These values are compared against the corresponding logarithms of stimulus ratios, $ln(\mathbf{S}_i/185)$. Because $\theta_i^*A = ln(\mathbf{S}_i/185)A$, the unknown Slope can be estimated. The Slope proves to be 26.55 and the linear fit is quite extraordinary. Thus there is little question about the applicability of Fechner's Law with respect to the estimates of θ_i^*.

The quantity of sensation is measured by θ^*A. The value of sensation, θ_i^*A, should equal $ln(\mathbf{S}_i/\mathbf{S}_0)A$. This function of $(\mathbf{S}_i/\mathbf{S}_0)^A$ is shown in Fig. 15.7 for Guilford's data. For simplicity the average θ_i^*A values are reduced by subtracting the θ^*A values for the lightest, 185 g, standard, as if this standard applied to all comparisons. The graph shows the estimated Psychophysical Function.

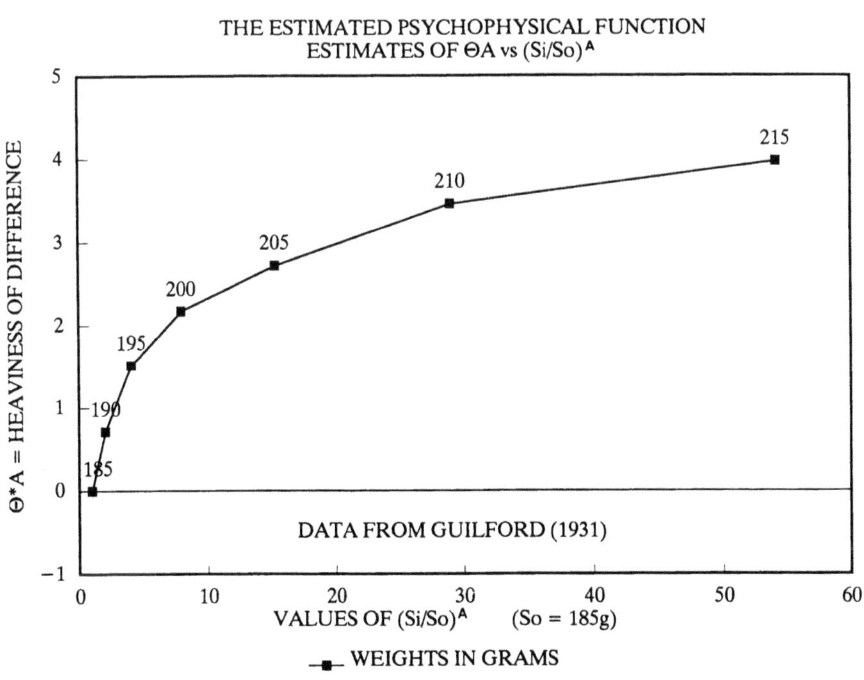

FIG. 15.7. The estimated Psychophysical Function for lifted weights using the lightest, 185 g, weight as a standard.

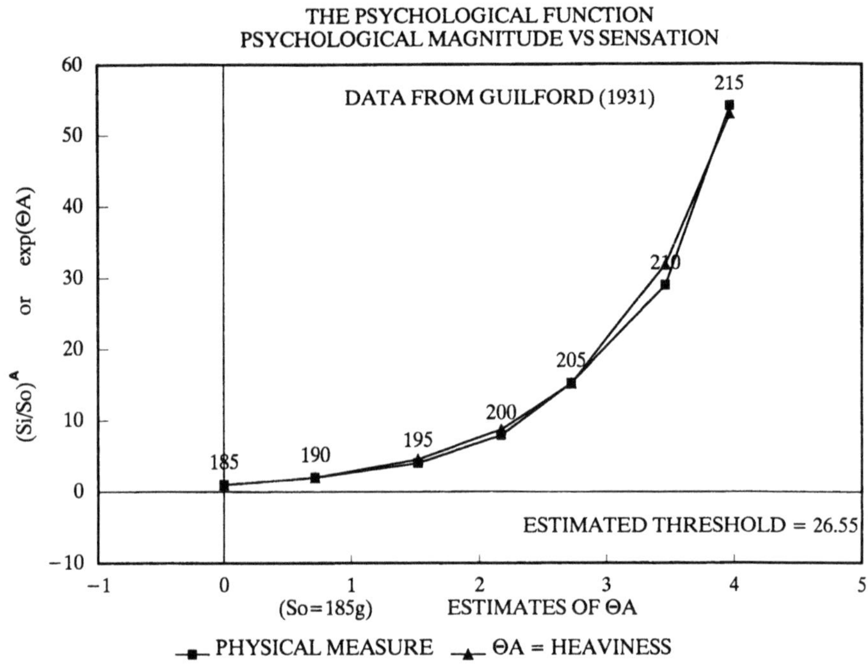

FIG. 15.8. The Psychological Function, Ψ, shows the feelings of stimuli that also give rise to sensation. The ordinate represents the subject's feeling. One function derives from sensation magnitudes determined from response probabilities. The other comes directly from the power function of stimulus magnitudes.

MEASURING FEELING

Plotting feeling against the magnitude of the sensation θ^*A shows what is felt in terms of heaviness. This is the Psychological Function, Ψ. From the subject's point of view feeling depends upon sensation. From the experimenter's point of view, feeling is a power function of the stimulus magnitude.

Guilford's (1931) data provide the opportunity to compare two different functions generating feeling. The first function is the Psychological Function illustrated in Fig. 15.8 for Guilford's weights. The Psychological Function shows the values of feeling, $\exp(\theta_{ij}^* A)$, derived from the difference in sensations obtained from the estimates of sensation for each stimulus, $\theta_{ij}^* A = \theta_i^* A - \theta_j^* A$, using the column averages of Table 15.2. These estimates of feeling increase as an exponential function of sensation, θ^*A. The other function can be determined because the stimuli have physical values and, therefore, there is a physical measure of their ratio. For this function the values on the ordinate of Fig. 15.8 equal $(\mathbf{S}_i/\mathbf{S}_{185})^A$ where A was estimated to be 26.55. These values increase as a power function of stimulus magnitude $\mathbf{S}_i$, scaled in units of 185 g. These two remarkably identical relations reveal that the feeling ascribed by the subject to the stimuli is not a power function of the physical stimulus magnitudes but an exponential function of heaviness.

This Psychological Function solves Fechner's problem for it allows determination of the feeling associated with a stimulus that has no physical measure. That measure is

$$\Psi(\theta^*A) = e^{\theta^*A}. \qquad (15.6)$$

Apparently the amount of feeling can be calculated from knowledge of an individual's Weber Constant, because the reciprocal of the Weber Constant is an estimate of A. Further the Weber Constant may be determined quickly by adaptive methods (Cornsweet, 1971). More important, when response probabilities permit estimation of θ^*A, estimates of the feelings created by the stimuli are easily obtained, regardless of whether the stimuli have physical magnitudes. The stimuli may be paintings, musical recordings, handwriting samples, essentially all those psychological things that involve feelings but are hidden from direct observation.

THURSTONE'S SERIOUSNESS OF CRIMES

The classic proportions in this regard are from the seriousness-of-crime judgments that caused Thurstone (1927a) to create the Law of Comparative Judgment. Analyzing again the 19 × 19 set of response proportions obtained by Thurstone produced the column-average estimates of θ^*_{ij} given in the inset of Fig. 15.9. Using these values the complete paired-comparison matrix of 361 response probabilities was regenerated. The number of independent probabilities is less than 361 because the main diagonal is necessarily 0.50 and the mirror-image comparisons have probabilities equal to 1 minus the response probability of the mirror comparison. In total, then, there are $(19 \times 19 - 19)/2 = 171$ independent response probabilities. The chi-squared goodness-of-fit statistic yields a value of 101.8, which, with $171 - 19 = 152$ degrees of freedom, is quite acceptable.

Much more than just a good fit of theory to data comes from this analysis. Each column estimate of θ^*_jA allows for the computation of a response probability. That value is the probability, P_{Aj}, of choosing the crime to be more serious than the Adaptation-Level crime, **S**. Specifically,

$$P_{Aj} = \frac{1}{1 + e^{-\theta^*_jA}}. \qquad (15.7)$$

Note that $\theta^*_jA = (\Delta s/\mathbf{S})A$ is also a measure of the discrepancy between the seriousness of the Adaptation-Level crime **S** and the seriousness of the crime being judged. Thus plotting the value of P_{Aj} generated by each column estimate of θ^*_jA against $\Delta s/\mathbf{S}$ produces a Derived Psychometric Function. For these data,

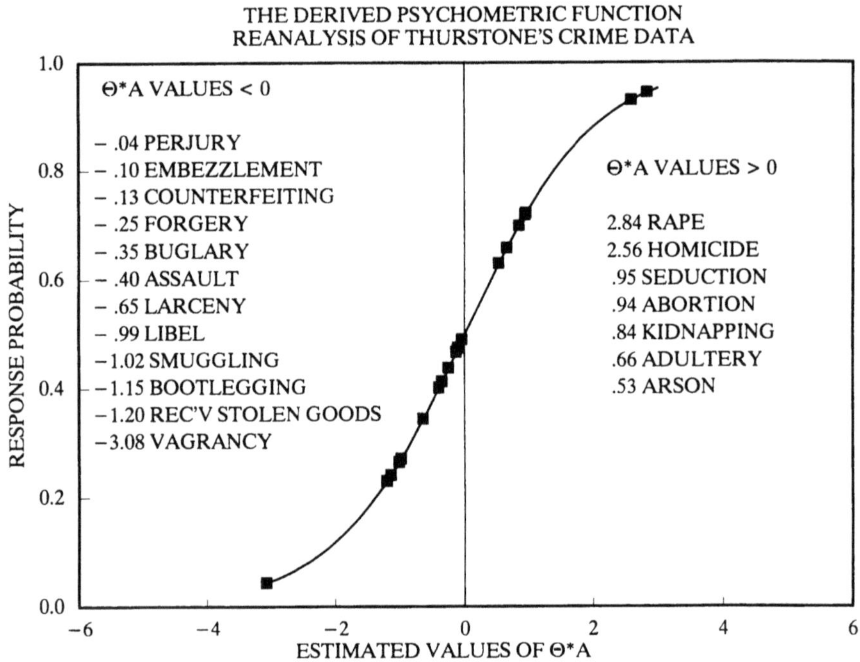

FIG. 15.9. The Derived Psychometric Function for the seriousness of crimes based on Thurstone's (1931) data.

the Derived Psychometric Function shown in Fig. 15.9 captures all the information about the seriousness of crimes. The majority of crimes have θ_j^*A values within 1 log unit of the average. Of the three extreme cases, Vagrancy is several units below the average and both Rape and Homicide are several units above.

These measures depend upon the ensemble of Poisson intensities to determine **S**, their geometric mean, or Adaptation Level. There is no dispersion or error as was assumed by Thurstone in the Law of Comparative Judgment. Thus the abscissa provides units of sensation with respect to the median crime.

Sensation, however, is not feeling. To measure the feeling, the felt seriousness of these crimes, requires calculation of the Psychological Function. Using Vagrancy as a baseline and subtracting the estimated seriousness, θ_V^*A for this crime from that for the j-th crime yields the value,

$$\theta_j^*A - \theta_V^*A = \theta_{jV}^*A$$
$$= [ln(\alpha_j/\alpha_V)]A.$$

This value depends upon on the ratio of Poisson intensities for Crime j and Vagrancy. The feeling evoked by Crime j with respect to Vagrancy follows by exponentiating θ_{jV}^*A, to obtain,

$$e^{\theta^*_{jV}A} = e^{[ln(\alpha_j/\alpha_V)]A}$$
$$= (\alpha_j/\alpha_V)^A \ .$$

The values α_j and α_V are like physical measures of stimuli. They are, in fact, the average values of Poisson waves. If they arose from physical stimuli they would be treated as the result of a similarity transformation of stimulus magnitude. Dividing the intensity parameter for Bootlegging by the intensity parameter for Vagrancy is similar to dividing a weight of 195 g by the minimum weight, 185 g. Thus, the ratio of α_j to α_V is similar to a measure based on physical stimuli. The values of $(\alpha_j/\alpha_V)A$ are dimensionless, they may arise from physical stimuli or other sources, but to the subject $(\alpha_j/\alpha_V)A$ is a feeling of how much more serious, for want of a better word, Bootlegging is than Vagrancy.

The Psychological Function for crime appears in Fig. 15.10. The ordinate scales the feeling of seriousness of crime with respect to Vagrancy, which has a unit value. Small crimes are those bounded by Bootlegging and Perjury. Major increments in the magnitude of crime begin with Arson and continue through Seduction (statutory rape), Homicide, and Rape.

The relation between the feeling of seriousness and the sensation evoked by a

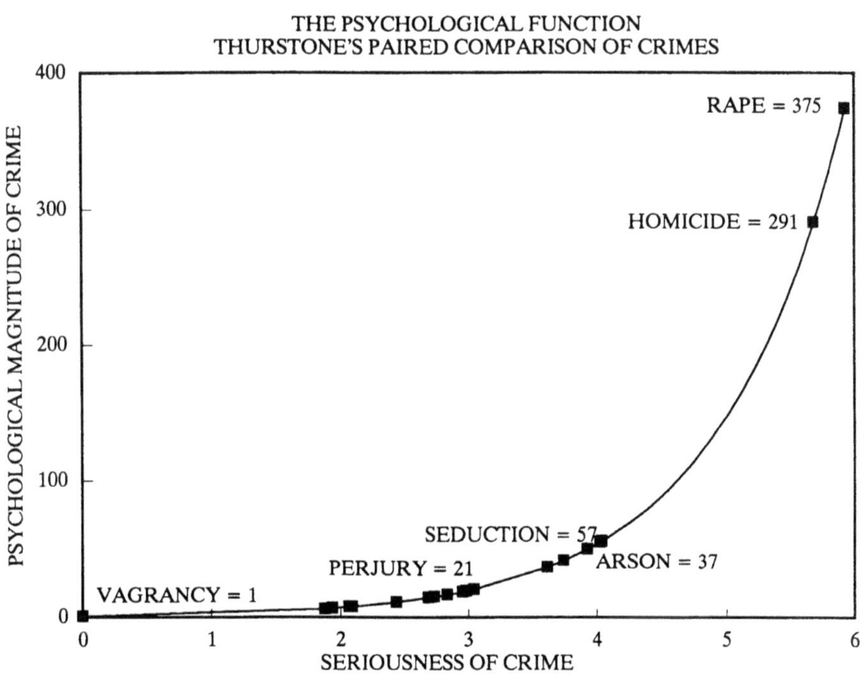

FIG. 15.10. The Psychological Function, Ψ, for the magnitude of crime. The abscissa shows measures of sensation, the ordinate shows the magnitude of feeling about the crime.

crime solves the problem faced by Fechner in his studies of aesthetics. The sensations are measured by values of θ^*A, generated by wave differences. Their values depend upon comparing one stimulus to some other stimulus in the ensemble. Feeling depends upon sensation. Although the feeling is actually a function of sensation, it occurs after the sensation makes the subject aware of the presence of the stimulus. Feelings themselves then become stimuli that are subject to the same mechanisms of comparison and judgment as the stimuli that gave rise to them.

These examples show how to measure feelings—the hidden psychological elements of mental life. The context of paired comparisons provides for new theoretical results that show how sensation is the foundation of response probability. Therefore, response probabilities provide the basis for estimates of sensation. From sensations come estimates of feeling, the measure of awareness. The Psychological Function defines the relation between sensation and feeling and provides a measure of the amount of feeling. The techniques are applicable generally to the measurement of concepts that generate feelings heretofore beyond the reach of psychological measurement. Many feelings may be common to all mental life and transcend species. Perhaps the measures developed here will not only find application to clinical evaluation but also provide for a better understanding of other sentient creatures.

16 Bias, Responsiveness, and Strategy

RESPONSE TIMES AND PROBABILITIES

Urban's (1908) Method of Just Perceptible Differences showed response proportions to increase gradually as the difference between a standard and comparison stimulus increase. The Psychometric Functions defined by these changes in response probability became a fundamental tool for the analysis of discriminability, especially for determining 75% thresholds and the Interval of Uncertainty. Yet, attacks by Thomson (1913) and Culler (1927) on Urban's frankly behavioristic psychophysics demonstrated errors in the mathematical formulations of this atheoretical approach. Advancing a more theoretical analysis, Thomson (1920) proposed that Psychometric Functions depend upon stimulus difference, response thresholds, and the number of sampled stimulus elements. Urban's Psychometric Function became the victim of increasing criticism and disuse, but the coming of Thurstone's Law of Comparative Judgment overshadowed both the criticism of Urban and Thomson's insightful achievement.

Link (1978a) proposed that Psychometric Functions depend upon three parameters; *discriminability*, as measured by θ^*, *response bias*, as measured by B, and *resistance to respond*, as measured by A. Wave Discrimination Theory extends this previous analysis by demonstrating how these three parameters influence the Psychometric Function's shape and position. The Psychometric Function for responding *larger* depends upon differences between the waves for standard and comparison stimuli, the accumulation of these differences to overcome a resistance to respond, and the amount of response bias. The probability of responding *larger* (see chap. 11, Equation 11.1) is

$$P_L = \frac{e^{\theta^*(A+B)} - 1}{e^{\theta^*(A+B)} - e^{-\theta^*(A-B)}}. \qquad (16.1)$$

This probability is easily separated into components that represent the probability that the threshold at A is exceeded given no response bias, $B = 0$, plus another term that depends on the amount of response bias and the response thresholds. Multiplying the numerator and denominator by $e^{-\theta^* B}$ and then adding and subtracting 1 in the numerator makes the two terms apparent. In this format the probability of responding *larger* is, (Link, 1978b)

$$P_L = \frac{e^{\theta^* A} - 1}{e^{\theta^* A} - e^{-\theta^* A}} + \frac{1 - e^{-\theta^* B}}{e^{\theta^* A} - e^{-\theta^* A}} \qquad (16.2)$$

$$= \text{Probability of } R_L \text{ given no Bias} + \text{Change due to Bias.}$$

The first term is the probability of responding *larger* given no bias at all. The second term is a measure of the advantage to responding *larger* given a response bias of size B. When the advantage is positive the bias is toward responding

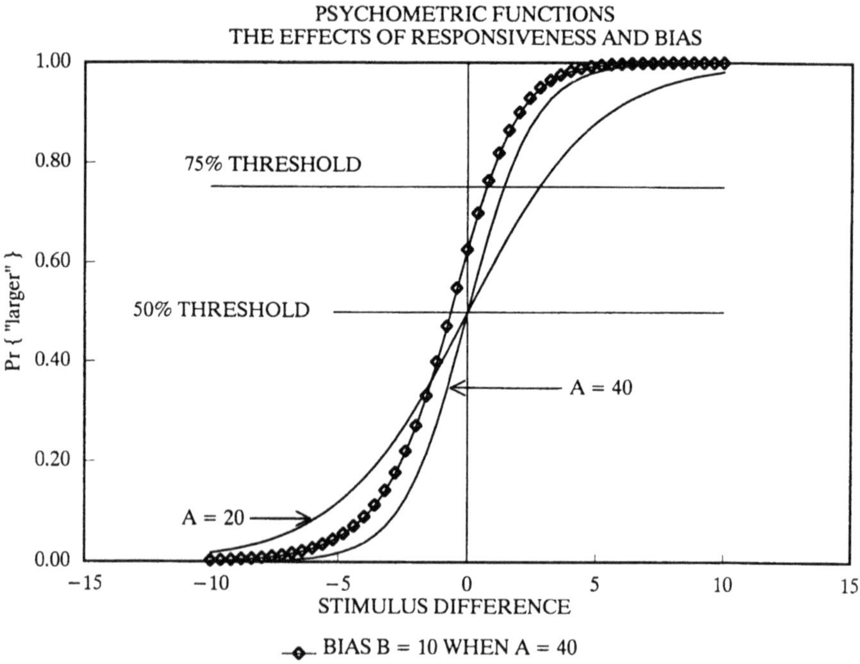

FIG. 16.1. Three Psychometric Functions illustrate the effects of changes in discriminability, resistance to respond, and response bias on response probabilities.

larger, that is, $B > 0$, and the value of P_L increases. When the bias is toward responding *smaller*, $B < 0$, the probability of responding R_L decreases.

The effects of these parameters are illustrated in Fig. 16.1. The abscissa is scaled in units of difference from a standard stimulus $\mathbf{S}_B$. For the purpose of illustration the image contains three functions. The standardized Psychometric Function is symmetric with respect to the value of the standard and assumes the value 0.731 when θ^*A equals 1.0. When A increases, the subject appears to show greater discriminability, as the increased slope of the Psychometric Function indicates. The discriminability, as determined by θ^*, actually remains fixed but the subject's response threshold increases thereby increasing correct response probabilities. The third Psychometric Function illustrates the effect of positive response bias on the standardized Psychometric Function. The subject tends to respond *larger* more readily than *smaller* thereby shifting the 50% value of the Psychometric Function to the left, away from the location of the standard. Examples of these effects drawn from experimental data of Brown (1910), Guilford (1931), and Link and Tindall (1971) appear in the preceding chapters.

Decision times also depend upon three parameters as shown in Equation 16.3. The parameters representing responsiveness and response bias remain unchanged, but the effective discriminability parameter changes from θ to μ, the average difference between amplitudes of the standard and comparison waves. Increases in discriminability result in smaller decision times because μ is a divisor of both resistance and bias. Increasing the resistance to respond increases decision time. Reducing the resistance to respond reduces decision time. Finally, response bias shifts the peak of the Chronometric Function.

$$E\{\boldsymbol{DT}\} = \frac{A}{\mu}(2P_L - 1) - \frac{B}{\mu}. \tag{16.3}$$

Figure 16.2 illustrates these effects. The Chronometric Function for unbiased responses, corresponding to the unbiased Psychometric Function in Fig. 16.1, is symmetric with respect to the location of the experimenter's standard, $\mathbf{S}_B$. When the resistance to respond increases the mean decision time also increases. With a response bias equal to that generating the biased Psychometric Function in Fig. 16.1 there is a noticeable shift in the position of the maximum mean decision time. Mean decision times compatible with these interpretations also occur throughout earlier chapters.

With three parameters to define a response probability, two of which are under the voluntary control of the subject, a bewildering variety of results may occur. Indeed, each point in the three dimensional parameter space shown in Fig. 16.3 generates a response probability. The line labelled $A_{1,B=0}$ is the locus of parameter values that define an unbiased ($B = 0$) Psychometric Function with resistance to respond A_1. The line labelled $A_{2,B=0}$ is the locus of parameter values that define an unbiased ($B = 0$) Psychometric Function with resistance to respond A_2. The line labelled $A_{1,B=1}$ generates a positively biased ($B > 0$) Psychometric

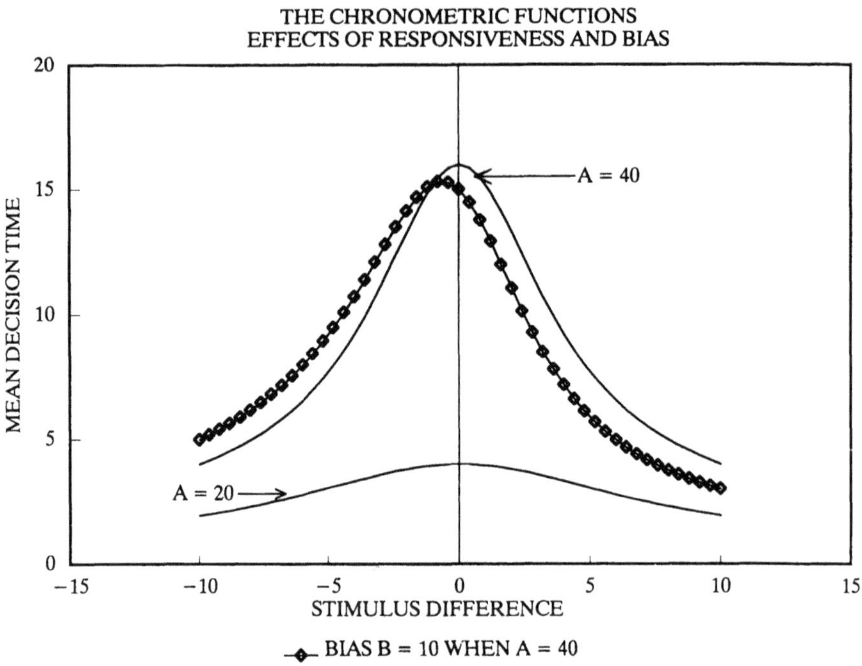

FIG. 16.2. Three Chronometric Functions show effects of changes in discriminability, resistance to respond, and response bias on response time.

Function with resistance A_1. At the other extreme of bias, the line labelled $A_{1,B=-1}$ corresponds to a biased ($B < 0$) Psychometric Function for the same range of discriminability and resistance.

An *Isoprobability Curve* appears as a solid black line in Fig. 16.3. This curve represents the many cases where a stimulus is judged to be just noticeably different from a standard. The Isoprobability Curve labelled $p = 0.731$ defines all cases where unbiased responding leads to a *jnd*. Each value of θ^* corresponds to wave differences between two stimuli. Thus a large range of stimuli and associated resistances produce a *jnd*. If the subject adopts a small resistance to respond the experimenter must present larger stimulus differences in order for correct discriminations to occur 73% of the time, that is, to produce a *jnd*.

The well-established empirical fact that at one or another time different stimulus magnitudes are required to generate a *jnd* imperiled the measurement of psychophysical thresholds. Many futile attempts to reduce the variability of threshold measures to a respectable level cast the very measurement of thresholds into disrepute. Experimental psychologists such as S. S. George (1917) suggested that these difficulties arose from the subject's attitude; that somehow, attitude affected performance. However, George's introspective analysis produced no measurements of the suspected influence of attitude.

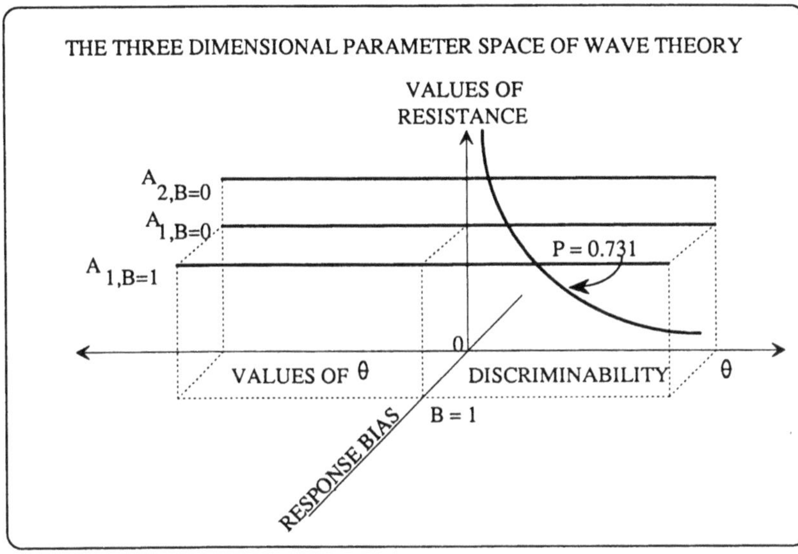

FIG. 16.3. Within the parameter space of Wave Discrimination Theory are lines characterizing Psychometric Functions and just noticeable differences.

With respect to response bias Thomson's (1920) three-choice theory proposed that variations in response thresholds accounted for many variations in Psychometric Functions. The theory anticipated by nearly 30 years Blackwell's introduction of response bias and the decision thresholds of Signal Detection Theory. Later, Shipley, et al. (1945), studying color preferences, discovered two different factors that contributed to the size of choice response times. One factor, affective distance, accounted for 74% of response time variance, whereas 16% of the variance resulted from "a positive value of the stimulus factor," actually another form of response bias. George's introspections and Shipley's correlational analysis foreshadowed Blackwell's introduction of response bias as a subjective component of performance. But the most significant analysis of this subject-controlled factor awaited the coming of Signal Detection Theory.

THE TWO-STIMULUS DISCRIMINATION EXPERIMENT

The outstanding achievement of Signal Detection Theory comes from clarifying the distinction between objective- (stimulus) and subjective- (response) performance factors. The stimulus factor corresponds to the difference between the means of two Gaussian random variables, a measure that owes much to both Fechner and Thurstone. The response factor corresponds to a bias toward one or the other response. In Signal Detection Theory these two variables determine the probability of choosing to respond *Yes*, signal present, or *No*, signal absent.

Because no sensory threshold existed to be measured, differences between sensory signals played second fiddle to demonstrations of response bias. To examine the theory many experiments promoted only changes in response bias and only a single stimulus difference occurred. To obtain sufficient data to estimate both the stimulus difference, d', and the response bias, β, large numbers of responses became necessary. The large number of trials yielded quite stable estimates of response probabilities. Yet, the remarkably stable response probabilities led to a theoretical enigma.

Chapters 7 and 8 show various Receiver Operating Characteristics obtained by use of this two-stimulus two-response experimental method. The surprising feature of the experiments is the remarkable correspondence between the various stimuli and the amazing disparity of experimental results. The ROCs which are to provide a more sensitive analysis of performance than the Psychometric Function are replicable, yet disturbingly different. How can experiments so similar in design, employing similar, if not identical, stimuli generate such diverse results?

Wave Discrimination Theory shows how response strategies rather than different theories of discrimination define the various forms of Receiver Operating Characteristics[1] Previously these different shapes identified a particular theory, but the ROCs predicted from the major psychophysical theories, High Threshold Theory, Low Threshold Theory, Signal Detection Theory and Variable Sensitivity Theory, are most likely the consequence of a single source.

To model the effects observed in these two-stimulus two-response experiments, Wave Discrimination Theory generalizes the original idea of Fechner, that a comparison stimulus is judged against the mean value of the standard and comparison values. Instead, stimulus waves defined as symmetric (Appendix C) produce mirror-symmetric difference waves. Mirror symmetry, as in Fechner's theory or Signal Detection Theory's Ideal Observer Theory (Green & Swets, 1966) leads to surprizing relations between Hit and False Alarm probabilities.

BIAS AND RESISTANCE IN THE ROC SPACE

When the wave difference accumulated under a presentation of stimulus $\mathbf{S}_\mathrm{A}$ is symmetric to the accumulation under a presentation of stimulus $\mathbf{S}_\mathrm{B}$ a startling relation exists between the probabilities of responding *larger* given stimuli $\mathbf{S}_\mathrm{A}$ and $\mathbf{S}_\mathrm{B}$. When stimulus $\mathbf{S}_\mathrm{A}$ occurs the probability of responding *larger* is

$$Pr\{larger|\mathbf{S}_\mathrm{A}\} = \frac{e^{\theta^*(A+B)} - 1}{e^{\theta^*(A+B)} - e^{-\theta^*(A-B)}}. \qquad (16.4a)$$

Because the only difference which occurs when presenting symmetric stimulus $\mathbf{S}_\mathrm{B}$ is the change in discriminability from plus to minus, $+\theta^*$ changes to $-\theta^*$.

---

[1]The first demonstration that sequential theories generate the ROC curves of Signal Detection Theory, High Threshold Theory, and Low Threshold Theory is found in Link and Heath (1975).

The probability of responding (incorrectly) *larger* to stimulus $\mathbf{S}_\mathrm{B}$ is obtained by simply changing the sign of θ^*.

$$Pr\{larger|\mathbf{S}_\mathrm{B}\} = \frac{e^{-\theta^*(A+B)} - 1}{e^{-\theta^*(A+B)} - e^{+\theta^*(A-B)}}. \tag{16.4b}$$

In the parlance of Signal Detection Theory these relations define the

$$Pr\{\text{Hit}\} = Pr\{larger \mid \mathbf{S}_\mathrm{A}\}$$

and the

$$Pr\{\text{False Alarm}\} = Pr\{larger \mid \mathbf{S}_\mathrm{B}\}.$$

The ratio of these seemingly complex probabilities yields a surprizing, fundamental result,

$$\frac{Pr\{\text{Hit}\}}{Pr\{\text{False Alarm}\}} = e^{\theta^*(A-B)}. \tag{16.5}$$

In terms of the theory of sensation presented in chap. 13, the value $\theta^*(A - B)$ is the sensation experienced when response R_A occurs following a $\mathbf{S}_\mathrm{A}$ presentation. The corresponding feeling is $e^{\theta^*(A-B)}$. Thus, the values of A and B, which determine the ratio of the probability of a Hit to a False Alarm, also generate a feeling that is under the subject's voluntary control. By adjusting the feeling generated by stimulus $\mathbf{S}_\mathrm{A}$ the subject can generate for the experimenter a point in the ROC space. The point is defined by a relation between Hits and False Alarms:

$$Pr\{\text{Hit}\} = e^{\theta^*(A-B)}Pr\{\text{False Alarm}\}. \tag{16.6}$$

This functional relation describes the many different ROC results attributed to a variety of quite distinct theories. Although all three parameters, θ^*, A, and B, determine performance within the ROC space, the major determinant of performance is the subject's selection of a resistance to respond and a response bias based on feeling. The way in which these parameters change creates the remarkably different performances shown in the following figures.

First, when there is no discrimination the value of $\theta^* = 0$ and Equation 16.6 reduces to

$$Pr\{\text{Hit}\} = Pr\{\text{False Alarm}\}. \tag{16.7}$$

As shown in Fig. 16.4, this equation defines the straight line beginning at the origin (0, 0) and continuing with Slope 1.0 to the upper corner of the ROC space at the point (1, 1). The locus of points generating purely biased responses is defined by the area within the Wave Discrimination Theory parameter space where $-A < B < A$ and $\theta^* = 0$.

According to Equation 16.6, guessing occurs in yet another way. When the resistance to respond is zero the response bias must also equal zero and Equation 16.6 reduces again to Equation 16.7. In this latter case the value of A equals zero

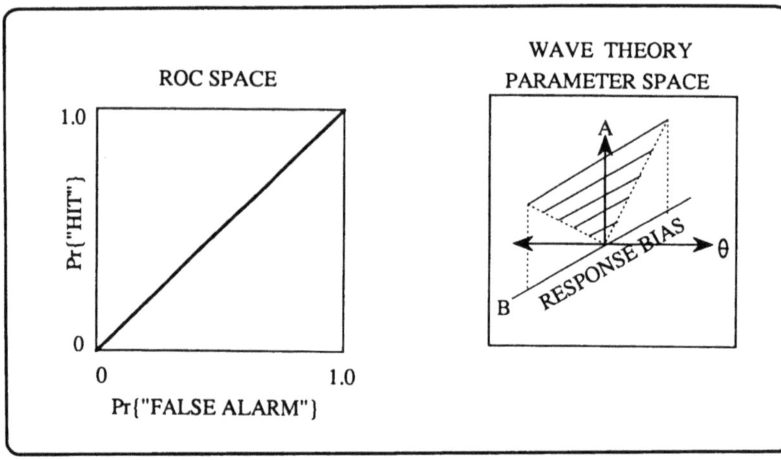

FIG. 16.4. The ROC space shows guessing response probabilities derived from purely biased responding as shown in the WDT parameter space.

and the first accumulation of a wave difference leads to a response. Such responses are fast guesses in the sense of Ollman (1966) and Yellott (1967, 1971). This agrees with the intuition that very fast responses based on no discriminative information must be guesses.

When the value of B is fixed, as shown for the experiments of D. M. Johnson (1939) and Link and Tindall (1971), changes in performance result from changes in stimulus difference and response-time deadlines. Response bias in these studies is essentially zero. Therefore, the expression for Hits versus False Alarms may be written as

$$Pr\{\text{Hit}\} = e^{\theta^* A} Pr\{\text{False Alarm}\}. \tag{16.8}$$

Because θ^* and A multiply each other changes in either θ^* or A affect the ROC in the same way. The effects of discriminability and resistance on response probabilities are indistinguishable. Increases in d' leading to better discriminability are often interpreted as evidence favoring the Signal Detection Theory of discrimination. However, changes in the resistance to respond produce identical operating characteristics. Thus evidence taken as support for changes in discriminability may be evidence in favor of changes in resistance to respond.

In Fig. 16.5 different values of response bias ranging to A from $-A$ sweep out the ROC. At the upper right-hand corner the bias equals A. There is actually no discrimination and the subject always responds that a signal is present. The lower-left corner shows the effect of bias toward the lower-response threshold: No signal is present, and the subject always responds *No*. Between these extremes the ROC passes smoothly through a set of performance values that, in practice, are indistinguishable from the predictions of Signal Detection Theory.

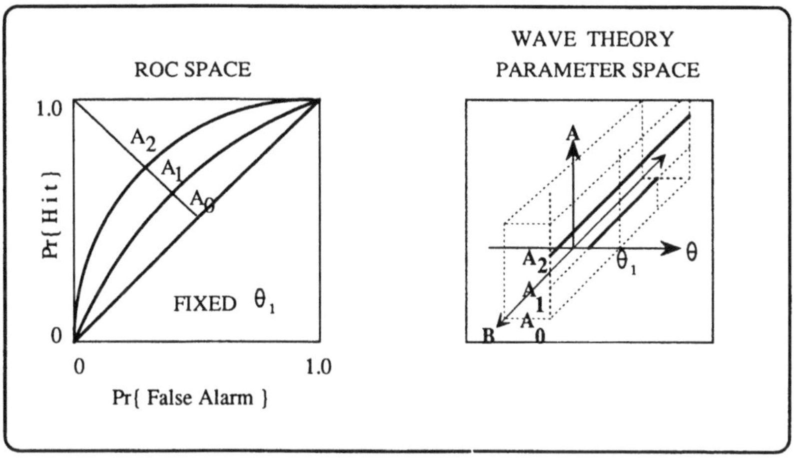

FIG. 16.5. The left-hand panel shows ROCs derived from the Wave Theory parameters shown in the right-hand panel. Changes in response bias produce changes for a fixed discriminability.

The three operating characteristics appearing in the ROC space appear to indicate better discriminability. However, the values in the parameter space show the value of θ^* to remain constant. Only the resistance to respond changes, from A_0 (guessing) to A_1 (moderate performance) to A_2 (good performance).

A major difference between Wave Discrimination Theory and Signal Detection Theory arises in the case of speeded responses where the subject is required to beat a response-time deadline. In these experiments (e.g., Link & Tindall, 1971), the subject usually beats the time deadline. Of course, all the results, including those cases where the subject does not beat the deadline, and errors are included in the analysis. As the value of A increases the ROC, as shown in Fig. 16.5, exhibits a greater curvature characteristic of greater discriminability. The change in performance is due to compliance with the experimenter's instructions, not with a change in discriminability. The subject's discriminability does not change at all. Instead the subject, responding to the experimenter's implicit or explicit instructions, alters the resistance to respond.

These changes in the resistance to respond may be easily mistaken for changes in discriminability. As a practical matter, changes in resistance to respond correspond to changes in vigilance. Fatigue may not cloud sensory processing at all, but may instead produce decreases in the subject's resistance to respond. A less vigilant subject exhibits a form of speed-accuracy tradeoff. The subject appears less discriminating, makes more errors, and faster responses.

Yet changes in discriminability cause similar changes in the ROC. Three examples corresponding to three levels of discriminability $\theta_0 < \theta_1 < \theta_2$ appear on the left of Fig. 16.6. Points along the ROC derive from different response biases, as shown on the right side of Fig. 16.6. The first ROC corresponds to a

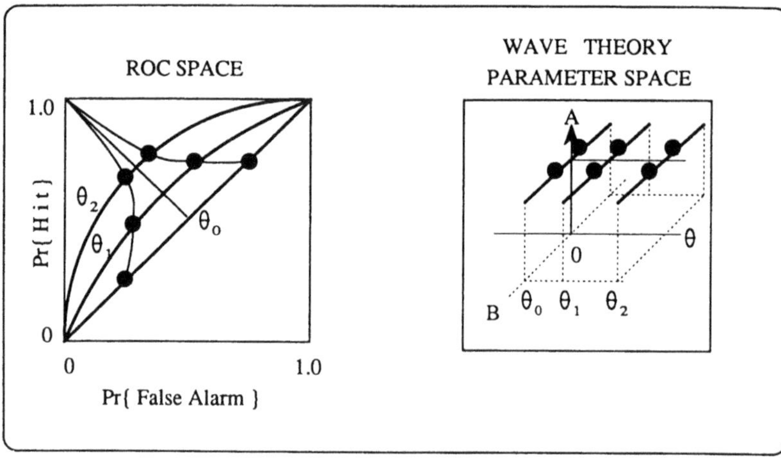

FIG. 16.6. The increasing curvature of these ROCs results from changes in θ^*, the WDT measure of discriminability. The points defining a curve result from different response biases. The resistance to respond, A, is constant, as shown in the accompanying parameter space.

value of $\theta_0^* A$ equal to 0 because the value of θ_0^* equals 0. Changes in B change response proportions but no discrimination occurs. In the second case θ_1^* yields a slightly bowed ROC. In the third case θ_2^* is large and the ROC is quite curved. The six hypothetical data points represented by circles and connected by solid lines show the effect of response bias. Equivalent biases toward the *larger* or *smaller* response appear on the opposite sides of the main diagonal. Lines connecting these points illustrate constant bias across changes in discriminability. These are *Isobias Curves* (Luce, 1960). No bias at all occurs on the diagonal connecting the upper left-hand corner to the figure's center.

If both discriminability and resistance to respond change, θ^* and A both change; consequently there should be an ordering of points in the ROC space that appear to correspond to changes in discriminability but are due to the joint influence of discriminability and the resistance to respond. Such results are shown in Fig. 14.3 for data drawn from the experiment of Link and Tindall (1971), where the various points in the ROC space may be taken as points on different ROCs showing apparently different amounts of discriminability. As shown in Fig. 14.4 these data arise from the subject controlling performance by manipulating the resistance to respond while the experimenter manipulates differences between stimuli. Quite clearly the ROC space is not a useful method of distinguishing between changes in performance due to increases in discriminability versus changes in the resistance to respond.

The failure of the ROC space to distinguish between improvements in performance due to changes in discriminability or changes in the resistance to respond also applies to the linear operating characteristics of Blackwell's High Threshold

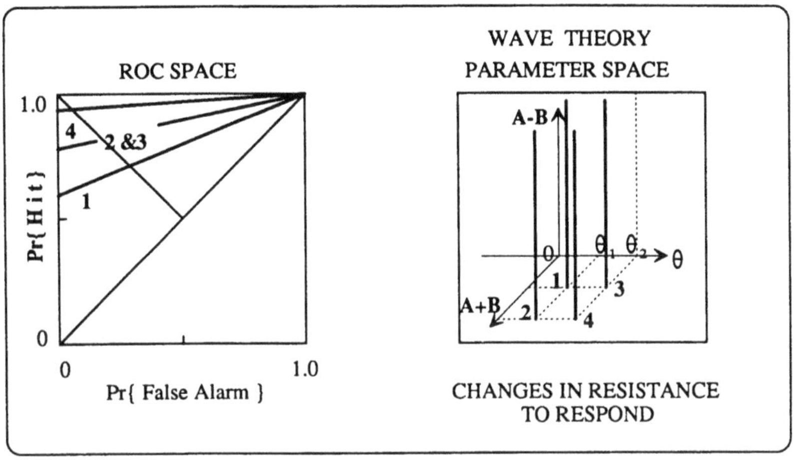

FIG. 16.7. The ROC space reveals linear response characteristics for Blackwell's High Threshold Theory.

Theory. Blackwell's linear ROCs come from observing that Equation 16.6 may be rewritten as,

$$Pr\{\text{Hit}\} = Pr\{\text{False Alarm}\}e^{-\theta^*(A+B)} + (1 - e^{-\theta^*(A+B)}). \qquad (16.9)$$

This equation expresses Hit and False Alarm probabilities as a linear function with Slope $e^{-\theta^*(A+B)}$ and Intercept $1-e^{-\theta^*(A+B)}$. For a fixed pair of stimuli and a fixed threshold $(A + B)$ for responding *No*, changes in the Hit and False Alarm probabilities generate a straight-line operating characteristic. These changes in probability must result from changes in the value of $A - B$, the informative value that is needed to respond *Yes*.

Four examples of parameter sets that generate such operating characteristics appear in Fig. 16.7. To ease the discussion of how Wave Discrimination Theory predicts straight, as well as curved, operating characteristics, the parameter space takes as dimensions, Discriminability as measured by θ^*, the *Yes* response threshold $(A - B)$, and the *No* threshold, $(A + B)$. For the purpose of illustration two discriminabilities, $\theta_1 = 0.03$ and $\theta_2 = 0.06$ and two *No* response thresholds equaling 30 and 60 are shown in Fig. 16.7. The values of the *Yes* response threshold $(A - B)$ vary upward as shown by the lines labelled 1, 2, 3, and 4 on the figure's right-hand side. The combinations of parameters produce four linear operating characteristics.

A value of $A + B = 30$ results in poorer discrimination than a value of $A + B = 60$. The two operating characteristics that illustrate this change in performance are 1 and 2. The increase in the probabilities of a Hit from 1 to 2 suggests an increase in discriminability. Actually, the subject simply requires a greater informative value before responding *No*. This consequence is obvious

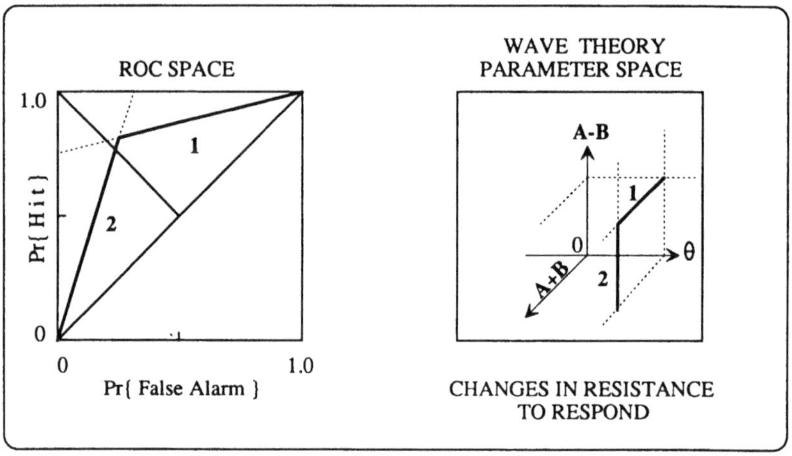

FIG. 16.8. The dog-leg operating characteristic of Low Threshold Theory may be the signature of a response strategy for manipulating response bias and the resistance to respond.

when response times are available, but with only ROC results, the experimenter would draw the (incorrect) conclusion that the improvement in performance resulted from greater discriminability.

Some changes in discriminability do appear as improvements in performance but are indistinguishable from improvements due to increases in the resistance to respond. The two sets of points labelled 3 and 2 in the parameter space correspond to the same operating characteristic in the ROC space in spite of their entirely different parameters. For the operating characteristic labelled 3 the value of $A + B = 30$ whereas the value of $\theta_2^* = 0.06$. For the operating characteristic labelled 2 the value of $A + B = 60$ and θ_1^* equals 0.03. A single operating characteristic cannot reveal the tradeoff between θ^* and the response thresholds.

Even more surprising is the operating characteristic in Fig. 16.8, which at first sight confirms predictions of Luce's (1960) Low Threshold Theory. However, this dog-leg operating characteristic derives from a response strategy that in one case emphasizes control of the resistance to respond *Yes*, as in the case of High Threshold Theory, and in the other case emphasizes the control of the resistance to respond *No*. The previous example for High Threshold Theory illustrates how the upper leg of the function in Fig. 16.8 occurs. The upper line need not extend to the point where the probability of a False Alarm equals zero. Rather, at some maximum value for the resistance to respond *No* the subject's response strategy shifts to varying the resistance to respond *Yes*. Because the resistance to respond *Yes* is constant but resistance to respond *No*, $A + B$, varies, another linear function occurs in the ROC space. This function defines the lower part of the dog-leg ROC. With this emphasis the probability of a Hit is,

$$Pr\{\text{Hit}\} = e^{\theta^*(A-B)} Pr\{\text{False Alarm}\}. \qquad (16.10)$$

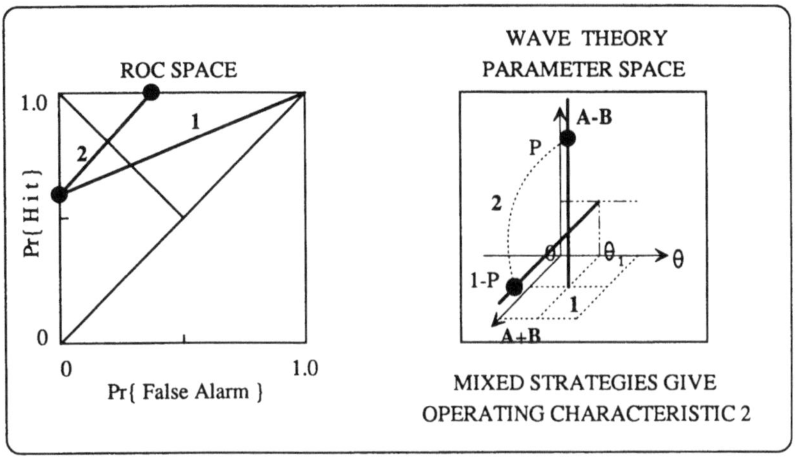

FIG. 16.9. Atkinson's Variable Sensitivity Theory predicts linear ROCs such as these.

When the maximum resistances to respond *Yes* and *No* are equal, the dog leg bends at the unbiased performance diagonal of the ROC space. When the maximum resistance to respond *Yes* differs from the maximum for *No* an asymmetric dog leg occurs. If the maximum resistance to respond *Yes* exceeds the maximum for *No* then the lower asymmetric ROC results.

The last example of many possible response strategies that generate unique performance functions in the ROC space appears in Fig. 16.9. The two straight lines derive from the data shown in Fig. 8.7 which support Atkinson's Variable Sensitivity Theory. The straight line ROC terminating at the upper right-hand corner of the ROC space occurs in the same way as in High or Low Threshold Theories, but the line parallel to the guessing axis, with intercept equal to $(1 - e^{-\theta*(A+B)})$, derives from a mixture of two opposite response tendencies.

The subject can emphasize the capture of noise trials by fixing the value of $A + B$ and increasing the value of $A - B$ to an arbitrarily high value. The left-hand point $[0, (1 - e^{-\theta*(A+B)})]$ corresponds to a fixed value of $A + B$ and a large, essentially infinite value for $A - B$. Theoretically an infinite value for $A - B$ makes this performance unrealizable because the subject may never respond to the presentation of a signal. However a large finite value for $A - B$ produces a realizable response probability. This pair of resistances to respond defines a single point near the edge of the ROC space.

To emphasize the capture of signal trials the resistances to respond, including response bias, are reversed. The resistance to respond *Yes*, $A - B$, remains fixed but the resistance to respond *No*, $A + B$, assumes an arbitrarily high value. As $A + B$ increases the point in the ROC space approaches the point $[(1 - e^{\theta*(A+B)}, 1)]$. This point is an extreme upper value for the lower segment of the dog leg in Fig. 16.8.

These two extreme ROC points represent two different viewpoints about how to meet the experimenter's performance demands. The subject may be induced through payoffs, instructions, presentation probabilities, and unknown factors to adopt one point of view with probability p, and another point of view with probability $1 - p$. Thus the data from a single experimental session consists of a mixture of these two different viewpoints. The result is an ROC point for a single experimental session that is a mixture of the two points at the edges of the ROC space. The probability of a Hit is a mixture

$$Pr\{\text{Hit}\} = p[(1 - e^{-\theta^*(A+B)})] + (1 - p)[1],$$
$$= 1 - pe^{-\theta^*(A+B)}, \quad (16.11)$$

whereas the probability of a False Alarm is a mixture of the form,

$$Pr\{\text{False Alarm}\} = p[0] + (1 - p)[e^{-\theta^*(A-B)}]$$
$$= (1 - p)[e^{-\theta^*(A-B)}]. \quad (16.12)$$

Writing the probability of a Hit as a function of the probability of a False Alarm yields,

$$Pr\{\text{Hit}\} = Pr\{\text{False Alarm}\} + (1 - e^{-\theta^*(A+B)}). \quad (16.13)$$

Thus, the mixing probability p disappears. To the experimenter the probability of a Hit appears to be a linear function of the probability of a False Alarm with Slope 1.0 and Intercept $[1 - e^{-\theta^*(A+B)}]$, a linear function such as that labelled 2 on the left side of Fig. 16.9.

The Wave Discrimination Theory parameter space provides a convenient method of summarizing these various response strategies. The three axes correspond to the parameters measuring discriminability θ^*, resistance to respond A, and response bias B. A point in this space represents a possible outcome for a triple of the three parameter values. A line in this space parallel to one axis represents the set of outcomes obtained when two parameter values are held constant but one varies. For example, holding discriminability and resistance to respond constant but allowing response bias, B, to vary to the value A from the value $-A$, generates the bow-shaped ROC suggested by Signal Detection Theory. For High Threshold Theory a different locus of points defines a straight line in the ROC space. Luce's Low Threshold Theory is characterized by two connecting lines. Finally Atkinson's Variable Sensitivity Theory is defined by a probability mixture of the two parameter values that define the subject's two response strategies.

The ROC space leaves much to be desired as a summary of performance. As a substitute the parameter space of Wave Discrimination Theory exposes immediately those changes and covariations between parameters that are characteristic of a response strategy. Lines within the parameter space that show changes in resistance to respond or response bias define subject-controlled response strategies. This account of the subject's reaction to the experimenter's inducements to perform makes clear that many different theories of discrimination are not necessary to account for the wide variety of experimentally observed ROCs.

THE NOREEN STUDIES[2]

Testing the predictions of Wave Discrimination Theory against other theories in the light of such a diverse array of performances requires the use of new experimental techniques. In this regard an important stepping stone in any scientific study is the creation of experimental designs that allow good parameter estimation. Link and Heath (1975) and Link (1975) showed how parameters are estimated from experiments using symmetric stimuli. Link (1985, 1990b) introduced the Method of Symmetric Stimuli to facilitate parameter estimation when two stimuli are symmetric. Azzarello (1985) used this method in the study of numerical comparisons.

For two symmetric wave differences, $\boldsymbol{W}_A - \boldsymbol{W}_R$ and $\boldsymbol{W}_B - \boldsymbol{W}_R$, the logarithm of the Hit to False Alarm probabilities is

$$ln\left[\frac{Pr\{larger|\mathbf{S}_A\}}{Pr\{larger|\mathbf{S}_B\}}\right] = \theta^*(A - B). \tag{16.14}$$

In a similar way the logarithm of the ratio of the probabilities of responding *smaller* given $\mathbf{S}_B$ and responding *smaller* given $\mathbf{S}_A$ gives

$$ln\left[\frac{Pr\{smaller|\mathbf{S}_B\}}{Pr\{smaller|\mathbf{S}_A\}}\right] = \theta^*(A + B). \tag{16.15}$$

Estimates of the unknown measure of discriminability θ^*, the resistance to respond A, and the response bias B are obtained by combinations of these logarithmic relations. First, adding Equation 16.6 to Equation 16.8 and dividing by 2 provides an estimate of θ^*A. Subtracting Equation 16.8 from Equation 16.6 and dividing by 2 estimates θ^*B, specifically,

$$\theta^*A = \left(\frac{1}{2}\right) ln\left[\frac{Pr\{larger|\mathbf{S}_A\}Pr\{smaller|\mathbf{S}_B\}}{Pr\{larger|\mathbf{S}_B\}Pr\{smaller|\mathbf{S}_A\}}\right] \tag{16.16}$$

and,

$$\theta^*B = \left(\frac{1}{2}\right) ln\left[\frac{Pr\{larger|\mathbf{S}_B\}Pr\{smaller|\mathbf{S}_B\}}{Pr\{larger|\mathbf{S}_A\}Pr\{smaller|\mathbf{S}_A\}}\right]. \tag{16.17}$$

[2]This research formed part of Dr. Noreen's PhD thesis at Rockefeller University.

These simple relations between response proportions and unknown parameters arise from the symmetry relations between difference waves. Although they are quite easily computed, the two independent response proportions can estimate only two unknown values. In Wave Discrimination Theory the two values are combinations of the three parameters that determine response probabilities. Thus, neither discriminability nor response bias are directly observable. Each parameter is multiplied by the resistance to respond. The fact that discriminability and response bias must covary with the resistance to respond raises disturbing implications for previous interpretations based on ROC analysis (Egan, 1976).

Noreen (1979) designed an ingenious experiment to contrast response-probability predictions of Wave Discrimination Theory against those of Signal Detection Theory. Noreen proved that the estimators of θ^*A and θ^*B defined by Equations 16.16 and 16.17 were also identical to estimators of parameters from other theories. For Signal Detection Theory Noreen assumed (as did Luce, 1959) that the hypothesized signal and noise distributions were Logistic, a very close approximation to a Gaussian distribution (see Appendix B). With this assumption the estimators in Equations 16.16 and 16.17 proved to be identical to those suggested by Luce (1959), Bush (1963), Wickelgren and Norman (1966), and Ogilvie and Creelman (1968), for estimating values of d' and β, the measures of sensitivity and bias in Signal Detection Theory.

According to Wave Discrimination Theory, the Signal Detection Theory estimates of discriminability must change as the value of A increases. The results of Link and Tindall (1971) show the accuracy of this prediction. More important, the estimate of response bias, θ^*B must increase whenever discriminability increases, even though the bias B remains constant! This prediction is at odds with the very foundations of several theories of detection and recognition. For example, the most primitive form of Signal Detection Theory must assume that bias and discriminability are unrelated, otherwise some form of relation between d' and β must exist. Also Signal Detection Theory makes no specific prediction concerning the ratio β/d'. However, for Wave Discrimination Theory the ratio $(\theta^*B)/(\theta^*A)$ should remain constant as the value of θ^* varies. Thus Wave Discrimination Theory predicts that estimates of θ^*B plotted against θ^*A for differing values of θ^* should generate a straight line with slope equal to the constant k $= (\theta^*B/\theta^*A) = B/A$.

Noreen's experiment tested this prediction in the context of memory for the frequency of presentation of CVC (consonant-vowel-consonant) trigrams such as QUH. To vary the presentation frequency, a series of 30 trigrams appeared sequentially at a fixed location on a computer controlled X-Y display screen within a 30-sec time period. Within each second a trigram appeared for 0.5 sec followed by a blank screen until the next second when the next trigram appeared for 0.5 sec followed by a blank 0.5-sec interval followed by the 28 remaining trigrams presented in the same manner.

TABLE 16.1
Estimates of θ^*A and θ^*B for Noreen's (1979) Study

		Pr{Top Correct}					
Parameter	*Frequency Difference*	*.3*	*.4*	*.5*	*.6*	*.7*	*Average*
θ^*A	1	.501	.548	.500	0.580	.559	.538
	2	.978	.917	.949	1.051	1.099	.999
	3	1.365	1.580	1.451	1.598	1.445	1.488
	Average	.948	1.015	.967	1.076	1.034	1.008
θ^*B	1	−.172	−.152	.018	.180	.200	.015
	2	−.346	−.282	.041	.317	.422	.031
	3	−.476	−.471	.040	.542	.587	.044
	Average	−.331	−.302	.033	.346	.403	.030

To alter discriminability of frequency, a trigram appeared either 1, 2, 3, or 4 times. Within a set of 30 presentations, 3 different trigrams appeared 4 times each, 3 different trigrams appeared 3 times each, 3 different trigrams appeared twice each, and 3 different trigrams appeared only once each. Thus a total of 12 different trigrams appeared during each 30-sec presentation phase. The difference in presentation frequency varied from one to three, which seemed large enough to demonstrate changes in the remembered frequency of presentation.

Immediately following the 30-sec presentation interval, subjects received 6 paired-comparison tests. Two trigrams were presented simultaneously, one above the other on the X-Y display screen until the subject responded by indicating which trigram (*top* or *bottom*) appeared with the greater frequency in the previous 30 trigrams. This test phase presented a total of 6 paired-comparison tests "constructed so that (i) each of the 12 CVCs from the presented list occurred in exactly one test pair, and (ii) each of the four levels of presentation frequency was paired with the other three levels exactly once" (Noreen, 1979 pp. 41–42).

The experimenter varied response bias by preceding each paired-comparison test with information designed to bias the subject's response for that one test. Following the method used by Link (1975), the subjects saw, prior to each paired comparison, the probability that the top member of the test pair was the more frequently presented trigram. To gather plenty of test data, five probabilities were used: 0.30, 0.40, 0.50, 0.60, and 0.70. Noreen tested with a total of 28,800 paired comparisons. Summary values of θ^*A and θ^*B obtained using the estimators in Equations 16.16 and 16.17 appear in Table 16.1

The experimental method must hold constant the value of A but vary the values of θ^* and B. Evidence favoring the constancy of A appears in Fig. 16.10. The values of θ^*A are shown as a function of the actual differences in frequency of CVC presentation. Increasing the frequency of presentation should increase θ^*. With a constant value for A the parameter estimate θ^*A should show the same

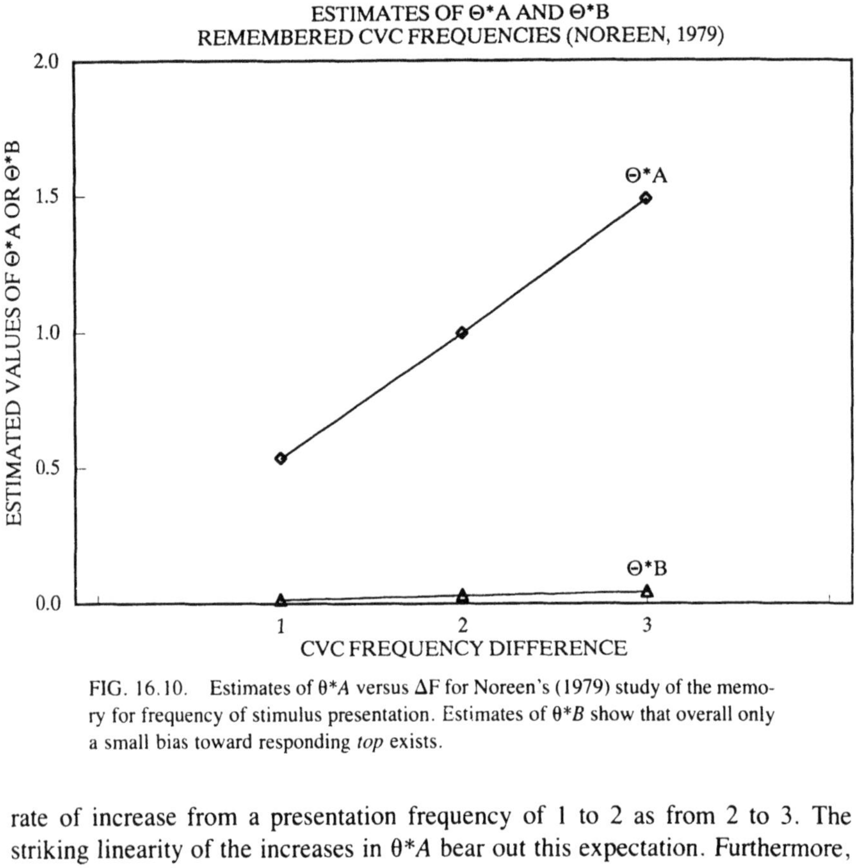

FIG. 16.10. Estimates of θ^*A versus ΔF for Noreen's (1979) study of the memory for frequency of stimulus presentation. Estimates of θ^*B show that overall only a small bias toward responding *top* exists.

rate of increase from a presentation frequency of 1 to 2 as from 2 to 3. The striking linearity of the increases in θ^*A bear out this expectation. Furthermore, the constant value for A suggests that the increases in θ^*A reflect the increases in presentation frequency. This suggests that A be estimated by using the equation $ln\theta^*A = ln\theta^* + lnA$.

Link (1975) answered the question of whether the presentation of stimulus likelihoods is sufficient to bias performance. Noreen provided greater confirmation. The response bias values θ^*B found at the bottom of Table 16.1 increase steadily as the likelihood increases that the upper member of the paired-comparison test pair will be the more likely of the comparisons. Also, as Fig. 16.10 reveals, the average estimates of θ^*B are essentially zero.

The major theoretical question is whether the ratio of θ^*B to θ^*A remains constant across increases in θ^*. To test this prediction values of θ^*B are plotted against values of θ^*A in Fig. 16.11. The result is perfectly compatible with theoretical expectations. For each bias condition (0.3, 0.4, 0.5, 0.6, 0.7) there are three ratios of θ^*B to θ^*A. As the bias increases, as it must according to the marginal estimates in Table 16.1, the slope of the line connecting the three points resulting from changes in presentation frequency must also increase.

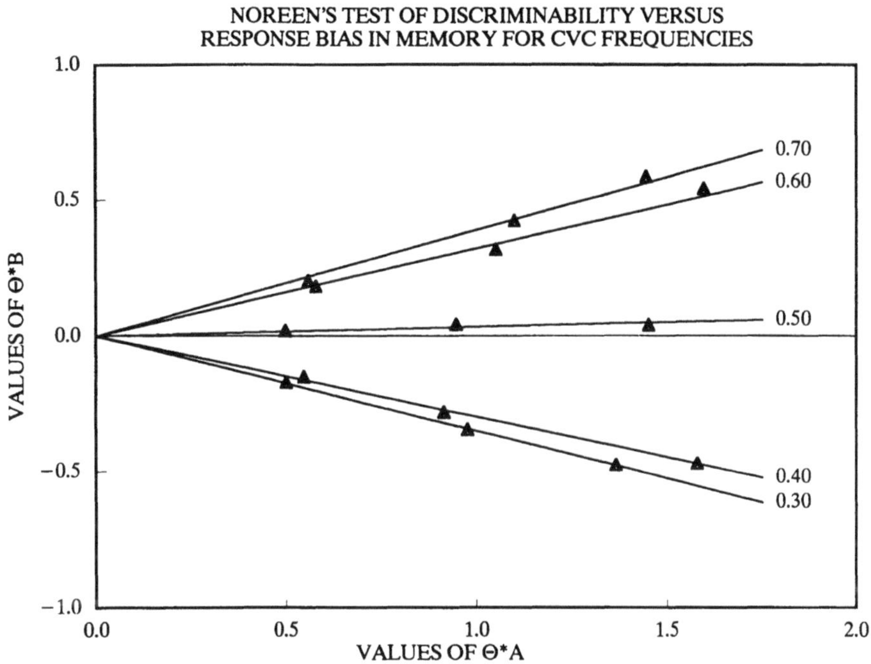

FIG. 16.11. Estimates of θ^*B versus θ^*A test the prediction that for a fixed response-bias B the ratio θ^*B/θ^*A remains constant across increases in discriminability. Response-bias conditions are indicated on the right. Lines are best fits constrained to zero Intercept.

Bias toward responding that the lower member of the test pair was presented more often is negative bias. The smallest negative bias occurs in the 0.3-bias condition. Here the subject is led to expect that the lower member of a test pair is the more frequent. The slope of the test line is the most negative, and the data points all fall on a single line. A similar result with slightly larger slope occurs for data from the 0.4-bias condition. In the equiprobable condition there is no suggested bias. The slope nearly equals zero, as it should. For bias toward responding that the upper member of a test pair is more frequent the slope of the test line is positive as shown by the lines for the 0.6- and 0.7-bias conditions.

Noreen's attention to experimental detail produced easily interpretable results showing the Theory of Wave Discrimination to apply in the context of memory experiments. The stimuli, which generate waves defining the memory for frequency, are compared, the differences accumulated, and a response threshold for the upper or lower member of the test pair exceeded. In summary the apparently mentalistic stimuli of frequency are subject to the same process of comparison and decision as are the sensory stimuli common to psychophysical experiments.

SEQUENCE EFFECTS[3]

During World War II the creation of powerful mathematical tools for the analysis of encoded secrets permitted the Allied Command to intercept and then capitalize on the enemy's supposedly secret plans. A major improvement in cryptanalysis came from Claude Shannon's (1948) invention of Information Theory, a theory about communication inspired in part by Norbert Wiener (cf. Kahn, 1967; Shannon & Weaver, 1949). According to this theory, communication occurs by transmitting a message through a communication channel to a receiver. The message consists of characters that are drawn from a finite alphabet but characters are concatenated in such a way that certain character sequences occur with probabilities greater than that expected on the basis of the alphabet's marginal probability distribution. Deviations of character sequence probabilities from that expected on the basis of statistically independent arrangements of characters indicate the existence of redundancy, a fundamental concept of Information Theory.

In creating a mathematical theory of communication Shannon created a quantitative measure of information that also provided a measure of redundancy. The greater the redundancy in the message the less the information value of the message. The less the redundancy the greater the amount of information. A message that was maximally redundant, that occurred with perfect certainty, contained no information at all. A completely random sequence of characters was not at all redundant, was maximally uncertain, and produced the maximum amount of information. Shannon defined the measure of this information as the number of bits transmitted per message.[4]

Psychologists (Miller & Frick, 1949) were quick to seize upon this measure as a means of assessing the information processing capacities of the human and animal information channel. Hick (1952) in one of a cascade of papers applying Information Theory to response time (Attneave, 1959; Garner, 1954; McGill, 1953, 1954; Miller, 1953) discussed changes in response time due to different probabilities for characters in the alphabet. According to Hick (1952):

> The analytical methods of information theory are applied to the data obtained in certain choice-reaction-time experiments. Two types of experiment were performed: (a) a conventional choice-reaction experiment, with various numbers of alternatives up to ten, and with a negligible proportion of errors, and (b) a ten-choice experiment in which the subjects deliberately reduced their reaction time by allowing themselves various proportions of errors. (p. 11)

---

[3]Some of these results were reported at various scientific meetings.

[4]A bit (binary integer) represents the information conveyed by a choice between two equally likely alternatives and equals $-\log_2(1/2)$. The name was suggested by J. W. Tukey.

> The principal finding is that the rate of gain of information is, on the average, constant with respect to time, within duration of one perceptual-motor act, and has a value of the order of five "bits" per second. (p. 11)
>
> The distribution of reaction times among the ten stimuli in the second experiment is shown to be related to the objective uncertainty as to which response will be given to each stimulus. The distribution of reaction times among the responses is also related to the same uncertainty. This is further evidence that information is intimately concerned with reaction time. (p. 11)

Hyman (1953) made clear the analogy between communication theory and the human participant in a choice experiment:

> In communication theory the amount of *information* which a message conveys is an increasing function of the number of possible messages from which that particular message could have been selected. The *S*'s reaction time seems to behave, under certain conditions, in a manner analogous to this definition of information. When a stimulus is chosen to which *S* must make a discriminatory response, his reaction time seems to be a monotonically increasing function of the number of possible stimuli from which the stimulus can be chosen. (p. 188)
>
> Thus the choice reaction-time experiment can be looked upon as a model of a communication system. The display represents a transmitter of information. Each alternative stimulus or *signal* represents a *message*; more information can be transmitted the greater the number of messages from which one can be chosen. The *channel* over which the signal is transmitted can be considered as the air space between the light and *S*, and might also include part of the *S*'s visual afferent system. The *S* also acts as a *receiver* or decoder in that at some point he decodes the signal into its message and reacts with the appropriate response (the *destination* of the information). (p. 188)

Hyman determined whether

> The regression of reaction time upon the amount of information is the same whether the amount of information per stimulus is varied by altering the number of equally probable alternatives, altering the relative frequency of occurrence of particular alternatives, or altering the sequential dependencies among occurrences of successive stimuli. (p. 189)

Through these three manipulations the amount of information varies over a range of 0 to 3 bits. Each of the three experimental methods used to alter the amount of information generated linear relations between reaction time and information. When the three methods were combined for each of the four experimental subjects, the correlations of 0.985, 0.953, 0.955, and 0.938 provided strong support for the Information Theory model of the human information processing system.

However, Hyman (1953) noted that

> high reaction times for the three conditions [within the sequential dependency conditions] wherein no immediate repetition of a stimulus was allowed raises some interesting questions. On the basis of information theory, the restriction "no repetition" lowers the amount of information per stimulus presentation since it effectively decreases the number of alternatives by one. Instead of lowering the subject's reaction time, however, this restriction caused his reaction time to be higher than if no such restriction had been imposed. (p. 193)

More extensive analysis of the data reveals the influence on reaction time of the number of other stimuli intervening between successive occurrences of a stimulus. More serious criticisms by Crossman (1953, 1956) pointed to the influence of errors on performance. At the time these criticisms paled when compared with the overall success of the Information Theory model.

Later Bertelson put a fly into the information theorists' ointment. In his extraordinary article entitled "Sequential redundancy and speed in a serial two-choice responding task" Bertelson (1961) argued:

> The task used is a self-paced 2-choice serial responding task, involving two neon lamps as signals [which were turned off by the subject's response] and two response keys in simple spatial correspondence with the signals. In Experiment 1, three conditions with respectively $^1/_4$, $^1/_2$ and $^3/_4$ alternations between successive signals were compared. . . . From the point of information theory, both conditions [$^1/_4$ and $^3/_4$] present the same amount of information per signal. (p. 90)

Given these experimental conditions the amount of information for the random condition equals 1 bit/sec, whereas for either the $^1/_4$ or $^3/_4$ alternating condition the amount of information is 0.81 bits/sec. If the amount of information governs the duration of a response, as Hick (1952) and Hyman (1953) argued, then because both the $^1/_4$ and $^3/_4$ conditions present the same amount of information, the same response time should occur. Furthermore a smaller amount of information, indicating less uncertainty, might be expected to reduce response time; therefore, the shortest response times should be those of the $^1/_4$ and $^3/_4$ alternating conditions and these two times should also prove to be equal. Using a measure of the total time to respond to 1000 stimulus presentations, Bertelson discovered that the total times for 1000 stimulus presentations within each condition were 445.0, 437.0, and 395.3 sec for the alternating, random, and repetition conditions, respectively. The results are hardly in keeping with the expectations of Information Theory.

Due to the type of experimental equipment available, the study measured only total time to complete 20 blocks of 50 trials each. Thus individual trials were not studied. Realizing the importance of the trial-by-trial analysis of response times

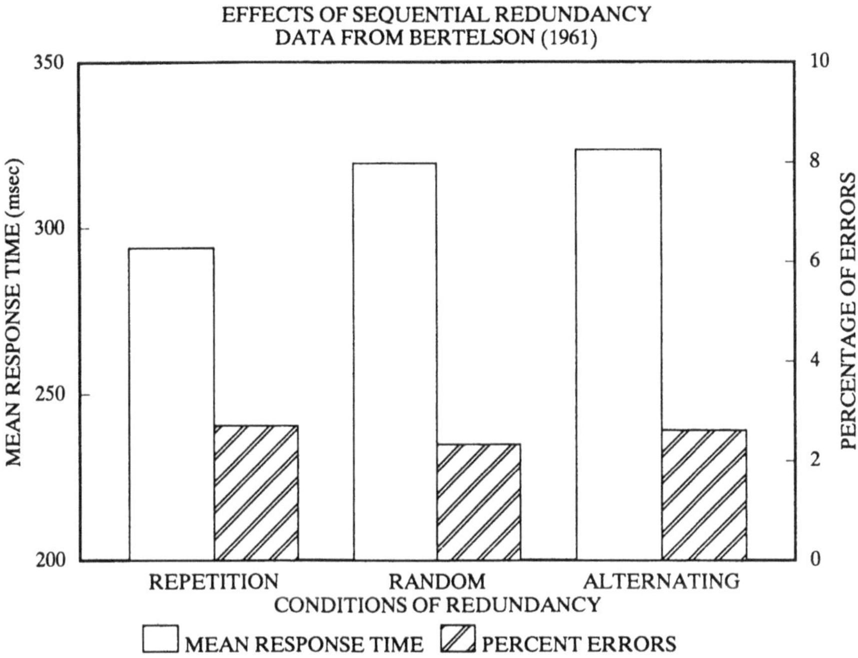

FIG. 16.12. Mean response times and error percentage for each experimental condition.

Bertelson replicated Experiment 1 in order to measure response times for every trial. Figure 16.12 shows the results. The mean response time in the random condition proves to be 320 msec and the mean response time in the alternating condition to be 324 msec. A slight difference at best. The more disturbing result occurred with the repetition condition where a stimulus repeats with probability $^3/_4$. As expected the mean response time of 296 msec is smaller than that of the random condition but is also substantially smaller than the mean response time in the alternation condition as in Experiment 1.

By examining a sequence of some 200 trials per subject per condition for each of the 12 subjects contributing to Fig. 16.12, Bertelson determined that mean response times depended critically on the stimulus-presentation sequence. The results shown in Fig. 16.13 illustrate this point. The rank of a stimulus repetition scales the abscissa. A zero rank means that the presented stimulus differs from the previously presented stimulus. A rank of one means that the stimulus repeats, a rank of two means the stimulus repeats twice, and so on. The possible number of repetitions depends upon the experimental condition. In the alternating condition the probability of a repetition is only $(^1/_4)(^1/_4)$, whereas the random condition produces a repetition with probability $(^1/_2)(^1/_2)$ and the repetition condition gives the highest repetition probability of $(^3/_4)(^3/_4)$.

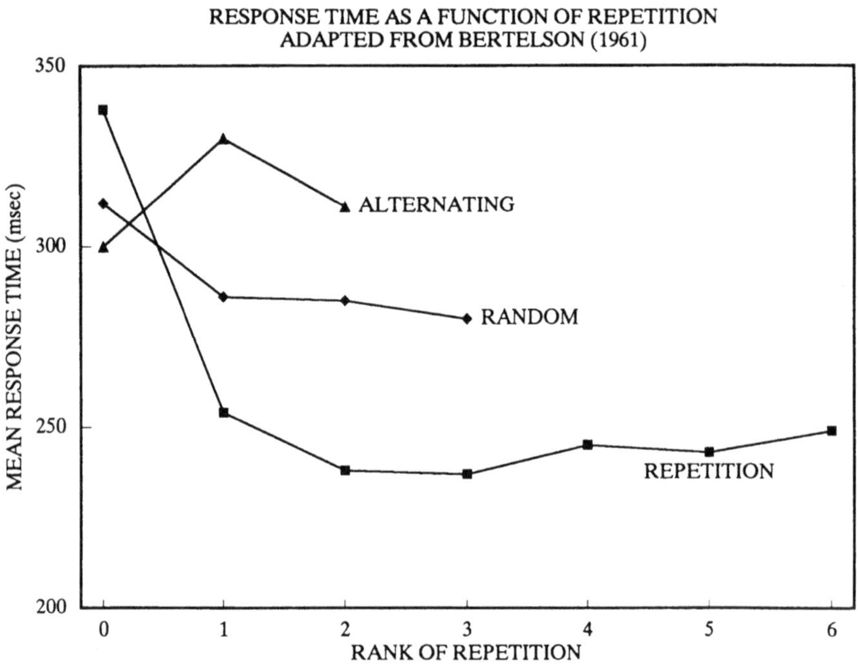

FIG. 16.13. Responses to repetitions of a stimulus reveal the effects of sequential redundancy.

The startling finding is that the repetition condition produces a steep decline in response time for the first repetition, whereas in the alternating condition the repetition of the same stimulus produces a sharp increase in response time. Also, remembering that a zero rank means that an alternation occurred, the mean zero-rank response time in the alternating condition is smaller than the mean response time for the repetition condition, yet in both instances alternations occurred. Bertelson's analysis of sequences of alternations showed that in the alternating condition, successive alternations substantially reduced response time, but in the repetition condition an alternation resulted in an increase in response time. On the surface the results deviate far from the expectations of Information Theory.

Such a major finding spurs the imagination to greater heights. To account for Bertelson's discovery Falmagne (1961) proposed an ingenious interpretation, and theory, for both the sequential effects and the reduction in response time due to different stimulus-presentation probabilities. The model proposed that the subject maintain a state of preparation (Mowrer, 1940) with respect to each stimulus. When a stimulus occurs the subject is either prepared or unprepared to respond. Response times depend upon which state of preparation meets the incoming stimulus. If prepared, the subject responds quickly with a mean response time of, say, 300 msec. If unprepared the subject responds to the same stimulus with a longer mean response time of, say, 500 msec. Luce (1986) provides an extensive review of this theory.

The probability of entering a state of preparation depends upon the sequence of stimulus presentations. Following the presentation of stimulus $\mathbf{S}_A$ the subject prepares for $\mathbf{S}_A$ with probability p. Thus after a sequence of repeated presentations of stimulus $\mathbf{S}_A$ the probability of being prepared for a presentation of $\mathbf{S}_A$ increases. Concurrently the mean response time to stimulus $\mathbf{S}_A$ decreases due to the increasingly high probability that the subject is prepared and can produce a mean response time of 300 msec. Subsequent presentations of stimulus $\mathbf{S}_A$ reduce average response time by increasing the mixture of quick, 300 msec, response times with slower responses caused by the unprepared state. A clever feature is the prediction that mean response time must increase as the number of equally likely to-be-presented stimuli increases, because the probability of preparation for any particular stimulus must decrease as the number of stimuli grows larger. The greater the lack of preparation, the more likely that the response time results from the slow unprepared state.

The formal mathematical model predicted changes in response time as a function of sequences of stimulus events such as repetitions or alternations. The theory suffered two flaws. First, Falmagne (1968) rejected the theory based upon a fixed-point theorem (cf. Wundt, 1896) that appeared to show inconsistencies between the assumption of a two-state process and mixtures of observed response-time distributions. Link (1982) suggested that the application of the fixed-point theorem did not apply to the histograms estimated by Falmagne and that the question of whether the theory failed remained open. Last, as early as 1952, Hick observed that errors played an important role in response-time experiments, and Rabbitt (1966, 1968) demonstrated that in 4- and 10-alternative choice tasks errors were as much as 150 msec faster than correct responses. The theory proposed by Falmagne left for later development a theory of how errors might occur.

Remington (1969) showed how response times change as a function of a history of preceding stimuli. Figure 16.14 illustrates the idea for a two-choice experiment. The critical point of Fig. 16.14 is that each node of the response-time tree shows the mean response time to stimulus $\mathbf{S}_A$ presented on trial n given the sequence of stimuli shown at each node. Audley (1973) suggested that these orderly changes in response time resulted from the subject either perseverating, that is, remaining set for the same stimulus, or changing set once an expectancy went unconfirmed. Theios and Smith (1972), Lupker and Theios (1975), and Theios and Walter (1974) developed and tested a more extensive Markov model incorporating these ideas. Their model illustrates the powerful impact of the original idea of Mowrer (1940), that expectancy of a stimulus accounts for changes in response time. The notion of preparation for a stimulus, believed to

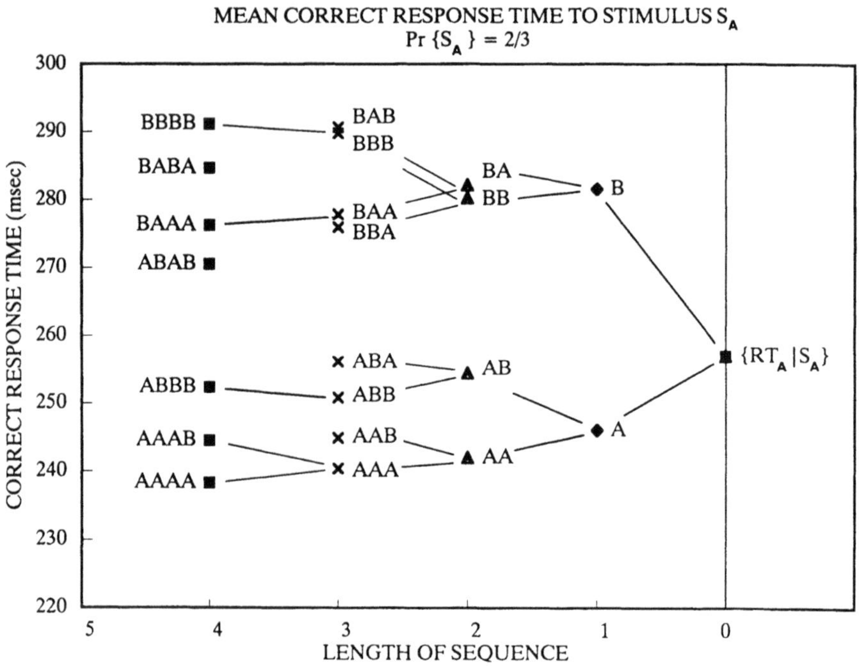

FIG. 16.14. Mean correct response times to stimulus S_A on trial *n* given various histories of previous stimulus presentations.

be the basis for these many results, creates the impression that preparation for a stimulus is a general feature of performance that must be included in cognitive models of the human information-processing system.

Many models developed for choice response tasks assume that errors are unwanted intrusions, experimental error of some sort, that simply plague the collection of otherwise pure data. Few previous studies provide details of error proportions or error response-time data. Often only a cursory comment, such as "the error rates are low" gave any indication that errors occurred at all. In the recognition literature, which contains most choice reaction time experiments, a commonly held belief pontificates that understanding changes in correct responses is difficult enough. Why analyze the small percentage of error data that make a formidable task more difficult. Thus, error rates and error response times are seldom reported, as if errors were the intrusions of a vindictive nature trying to frustrate the struggling scientist.

Yet, if error rates change as a function of experimental conditions then they are not just coincidental, but an inherent part of the choice process. Furthermore, these infrequent events are more informative, in an Information Theory sense, than the more frequent correct responses, and may therefore assist, rather than retard, the understanding of mental processes. As previous chapters show, errors are an inherent part of the decision-making mechanism. The correct and error response frequencies provide data necessary for measuring response biases and resistances to respond. The measurement of these effects permits the data to reveal the influence of a history of events responsible for a sequence effect.

There is no better starting point to illustrate the application of Wave Discrimination Theory to sequential effects than the data from a replication of an experiment by Falmagne, Cohen, and Dividi (1975) performed by Link in 1976. The replication improves estimates of low-probability events that are needed to determine values of theoretical parameters. The experimental details are straightforward.

Experimental Details

Arrows pointing to the left or right sides of an X-Y display screen served as stimuli. Subjects were to depress a left hand response key with the index finger of the left hand when the arrow pointing left appeared on the display. When the arrow pointing to the right appeared the subject was to depress the right hand response key with the right index finger. The stimulus remained visible until the subject responded but the next stimulus presentation occurred immediately after the depression of the choice key. The dominant stimulus, $\mathbf{S}_A$, occurred on a random 2/3 of the trials. The alternative stimulus appeared on the remaining trials. Three subjects were instructed to respond as quickly and as accurately as possible. As nearly as possible these conditions matched those of Falmagne et al. (1975).

Subjects practiced at this task for 2 blocks of 800 trials per day for a total of six weeks. From this extensive period an intermediate run of 30 blocks of 800 trials each provide the data base for the results presented below. The first 100 trials of each block were treated as warm-up trials and ignored. Each subject contributed 21,000 observations for a total of 63,000 trials. This is enough to yield reasonable estimates of several low probability events.

These three well practiced subjects are remarkably similar in performance characteristics. The mean correct response times are very fast, averaging 262 msec with a marginal error rate of 14%. The mean correct response time to the dominant stimulus, $\mathbf{S}_A$, shown on the right of Fig. 16.14, is 257 msec with an error rate of 5.7%. The alternative stimulus, $\mathbf{S}_B$, yields a mean correct response time of 295 msec and an astonishing error rate of 32%. These mean correct response times suggest some form of preparation that leads to longer response times for an infrequent event, $\mathbf{S}_B$, and a large probability of responding in error. On the other hand, the dominant stimulus, $\mathbf{S}_A$, generates fast response times and a low error rate. The result is at odds with the form of a speed-accuracy tradeoff that posits longer response times to be more accurate.

Sequences of stimuli dramatically influence both response time and error rate.

The measure of performance most often reported is the mean correct response time to the dominant stimulus, S_A. For each subject there are about 14,000 presentations of S_A. Of these, nearly 13,000 are correct responses. Across all subjects, S_A preceded another presentation of S_A on 26,700 trials. For these sequences the mean response time to S_A equals 246 msec. On 12,400 trials S_B preceded the presentation of S_A. For this sequence the mean correct response time to S_A equals 282 msec. These response times appear in Fig. 16.14, in a manner suggested by Remington (1969) which also illustrates the influence of other event sequences on response time. Only histories with at least 100 observations appear.

The labels in Fig. 16.14 refer to the sequence of stimulus events leading to the presentation of S_A on trial n. To be clear about the meaning of the figure, the data point plotted at -1 and 246 msec shows the mean response time to a presentation of stimulus S_A when on the preceding trial, S_A also occurred. Thus the sequence is of Length 1. To illustrate the principle of labelling the data points, the left-most letter indicates the stimulus presented on trial $n - 1$, just prior to trial n. The response time on trial n, given this history is the associated data point. For example, the sequence $B_{(n-1)}A_{(n-2)}$ has a response time to stimulus S_A (on trial n) of 282 msec. The sequence indicates that on trial $n - 2$ the subject responded to S_A, on trial $n - 1$ the subject responded to stimulus S_B, and then on trial n responded to stimulus S_A. This method of showing the stimulus history is similar to that used for Markov learning models. Lines connecting data points show how higher-order sequences, such as S_B, S_A, S_B and S_B, S_A, S_A combine to yield a lower-order sequence such as S_B, S_A.

In general the image shows results similar to those reported by Bertelson (1961). After a long sequence of infrequent events such as S_B, the response time to S_A is longer than for random mixtures of S_B and S_A stimuli. A run of four S_B stimuli followed by S_A produces the longest RT of 291 msec. When the preceding stimuli are the same as S_A, as in the case of four repetitions of S_A prior to the presentation on trial n, the response time is the lowest at 238 msec. These are remarkable quick choice reactions.

The picture that emerges from the figure appears orderly and is common in the choice reaction-time literature. Yet more needs to be reported to make sense of these seemingly orderly results. Typically the error data are missing completely, the mean correct response time appears and no indication of the error rates is given. If the error rates change systematically as a function of the preceding stimulus sequences then a successful theory must also account for the presence of these errors and the reason for the change in error rate. For example, given the sequence S_A, S_A, S_A, S_A shown at the bottom of Fig. 16.14 the probability of an error in responding to S_A is only 0.017, based on some 8,000 observations. The probability of an error to S_A given the upper-most sequence S_B, S_B, S_B, S_B, equals 0.171, based on 521 observations. This tenfold increase in error rate spans the range of error rates. In between these sequences are other sequences with

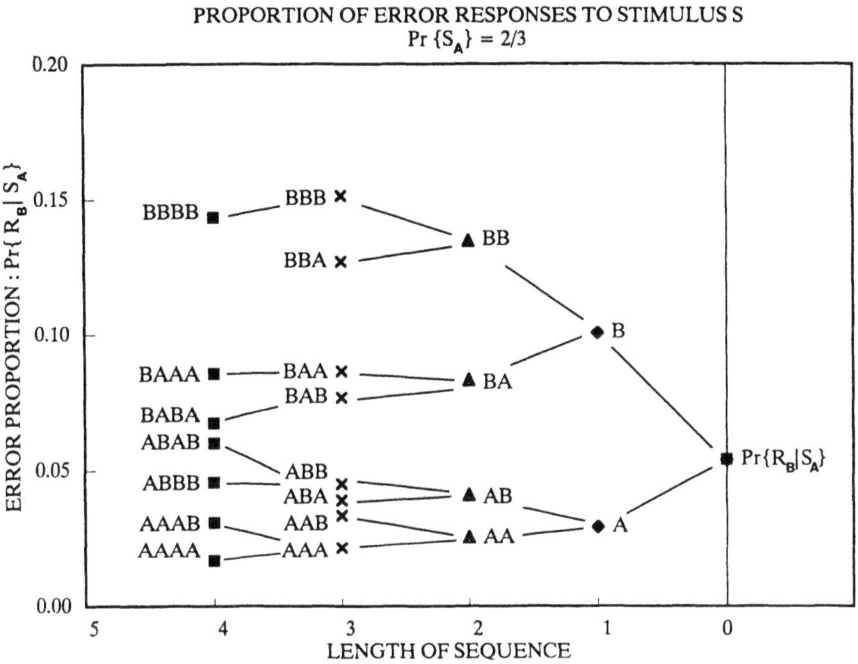

FIG. 16.15. Error proportions for various stimulus sequences.

error rates that form a treelike structure. Figure 16.15 shows the error probabilities for the responses to $\mathbf{S}_A$ shown in Fig. 16.14.

Because the experiment is like most other two-stimulus discrimination and/or recognition experiments there is ample reason to suggest that Wave Discrimination Theory applies to the discriminating process engaged by these subjects. The referent adopted by the subject is unknown but whatever its form the most plausible assumption concerning two stimuli is that they are symmetric with respect to each other. Thus a presentation of $\mathbf{S}_A$ drives the accumulation of wave differences toward the upper response threshold in the same manner as a presentation of $\mathbf{S}_B$ drives the process toward the lower response threshold. Stimulus-presentation probability may simply bias the subject toward one of the response thresholds. That is, all of these effects may simply be the result of response bias.

To investigate this possibility estimates of θ^*A and θ^*B were obtained for every stimulus sequence. These calculations assume that the mental state of the subject is influenced by the preceding history of events. The stimulus presented on trial n is not known in advance, thus a fixed history prepares the subject in the same way for a presentation of either $\mathbf{S}_A$ or $\mathbf{S}_B$. To estimate θ^*A when the sequence of events is $\mathbf{S}_{B,n-1}$, $\mathbf{S}_{A,n-2}$, $\mathbf{S}_{B,n-3}$, $\mathbf{S}_{A,n-4}$ requires the conditional probabilities of responses to $\mathbf{S}_{A,n}$ and $\mathbf{S}_{B,n}$. This is the same method used when estimating parameters for two stimuli, $\mathbf{S}_{A,n}$ and $\mathbf{S}_{B,n}$, without regard to the history. The ratios of probabilities conditioned on the same sequence yields,

$$e^{\theta^*(A-B)} = \frac{Pr\{R_A|\mathbf{S}_{A,n}, \mathbf{S}_{B,n-1}, \mathbf{S}_{A,n-2}, \mathbf{S}_{B,n-3}, \mathbf{S}_{A,n-4}\}}{Pr\{R_A|\mathbf{S}_{B,n}, \mathbf{S}_{B,n-1}, \mathbf{S}_{A,n-2}, \mathbf{S}_{B,n-3}, \mathbf{S}_{A,n-4}\}}$$

The value of $\theta^*(A - B)$ is obtained from

$$\theta^*(A - B) = ln\left[\frac{Pr\{R_A|\mathbf{S}_{A,n}, \mathbf{S}_{B,n-1}, \mathbf{S}_{A,n-2}, \mathbf{S}_{B,n-3}, \mathbf{S}_{A,n-4}\}}{Pr\{R_A|\mathbf{S}_{B,n}, \mathbf{S}_{B,n-1}, \mathbf{S}_{A,n-2}, \mathbf{S}_{B,n-3}, \mathbf{S}_{A,n-4}\}}\right]. \quad (16.18)$$

An estimate of $\theta^*(A + B)$ derives from

$$\theta^*(A + B) = ln\left[\frac{Pr\{R_B|\mathbf{S}_{B,n}, \mathbf{S}_{B,n-1}, \mathbf{S}_{A,n-2}, \mathbf{S}_{B,n-3}, \mathbf{S}_{A,n-4}\}}{Pr\{R_B|\mathbf{S}_{A,n}, \mathbf{S}_{B,n-1}, \mathbf{S}_{A,n-2}, \mathbf{S}_{B,n-3}, \mathbf{S}_{A,n-4}\}}\right]. \quad (16.19)$$

Adding the estimates together and dividing by two yields the previously unknown value of θ^*A. Subtracting Equation 16.18 from Equation 16.19 and dividing by two produces the estimate of θ^*B.

The simplest hypothesis regarding sequence effects is that they are the consequence of only response bias. Each history in the two-stimulus two-response experiment defines a point in the ROC space. If only response bias operates and response thresholds remain fixed at A and $-A$, then all points should lie on a single curved operating characteristic like those in Fig. 16.5.

Figure 16.16 shows the result of plotting the probability of responding correctly to the more dominant stimulus, $\mathbf{S}_A$, a Hit, versus the probability of responding incorrectly to the less dominant stimulus, $\mathbf{S}_B$, a False Alarm. The various points derive from histories leading to the response to stimulus $\mathbf{S}_A$ that give stable estimates of low frequency False Alarms. The shape of the ROC and the theoretical best fit based on a global estimate of θ^*A, suggest that changes in response bias are the basis for changes in performance.

A particularly informative analysis results from plotting estimates of response bias as measured by θ^*B as a function of the history of stimulus presentations. Figure 16.17 shows estimates of θ^*B obtained from sequences of $\mathbf{S}_A$ and $\mathbf{S}_B$ presentations plotted against the length of the stimulus sequence on which they are based. A particular history of stimulus presentations always reads toward the right. For example, the right-most points for the upper estimates of θ^*B are based on a single trial. One space to the left, at a depth of 1 the estimate is based on histories of a single preceding $\mathbf{S}_A$ or $\mathbf{S}_B$ presentation which occurred on trial $n - 1$. One more space to the left and the estimate is based on a history of depth 2. A parameter estimate in Fig. 16.17 is always determined by correct and error responses to $\mathbf{S}_A$ and $\mathbf{S}_B$ presented on trial n; thus there is a need to analyze errors and a need for sufficient numbers of trials to stabilize estimates of the error proportions.

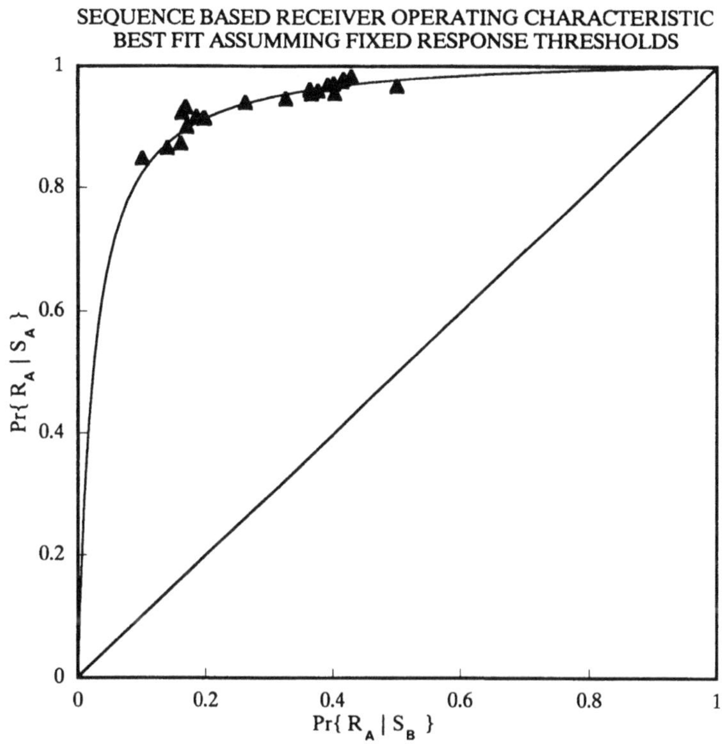

FIG. 16.16. The ROC space obtained from various stimulus histories.

Overall there is a large response bias, $\theta^*B = 0.734$, toward the response threshold for R_A. When the presentation of $\mathbf{S}_B$ on trial $n-1$ precedes a presentation of $\mathbf{S}_A$ on trial n, the bias is reduced to 0.22, but when the preceding stimulus is $\mathbf{S}_A$, the bias increases to 1.06. The change in bias is a measure of the preparation for the response on trial n. Notice that there is an independence of path revealed by sequences such as $\mathbf{S}_{B,n-1}$, $\mathbf{S}_{A,n-2}$, $\mathbf{S}_{B,n-3}$, $\mathbf{S}_{A,n-4}$, and $\mathbf{S}_{B,n-1}$, $\mathbf{S}_{B,n-2}$, $\mathbf{S}_{B,n-3}$, $\mathbf{S}_{B,n-4}$. The value of θ^*B appears to depend only on the last stimulus presented, that is, on the preparation for the immediately occurring trial. Thus the estimates of response bias change systematically depending on the most recent stimulus presentation. The alternations of bias are particularly evident in the set of histories $\{\mathbf{S}_A, \mathbf{S}_B, \mathbf{S}_A, \mathbf{S}_A, \mathbf{S}_B, \mathbf{S}_A, \mathbf{S}_B, \mathbf{S}_A, \mathbf{S}_B, \mathbf{S}_A\}$. The response bias on trial n evidently depends on the stimulus presented on trial $n-1$.

Knowing, now, that the response bias depends on the last stimulus presentation, suggests a reexamination of the data that motivate studies of sequence effects. The mean correct response times to $\mathbf{S}_A$ presented in Fig. 16.18 are the same as those in Fig. 16.14 but are connected differently. Again the mean

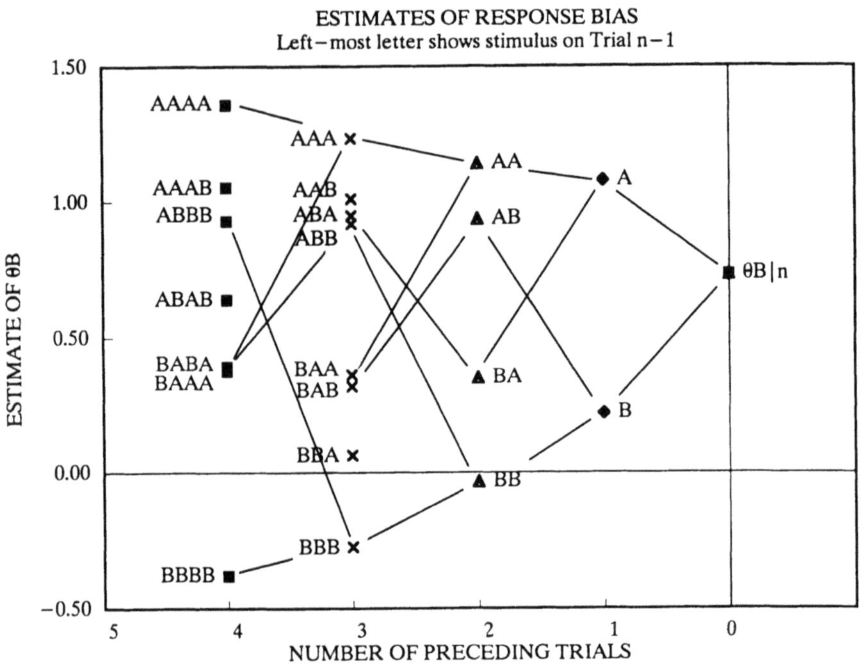

FIG. 16.17. Estimates of θ^*B show the influence of different stimulus-presentation histories on response biases.

response times are correct responses to stimulus $\mathbf{S}_{A,n}$. The graph reads from left to right depending on the depth of the history leading to the presentation of $\mathbf{S}_A$ on trial *n*. The mean response time to $\mathbf{S}_A$ is at the far right with depth 0. When $\mathbf{S}_B$ precedes $\mathbf{S}_A$, the depth is 1, the bias shown in Fig. 16.17, is 0.22, and the mean response time is 282 msec. When $\mathbf{S}_A$ precedes $\mathbf{S}_A$, the depth is 1, the bias is 1.06, and the mean response time is 246 msec. The changes in response bias are reflected by the response times.

To distinguish easily the factor determining performance, histories with $\mathbf{S}_A$ on trial $n-1$ are shown as squares, whereas histories including $\mathbf{S}_B$ on trial $n-1$ are shown as diamonds. To reduce the clutter not all sequences are presented in Fig. 16.18. Of those that are presented, several important sequences deserve special attention. The presentation of $\mathbf{S}_A$ on trial $n-2$ followed by $\mathbf{S}_B$ on trial $n-1$ leads to a response time of 282 msec to $\mathbf{S}_A$ on trial *n*. This time is essentially the same as the 281 msec when only $\mathbf{S}_B$ preceded $\mathbf{S}_A$. Also if a $\mathbf{S}_B$ stimulus occurs twice, $\mathbf{S}_{B,n-1}$, $\mathbf{S}_{B,n-2}$ and is followed by $\mathbf{S}_A$ the mean RT is 280 msec. Thus the response time seems to depend only upon the stimulus presented just prior to the presentation of $\mathbf{S}_A$ on trial *n*. Any number of $\mathbf{S}_A$ presentations followed by a single $\mathbf{S}_B$ presentation immediately before the $\mathbf{S}_A$ presentation on trial *n* leads to a

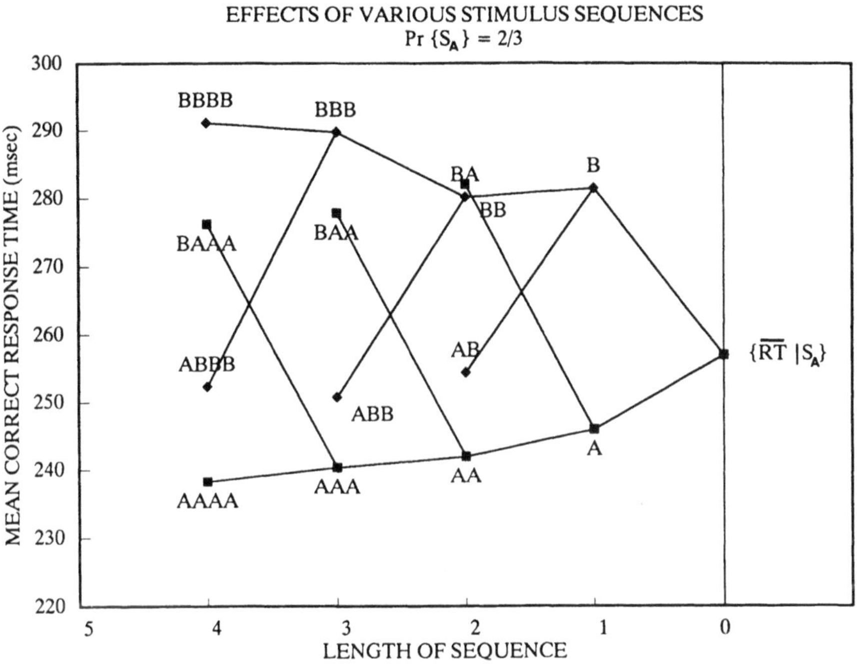

FIG. 16.18. Mean response times to S_A given various histories. The graph reads from left to right. Filled squares indicate an S_A presentation, diamonds indicate S_B on trial $n-1$.

mean response time to S_A of about 280 msec. Similarly a presentation of S_A yields a mean response time of 254 msec, whereas the sequence $S_{A,n-1}$, $S_{B,n-2}$ yields a mean response time to $S_{A,n}$ of 256 msec. A longer sequence such as $S_{A,n-1}$, $S_{B,n-2}$, $S_{A,n-3}$, $S_{B,n-4}$ gives a response time to $S_{A,n}$ of 254 msec. In other words the preceding S_A presentation leads to mean response times of 255 msec independent of path.

Two particularly revealing histories show that alternations affect performance on the basis of the stimulus preceding the presentation of $S_{A,n}$. The history $S_{B,n-1}$, $S_{A,n-2}$, $S_{B,n-3}$, $S_{A,n-4}$ yields a response time to $S_{A,n}$ of 283 msec, essentially the same as for any history with S_B on trial $n-1$. The other history is $S_{A,n-1}$, $S_{B,n-2}$, $S_{B,n-3}$, $S_{B,n-4}$. Here, three S_B presentations followed by a S_A precede S_A on trial n. The mean response time is 252 msec, the same as for the other histories where S_A precedes S_A on trial n.

If response bias causes changes in correct response times it must also cause changes in error response times. Although a full analysis of errors is beyond the scope of this discussion, there is sufficient curiosity about the behavior of errors to warrant their examination. Figure 16.19 shows changes in mean response

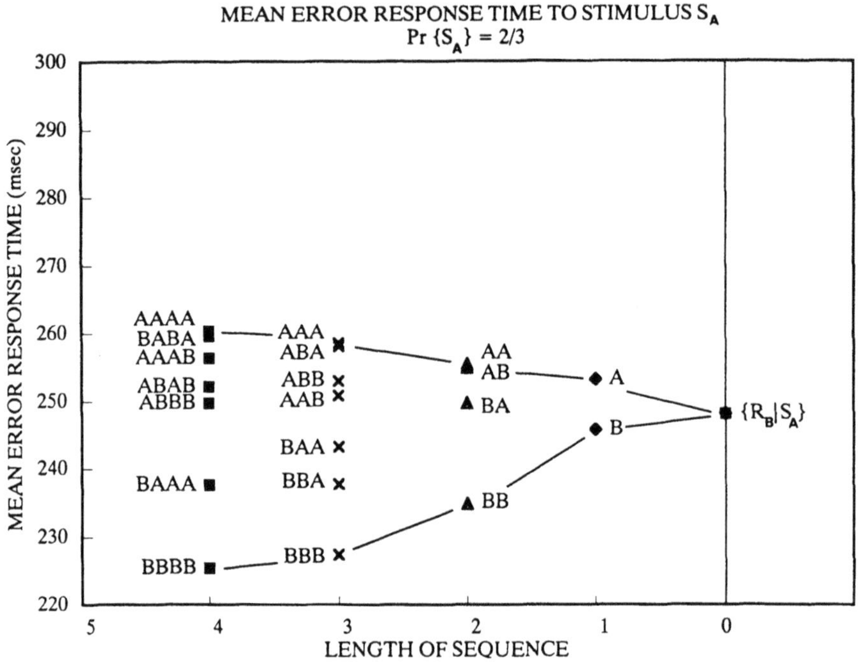

FIG. 16.19. Mean error response times to stimulus S_A on trial n depend upon the bias created by stimulus presentation histories.

times for errors that occur when $\mathbf{S}_A$ is presented. Again these mean error times are to stimulus $\mathbf{S}_A$ presented on trial n, but dependent upon various stimulus histories. When the subject is biased toward response R_A, and $\mathbf{S}_A$ occurs, the mean correct response time is small but the mean error response time, that is responding R_B to $\mathbf{S}_A$, is large when compared to other error response times. Alternatively, when the bias is toward response R_B, as it is after $\mathbf{S}_B$ stimulus presentations, the mean correct response time to stimulus $\mathbf{S}_A$ presented on trial n is relatively large, whereas the error response time, the time to respond R_B, is quite fast, certainly the fastest of the mean response times. As response bias changes due to changes in which stimulus occurs, the error response time also changes. The influence of response bias on the mean error response time is considerable, although the range of the mean error times is limited to about 40 msec.

These various relations among response thresholds and response bias are summarized in Fig. 16.20. Because discriminability is fixed, the value of θ^* may be taken as a constant multiplier of A and B. Thus the estimates of A and B may be used to determine scaled values for response thresholds and the amount of response bias. The upper estimates of $\theta^* A$ derive from sequences such as $\mathbf{S}_{A,n}$,

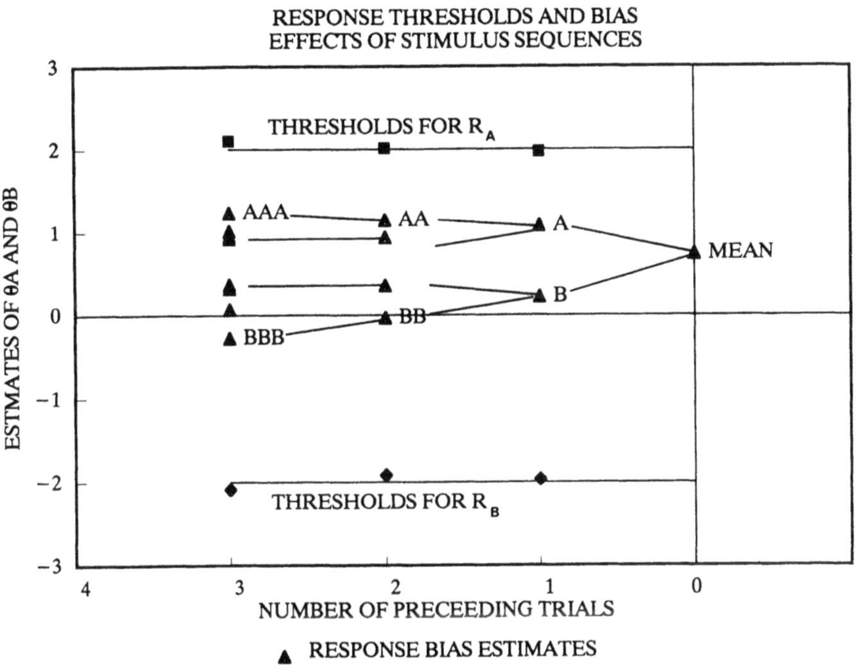

FIG. 16.20. Estimates of θ^*A and $-\theta^*A$ based on independent estimators produced the response-threshold values for R_A and R_B. The amount of response bias varies between these response thresholds depending upon the various stimulus-presentation histories.

$\mathbf{S}_{A,n-1}$, $\mathbf{S}_{A,n-2}$, $\mathbf{S}_{A,n-3}$, $\mathbf{S}_{A,n-4}$, and $\mathbf{S}_{B,n}$, $\mathbf{S}_{A,n-1}$, $\mathbf{S}_{A,n-2}$, $\mathbf{S}_{A,n-3}$, $\mathbf{S}_{A,n-4}$ by applying Equations 16.18 and 16.19. The lower set of estimates of θ^*A derive from sequences for the less frequently occurring stimulus. In this case new sequences and response probabilities occur by simply exchanging values and B for A and A for B in Equations 16.18 and 16.19. Presumably the estimates of θ^*A, whether from the sequences of $\mathbf{S}_A$ or $\mathbf{S}_B$ trials should agree. As is evident in Fig. 16.20, the two independent estimations produce comparable results, the lower estimates of θ^*A based on $\mathbf{S}_B$ sequences are essentially the same (except for sign) as the upper estimates based on $\mathbf{S}_A$ sequences.

A clear picture of the influence of response bias emerges. Estimates of θ^*B change systematically across the histories leading to the presentation of $\mathbf{S}_A$ on trial n. The response threshold remains fixed. Discriminability is fixed. Only response bias causes the many changes in performance that create the treelike structures for correct and error response time and the proportion of errors. Responses to stimulus $\mathbf{S}_A$, which occurs with probability $2/3$, are generally quick because the response bias is toward the R_A response threshold. The influence of the last stimulus is to move the bias toward, or away, from a mean value of

TABLE 16.2
Stimulus-Response Event Matrix

	Response Frequency		*Conditional Response Probability*	
Stimulus	R_A	R_B	R_A	R_B
$\mathbf{S}_A$	39100	2226	.945	.055
$\mathbf{S}_B$	6716	13961	.325	.675

$\theta^*B = 0.734$ by equal amounts. The particular stimulus, $\mathbf{S}_A$ or $\mathbf{S}_B$ simply governs the direction of the next response bias. Error proportions increase as the response bias moves nearer the R_B response threshold, because the stochastic properties of the accumulation of wave differences result in more stochastic paths reaching the lower response threshold first.

Surely the most parsimonious account of these many results is in terms of response bias. Response thresholds remain fixed throughout the experiment but the stimulus presented last determines response bias. The effect of response bias on response time is shown by this new analysis of RT tree structures. The values of response times in the trees are determined in the same manner as those of Remington (1969) and Kirby (1980). The conclusions drawn from this analysis are entirely different.

The control of response bias depends upon the subject's feeling about the stimulus presented by the experimenter. How the subject selects the response bias levels used in this experiment is evident in calculations related to the stimulus-response probability matrix of Table 16.2.

The probability of responding in error to stimulus $\mathbf{S}_B$, the less frequent stimulus, essentially equals the presentation probability for stimulus $\mathbf{S}_B$, which is 0.33. Remembering that

$$Pr\{\text{Hit}\} = e^{\theta^*(A-B)} Pr\{\text{False Alarm}\}$$

gives, in this context, the ratio

$$\frac{Pr\{R_A|\mathbf{S}_B\}}{Pr\{R_A|\mathbf{S}_A\}} = e^{-\theta^*(A-B)}$$

$$= \frac{e^{\theta^* B}}{e^{\theta^* A}}.$$

From the Theory of Feeling (chap. 15) the right-hand value is a ratio of feelings, the psychological magnitude of $\mathbf{S}_A$ with a response threshold equal to B, relative to the psychological magnitude of $\mathbf{S}_A$ with response threshold A. The left-hand side is the resulting ratio of response probabilities. By choosing the resistance to respond and response bias based on the apparent psychological magnitude of $\mathbf{S}_A$, the subject controls the probability of responding R_A to $\mathbf{S}_A$ and $\mathbf{S}_B$ presentations. In this case the response bias and resistance to respond are set so that the probability of an error to stimulus $\mathbf{S}_B$ matches the presentation probability for $\mathbf{S}_B$.

In this fast-paced experimental paradigm the next stimulus appears within 1 msec of the response, and the mean response time is only 262 msec. This remarkably fast response time suggests that the subject prepares for the next trial even prior to the physical depression of the response key. In this regard the experiment is quite similar to the prediction experiments used to study probability matching, and the same form of analysis is equally justified in that context. The subject's prediction of which stimulus will occur is not complicated by cognitive strategies focused on errors or gamblers' fallacies, for these require mental analysis after the response occurs. The data show that following $\mathbf{S}_B$, the response bias moves toward R_B. Following $\mathbf{S}_A$, the bias is toward R_A. The amount of bias and resistance to respond depends on psychological magnitudes, feelings, that occur in the presence of the two stimuli. For these very well practiced subjects only the immediately preceding stimulus controls the direction of response bias. Wave Discrimination Theory predicts that for fixed values of discriminability and resistance to respond, mean response times depend upon response bias. The response times support this prediction, showing that the mean correct and error response times to $\mathbf{S}_A$ depend on the bias produced by the last stimulus presentation.

This discussion of sequence effects began with the discovery that mean response times appeared to be linearly related to the amount of information contained in the signal to which the subject responded. The Information Theory model was of a communication channel through which a signal travelled to be ultimately decoded. Information Theory provided a general framework for the analysis of human information processing that ultimately proved to be incomplete because the theory assumes the subject to be only a medium through which the signal is transmitted, rather than an intimate part of the process of transmission. Information Theory analysis of human performance often fails because factors such as response bias are a part of human performance but are not part of the theory. Nevertheless, the general scheme of analysis, the idea of information flowing over time, is a fundamental contribution to modelling human performance. Wave Theory owes much to the Cybernetics created by Wiener (1948) and the Information Theory of Shannon (1949).

17 Wave Similarity Theory

As many introductory art history students are quick to explain, the art of Titian is similar to the art of Caravaggio. From such judgments of similarity, or even equality, comes the most baffling of psychophysical discoveries. The discovery is that these quite difficult judgments of *equality* or *sameness* occur with an ease and certainty that can only amaze the psychophysical theorist. The bafflement arises because these judgments of *sameness* or *equality* somehow compare stimuli that are not measured on a physical scale. How can stimuli without a physical measure be equated?

Initially this enigma promoted efforts to cast judgments of *equality* in the same mold as the bipolar judgments of *larger-smaller*, *greater-lesser*, or *more* or *less*. For example, Fechner, recognizing the existence of the feeling of equality, speculated about an interval of uncertainty within which two physically different stimuli felt the same, but then sought to minimize the influence of this interval by splitting any *equal* responses equally between the categories of Right or Wrong responses. Urban (1908), capitalizing on Fechner's discarded *equal* judgments, touted *equal* judgments as a more appropriate measure of discrimination. If two things are equal then they are indiscriminable, neither larger nor smaller. Thus, argued Urban, the range of indiscriminability measured by the Interval of Uncertainty measures discriminability. The smaller the Interval of Uncertainty, the better the subject's discrimination.

Urban's assertion that the Interval of Uncertainty measured discriminability met stiff opposition. Brown (1910) brought the question to the empirical arena. When measured outside the context of three-choice judgments (*smaller*, *equal*, or *larger*) the Psychometric Equal Function differed from that expected on the basis of previously obtained three-choice results (Part I, chaps. 3 and 5). Furthermore the two-choice probabilities of a separately run bipolar discrimination experiment gave no hint of the form taken by the separately obtained Psychometric Equal Function.

Alarmed at the inconsistency of subjects' *equal* judgments, Fernberger (1913, 1914, 1920, 1926, 1930, 1931) sought out and exposed the role of psychological factors that influenced such performance. At just the time von Békésy (1930) created Neural Quantum Theory to account for his peculiar Equal-Loudness Psychometric Function, Fernberger (1931) suggested that the judgments of *sameness*, or *equality*, differ from judgments of *larger* or *smaller* in ways that require extensive research.

Then Stevens, Volkmann, and Morgan (1941) adopted von Békésy's theory and claimed that the Psychometric Function for *different* judgments created from *equal* judgments was quite different from the smoothly changing functions that characterized judgments of *greater* or *smaller*. In a canny attempt to make judgments of *equality* increase in a manner like the Psychometric Function for *larger* judgments, Stevens plotted $1 - Pr\{equal\}$ as a form of *different* response. The subject never made a judgment of *different*, only of *equality*. Thus, the claim that these results reflect the subject's ability to judge differences is only a claim, not a demonstrated fact.

Nonetheless, Stevens (1975) showed that $1 - Pr\{equal\}$ equalled zero for comparisons near the standard and steadily increased as the comparison stimulus deviated farther from the standard. To add a mirror to the smoke, Stevens plotted the proportions, $1 - Pr\{equal\}$ as a function of absolute stimulus difference, thereby completely obscuring whether the comparison was *larger* or *smaller* than the standard, and suggesting that only the absolute difference from the standard mattered. The peculiar linearly increasing response proportions vigorously pursued by Stevens became a signature of Neural Quantum Theory.

To derive a linear Psychometric Function, Neural Quantum Theory posits that the standard $\mathbf{S}_R$ must activate N_R neural elements and the comparison stimulus $\mathbf{S}_A$ must activate an equal number, $N_A = N_R$, units in order for the response *equal* to occur. This seems too restrictive, if not entirely improbable, because these suprathreshold stimuli activate many neural units. The joint probability of activating an equal number seems astonishingly small. The need for an equal number of stimulated receptors quickly leads to theoretical problems that force the theory to postulate the linear Psychometric Function that it needs to predict.

By assuming that the probability of activating the last neural unit derives from a uniform probability distribution, the theory assumes its prediction. To follow von Békésy's argument, increases in stimulus intensity increase the mean value of this uniform distribution, thereby projecting more of it (see Fig. 5.1) onto the next in a sequence of unactivated neural units. According to the theory, the amount of this projection determines the probability that the last unactivated unit will activate. As the intensity of the comparison approaches that of the standard, the uniform probability distribution creeps across the last neural unit needed to

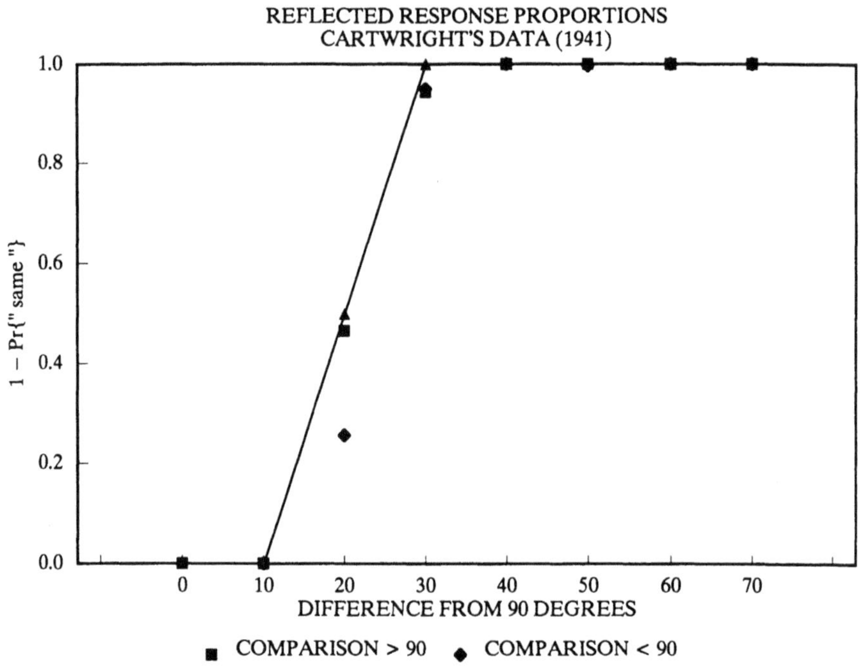

FIG. 17.1. Cartwright's *Pr{same}* data show the linear increase suggested by Neural Quantum Theory after being reflected about the midpoint of the range of standards and then converted to 1 − *Pr{same}*.

equate the number of units activated by the comparison. The creep, somehow, occurs steadily. Thus the probability of activating the last unit increases steadily, by assumption linearly, from zero to one. But this, then, is simply a restatement of the empirical result, because the uniform probability distribution produces the linearly increasing Psychometric Function.

The linearly increasing Psychometric Functions that the theory must account for are not at all unique to the sensory experiments of Stevens and von Békésy. A reanalysis of Cartwright's (1941b; chap. 5) data reveals a similar finding in a task requiring subjects to judge the size of an angle formed by two lines converging to form the symbol <. In a significant advance in methodology, Cartwright trained subjects to recognize a standard range of such angles. Then, by presenting comparison stimuli both inside and outside the range, Cartwright obtained a Psychometric Equal Function that showed a steep increase in the proportion of equal responses near the lower and upper bounds of the learned interval. Within the interval the proportion of *equal* responses remained nearly at 1.0 (Fig. 5.13).

Plotting these data according to the Stevens-von Békésy method reflects the proportions around the midpoint of the standard range. Then, plotting 1 − *Pr{same}* yields the linearly increasing Psychometric Function which is the signature of Neural Quantum Theory as in Fig. 17.1. There is a striking linearity to the proportions, enhanced by the straight line drawn through the majority of data points. Also, the slope of this straight line is about equal to two intervals of stimulus difference, as is found in Stevens' experiments. Of course, the subjects did not make the judgment *different*, rather the subject made the judgment *equal*.

The point, made perfectly clear by Bindra et al. (1968), is that the psychological mechanism necessary for the judgment of *equal* may be entirely different from that required for the judgments of *smaller-larger*, although Sorkin (1962) proposed that Signal Detection Theory gives a good account of *same-different* judgments. Trying to create *different* judgments out of *equal* judgments and then claiming, as did Stevens that theories of *smaller-larger* judgments are not sufficient to account for the results, simply obscures the issue of whether a new theory is required to account for *equal* judgments. As demonstrated in Part I, judgments of *equality* are not the obverse of judgments of *larger* or *smaller*. As Fernberger (1930) argued, these judgments require a theoretical treatment of their own.

The stimuli evoking these results are similar to those used to study bipolar judgments such as *louder* or *softer*, *larger* or *smaller*, *longer* or *shorter*, *heavier* or *lighter*, and so forth. Perhaps this is why Stevens transformed the measured *equality* into a measure of *differences* plotted as a function of stimulus difference. Yet the treatment strains credulity because the judgment made by the subject is not a *different* judgment similar in some way to the bipolar judgment required by Fechner and followers. In studies of *equality* there is a quality of similarity but no directionality of judgment.

WAVE EQUALITY THEORY

The goal of Wave Equality Theory is to account for such response measures as latency or proportion, and to measure the feelings that accompany such judgments. From the feelings expressed by subjects making *equal* judgments there are degrees of sameness or equality, just as there are degrees of difference or inequality. How much equality (sameness) or how much inequality (difference) is there? Are some things more equal than others? How are these feelings of equality to be measured?

Measuring the feeling of equality is made easier by introducing the concept of the *Region of Equality*. The Region of Equality defines the set of things that feel similar, either as acquired through experience, or defined by training the subject as in Cartwright's, Stevens', and von Békésy's experiments. When stimuli vary with respect to a single component the Region of Equality becomes an *Interval of Equality* defined by bounds set on the basis of previous experience or in compliance with the experimenter's explicit or implicit instructions. As Urban pointed out, these bounds may provide a basis for deciding whether some sub-

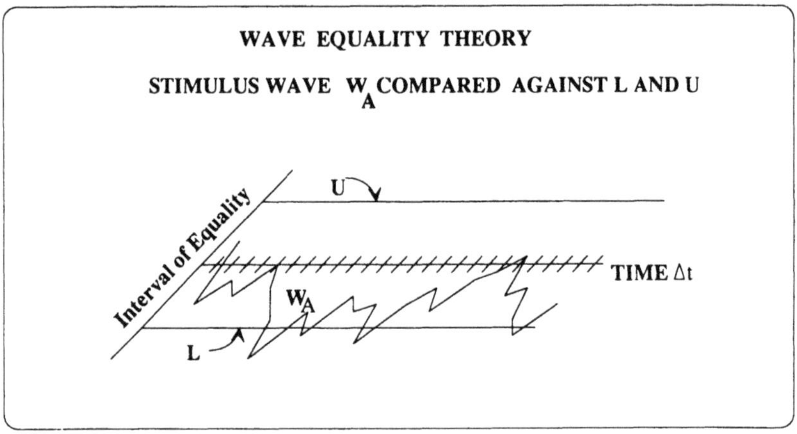

FIG. 17.2. Over time the stimulus wave for $\mathbf{S}_A$ is simultaneously compared against the upper and lower bounds of the Interval of Equality.

jects are more or less sensitive than others. A small Interval of Equality suggests a subject sensitive to *equality*, whereas a large interval characterizes a subject insensitive to differences.

How is the subject to decide whether a comparison stimulus lies within or outside the Interval of Equality? As in the previous theory of bipolar judgments such as *larger* or *smaller*, the stimulus activates a number of transceivers that generate a wave with an associated probability distribution. The stimuli are still waves, only the comparison mechanism changes to facilitate a judgment of *equality* or *difference*.

As illustrated in Fig. 17.2, the Interval of Equality is defined, during the epoch Δt, by a lower bound $\boldsymbol{L}(\Delta t)$ and an upper bound $\boldsymbol{U}(\Delta t)$. Comparison stimulus $\mathbf{S}_A$, characterized by a wave $\boldsymbol{W}_A(\Delta t)$, is compared against the bounds of the Interval of Equality. Both boundaries are postulated to enter simultaneously into the comparison, making this judgment more complicated than the judgment of *larger* or *smaller*. Furthermore, two waves result from the simultaneous comparison of the stimulus wave against each boundary. One wave measures the deviation of the stimulus from the lower boundary and produces a new random variable, $\boldsymbol{W}_A(\Delta t) - \boldsymbol{L}(\Delta t)$. The other comparison yields the difference between the upper boundary and the stimulus, that is $\boldsymbol{U}(\Delta t) - \boldsymbol{W}_A(\Delta t)$. Although $\boldsymbol{L}(\Delta t)$ and $\boldsymbol{U}(\Delta t)$ are boundaries they are also waves, thus the resulting comparative waves depend upon three random variables, $\boldsymbol{W}_A(\Delta t)$, $\boldsymbol{L}(\Delta t)$, and $\boldsymbol{U}(\Delta t)$.

How can the two measures of discrepancy lead to a judgment of equality? First, the result must measure *equality*. Second, it must have the property that when a value of the stimulus wave falls within the upper and lower bounds a qualitative difference exists from when another wave value falls outside the interval. A qualitative difference does exist when a wave value $\boldsymbol{W}_A(\Delta t)$ lies within the interval and the product of $\boldsymbol{W}_{\mathrm{A}}(\Delta t) - \boldsymbol{L}(\Delta t)$ and $\boldsymbol{U}(\Delta t) - \boldsymbol{W}_{\mathrm{A}}(\Delta t)$ is positive. When $\boldsymbol{W}_{\mathrm{A}}(\Delta t)$ lies outside the $(\boldsymbol{L}(\Delta t), \boldsymbol{U}(\Delta t))$ interval, the combination of $\boldsymbol{W}_{\mathrm{A}}(\Delta t) - \boldsymbol{L}(\Delta t)$ and $\boldsymbol{U}(\Delta t) - \boldsymbol{W}_{\mathrm{A}}(\Delta t)$ is negative. This qualitative distinction is captured by the sign of the wave.

These requirements for a measure of equality are met by the cross-product of the waves $\boldsymbol{W}_{\mathrm{A}}(\Delta t) - \boldsymbol{L}(\Delta t)$ and $\boldsymbol{U}(\Delta t) - \boldsymbol{W}_{\mathrm{A}}(\Delta t)$. During time epoch Δt these comparisons give a measure of *equality* equal to,

$$\boldsymbol{D}_{\mathrm{A}}(\Delta t) = [\boldsymbol{W}_{\mathrm{A}}(\Delta t) - \boldsymbol{L}(\Delta t)][\boldsymbol{U}(\Delta t) - \boldsymbol{W}_{\mathrm{A}}(\Delta t)]. \tag{17.1}$$

The value of $\boldsymbol{D}_{\mathrm{A}}$ (the notation Δt will be dropped in order to simplify the discussion) is positive when the stimulus wave is within the Interval of Equality because then $\boldsymbol{W}_{\mathrm{A}}$ is always larger than $\boldsymbol{L}$ but simultaneously smaller than $\boldsymbol{U}$. Otherwise $\boldsymbol{D}_{\mathrm{A}}$ is negative because either $\boldsymbol{W}_{\mathrm{A}}$ is smaller than $\boldsymbol{L}$ or larger than $\boldsymbol{U}$. The reversed polarity of the two difference-waves represents the use of inhibitory and excitatory waves to produce the resulting comparative value.

The theory of how a judgment of *equal* or *different* occurs uses the accumula-

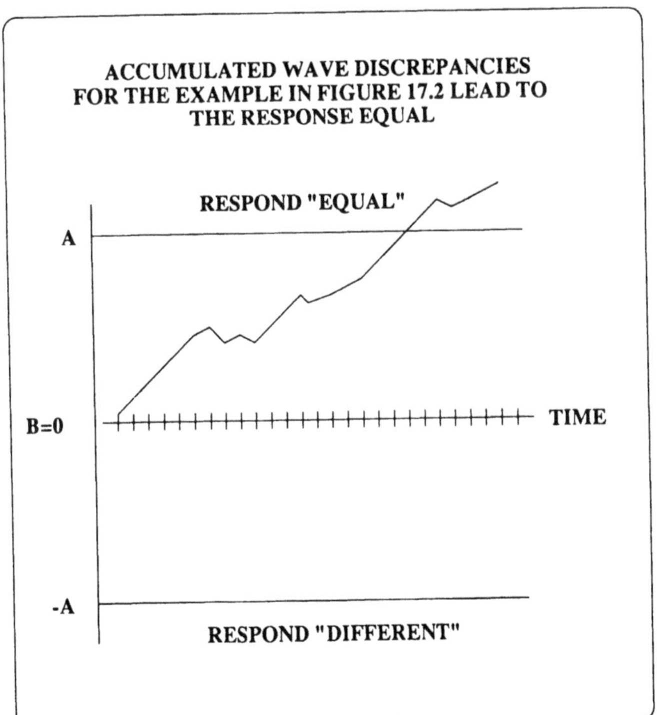

FIG. 17.3. The response thresholds at A and $-A$ and the response bias B are preset at the beginning of a trial. The accumulation of values of $\boldsymbol{D}_{\mathrm{A}}$ result in a response when a response threshold is exceeded.

tion mechanism used for judgments of *larger* or *smaller*. The value D_A is sampled and accumulated until a response threshold for *equal* or *different* is first reached. As shown in Fig. 17.3 the positive response threshold at A represents the Informative Value required to respond *equal*. The response *unequal* or *different* results if the response threshold at $-A$ is exceeded first. The bias toward responding *equal* or *different* is represented by the value B. In cases where no explicit *different* response is required the subject may still adopt a response threshold at $-A$ to decide *not equal* as in the case of Brown's (1910) study where the subject only responded when the comparison stimuli felt equal to a 96-g standard.

The expected value of D_A is the rate of drift toward the response thresholds. The rate of drift influences both response times and response probabilities. A rate of drift equal to zero produces chance performance, whereas positive or negative rates of drift result from informative values for equality or difference, respectively. The drift rate is,

$$E\{D_A\} = E\{(W_A - L)(U - W_A)\}$$
$$= E\{-W_A^2 + W_A(L + U) - LU\}. \quad (17.2)$$

Let the expected values for the boundaries be L for $E\{L\}$ and U for $E\{U\}$. Then, assuming that $L(\Delta t)$, $U(\Delta t)$, and $W_A(\Delta t)$ are independent random variables, a reasonable assumption given that the bounds are not dependent upon the stimulus, gives,

$$E\{D_A\} = -E\{W_A^2\} + (L + U)E\{W_A\} - LU\,.$$

Because $E\{W_A^2\} = Var\{W_A\} + E\{W_A\}^2$, a second form for the expected difference is,

$$E\{D_A\} = -Var\{W_A\} - E\{W_A\}^2 + E\{W_A\}(L + U) - LU. \quad (17.3)$$

In large part the expected value of D_A determines how the average response time changes as the average value of the stimulus changes with respect to the Interval of Equality. The expected value of D_A depends upon the expected value of W_A in such a way that when $E\{W_A\}$ is within the Interval of Equality the maximum is reached at the midpoint of the interval, as is shown in Fig. 17.4. The quadratic form of D_A also ensures that there are two values of $E\{W_A\}$ that cause $E\{D_A\}$ to equal zero. The stimuli producing zero drift rates are those that also produce the longest response times.

The expected value of D_A is also the cross-correlation or cross-covariance (Box & Jenkins, 1976; Uttal, 1975) of the differences from the edges of the Interval of Equality. Knowing the form of Equation 17.2 helps determine general characteristics of performance. Two predictions consistent with empirical facts are quickly revealed. Indifference occurs at the value of $E\{D_A\} = 0$. For those stimuli causing $E\{D_A\} = 0$, the probability of responding *equal* or *different* is

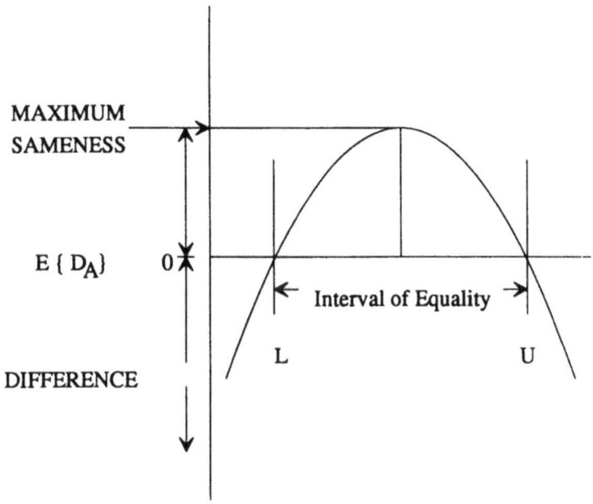

FIG. 17.4. The values of $E\{\boldsymbol{D}_A\}$ depend upon $\boldsymbol{W}_A$ and the bounds of the Interval of Equality. The value of $E\{\boldsymbol{D}_A\}$ reaches a maximum at or near the midpoint of the Interval of Equality. Zero values occur at the edges of the Interval of Equality.

0.50. This zero value occurs for values of the stimuli near L and U. More important, perhaps, is the observation that the value of $E\ \{\boldsymbol{D}_A\}$ approaches zero from either side of a boundary of the Interval of Equality. As $E\ \{\boldsymbol{D}_A\}$ approaches zero the mean response time must rise. Then, somewhere between L and U, $E\ \{\boldsymbol{D}_A\}$ reaches a local maximum, and response time reaches a local minimum. This point represents the stimulus that is most *equal*.

Figure 17.4 illustrates the change in $E\ \{\boldsymbol{D}_A\}$ as the expected value of the stimulus changes. When the expected stimulus value lies near either L or U, Equation 17.2 nears zero. There is no information available about equality or difference. This placement reduces Equation 17.2 to $-E\ \{\boldsymbol{W}_A^2\} + L^2$. Differentiating Equation 17.2 with respect to $E\ \{\boldsymbol{W}_A\}$ provides a maximum value for this function. The maximum is the largest possible value for *equality* judgments. Because this maximum is an upper bound on $E\ \{\boldsymbol{D}_A\}$ the maximum magnitude for *equality* is easily exceeded by the magnitude of *difference* for stimuli placed far outside the Interval of Equality.

Values of θ^* covary with values of $E\ \{\boldsymbol{D}_A\}$ so that as the comparison stimulus increases in value, $E\ \{\boldsymbol{D}_A\}$ and θ^* begin to decrease near the lower boundary of the Interval of Equality. At the boundary, $E\ \{\boldsymbol{D}_A\}$ equals zero and so does θ^*. Because the sensation produced by the stimulus equals $\theta^* A$ the sensation also equals zero. Beyond the lower boundary the values of $E\ \{\boldsymbol{D}_A\}$ and θ^* increase resulting in an increase in sensation until the upper boundary of the Interval of Equality is reached and the values of $E\ \{\boldsymbol{D}_A\}$ and θ^* again return to zero. Beyond the upper boundary both $E\ \{\boldsymbol{D}_A\}$ and θ^* become increasingly negative.

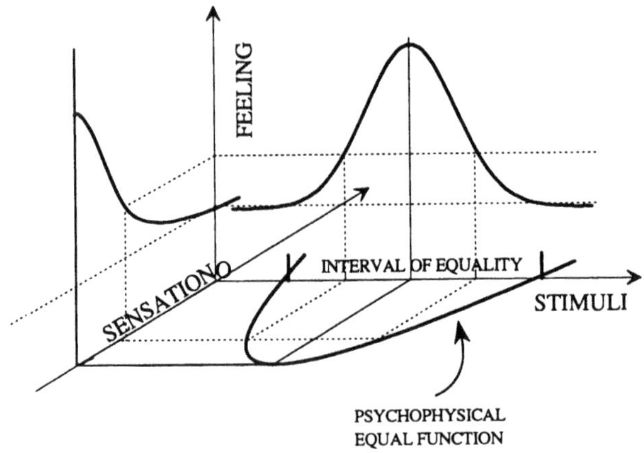

FIG. 17.5. Relations between sensation and feeling derived from Wave Equality Theory show the maximum feeling of equality to occur in the middle of the Interval of Equality.

Figure 17.5 illustrates the changes in sensation as a probe stimulus varies across the stimulus axis. The Psychological Equal Function shows how sensation changes. The feeling of equality changes differently from the sensation of equality. At the boundaries of the Interval of Equality the sensation is 0, and the subject chooses the responses *equal* or *different* with probability 0.50. This amount of feeling is a unit value. As the probe nears the midpoint of the Interval of Equality the feeling of equality nears its maximum. As the probe slides beyond the Interval of Equality the feeling of equality steadily declines. Apparently two distinct stimuli may give rise to the same feeling of equality. Therefore feelings of equality do not uniquely identify the stimuli that give rise to them.

The observation that the two different stimuli on opposite sides of the midpoint of the Interval of Equality give rise to the same feeling of equality resolves a formidable psychophysical enigma. Fechner first, and other psychophysicists later, discovered that when using the method of ascending and descending stimuli to determine a just noticeable difference, two different stimuli gave rise to the *jnd*. One of these stimuli defined a lower boundary for the *jnd* and the other, an upper boundary. Indeed Fechner (1860/1966) referred to "passing through an interval of equality" (p. 63) as the comparison stimuli continually increased away from the standard or decreased toward it.

Figure 17.5 suggests the reason for the existence of two stimuli that both provide a feeling of being just noticeably different from a standard. The task of the subject is to choose a stimulus just noticeably different from a standard, not choose the standard itself. To select a stimulus that "feels" just noticeably different, the subject chooses an Interval of Equality around a sensation that just differs from the standard. Evidence in previous chapters shows that this sensation has a unit value, and therefore is not just an arbitrary quantity, but instead one rooted in the mechanics of the sensory system.

The midpoint depends on a unit sensation value but the boundaries around the midpoint may appear more arbitrary. Yet even here there is an important constraint. The feeling that a comparison stimulus equals the (hypothetical) just noticeably different stimulus reaches a unit value at exactly the boundaries of the Interval of Equality surrounding the comparison stimulus that just differs from the experimenter's standard. At these edges the subject just feels the stimulus to be equal to the just noticeably different midpoint of the Interval of Equality. The edges are revealed by exactly the method that produced the theoretical enigma in the first place: increasing a comparison stimulus value away from the standard (and toward the Interval of Equality) or beginning with a very large comparison and decreasing its value toward the standard and the Interval of Equality.

S. S. Stevens' many experiments on matching different physical stimuli follow from similar considerations. The experimental subject is given a standard, which provides the midpoint for an Interval of Equality. The comparison stimulus, even in a different sense modality generates a sensation or feeling that can be matched against the boundaries of the Interval of Equality surrounding the sensation or feeling of the standard. By controlling the physical magnitude of the comparison stimulus the subject is able to comply with the experimenter's demand to "make the sound as loud as this light."

When more specific assumptions are made about the probability distribution for $\boldsymbol{W}_{\mathrm{A}}$ more specific results occur. For example, a Poisson input with parameter α generates an expected value equal to

$$\begin{aligned} E\,\{\boldsymbol{D}_{\mathrm{A}}\} &= -\alpha(1+\alpha) + (L+U)\alpha - LU \\ &= -\alpha^2 - \alpha(1-L-U) - LU\,. \end{aligned} \tag{17.4}$$

For fixed L and U this is a quadratic function of the comparison stimulus parameter α. Differentiating $E\,\{\boldsymbol{D}_{\mathrm{A}}\}$ with respect to α and setting the derivative to zero shows the function to reach its maximum when $\alpha = 0.5(L + U - 1)$, nearly at the midpoint of the Interval of Equality. At the maximum the expected difference has its largest value and is a function of the size of the Interval of Equality. That is,

$$\mathrm{Max}[E\,\{\boldsymbol{D}_{\mathrm{A}}\}] = 0.25(1 - L - U)^2 - LU\,. \tag{17.5}$$

These results suggest a picture of mean response time as a function of the comparison-stimulus values that corresponds to the results of Cartwright (1941b) and Johnson's Principle (chap. 5) that response time depends upon the distance of a stimulus from a category boundary. For unbiased responding, as illustrated in Fig. 17.5, the mean response time is largest near the boundaries of the Interval of Equality. Within the interval, the response time steadily decreases toward a minimum as the comparison stimulus nears the interval's midpoint. Within the interval the fastest mean response time occurs to the comparison stimulus nearest the midpoint. Yet if the range of stimuli is quite large the fastest response times may occur to comparison stimuli that lie outside the Interval of Equality and deviate further from its bounds than does the midpoint. These mean response times continually decrease as the comparison deviates farther from the boundaries.

Mean response times do not reveal whether the response made is *equal* or *different*. The picture in Fig. 17.4 suggests that within the Interval of Equality the probability of responding *equal* is greater than 0.50. This conclusion is forced by the fact that measures of discriminability, values of θ^*, have the same sign as values of $E\ \{\boldsymbol{D}_A\}$. According to Fig. 17.4, $E\ \{\boldsymbol{D}_A\}$ exceeds zero within the Interval of Equality. Therefore θ^* must also exceed zero within the Interval of Equality. When θ^* exceeds zero θ^*A exceeds zero and the probability of responding *equal*,

$$Pr\{equal \mid \text{Stimulus } \mathbf{S}_i\} = \frac{1}{1 + e^{-\theta_i A}}, \tag{17.6}$$

necessarily exceeds 0.50.

Thus, a reasonable expectation is that within the Interval of Equality the probability of responding *equal* exceeds the probability of responding *different*. Yet this seemingly reasonable expectation is apparently falsified by Brown's (1910) experiment investigating the Psychometric Equal Function (discussed in chap 3.). The comparison stimuli increased in weight from 84 g to 110 g with the mid-weight set at 96 g. Yet the probabilities of responding *same* were nearly always less than 0.50. Brown, regrettably, did not measure response times. Therefore it is not possible to determine whether the mean response times increase and decrease in the manner suggested in Fig. 17.6 even when response probabilities are less than 0.50.

Yet a remarkable conclusion remains to be drawn from these results. The probability of responding *same*, generally smaller than 0.50 even for stimuli quite near the middle weight of 96 g, does not exceed 0.50 because the comparison stimuli never probe the Interval of Equality. Although the test stimuli differ by only 1 g, the subject's Interval of Equality must be smaller than 1 g. Had the experimenter used a finer gradation of weight, substantial increases in the proportion of *same* responses above the 0.50 level may have occurred. The subject's very small Interval of Equality suggests a very fine sense of equality.

The subject's response proportions are available. Their logistic relation to values of θ^*A makes possible a reconstruction of the subject's view of *equality*. The sensation of *equality* is obtained by first estimating θ^*A values for each comparison stimulus. These estimates derive from the proportion of *same* judgments shown in Fig. 3.3. From Equation 17.6 the estimated value of θ_i^*A is computed from the relation $ln[Pr\{same\}/(1 - Pr\{same\})]$. The estimates of θ_i^*A

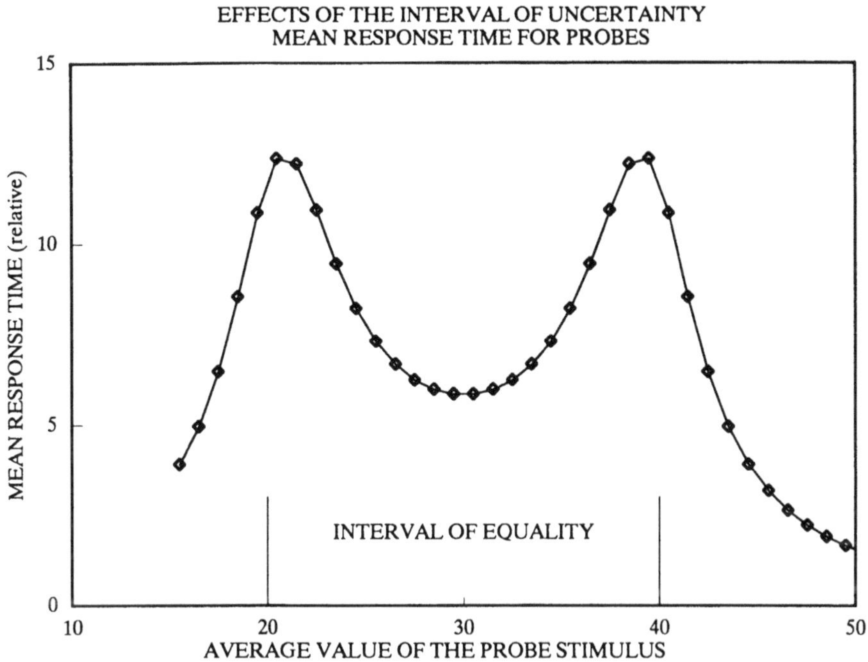

FIG. 17.6. Theoretical mean response times as a function of comparison stimuli that probe the Interval of Equality. The largest response times occur to stimuli near the borders of the Interval of Equality.

for each weight from 84 g to 110 g appear in Fig. 17.7. The parabolic shape formed by the θ_i^*A values is quite evident. The best estimate of the Point of Subjective Equality is 96.8 g, a value quite near the actual midpoint of 96 g. Generally, the values of θ_i^*A are less than zero, thus the comparison stimuli are not sufficiently refined to locate the bounds on the Interval of Equality.

This subject's feeling of equality may now be resurrected by exponentiating the estimated values of θ^*A to obtain the measure of the feeling of equality. Fig. 17.8 shows the best estimates of this subject's feeling of equality for the comparison stimuli. These estimates of the feeling of equality experienced by the subject are based on the best fit to the values of θ^*A in Fig. 17.7. The maximum occurs at 97 g but the decline in the feeling of equality, as the comparison stimuli deviate from the 96 g midpoint, is astonishing. This subject, very well practiced in the task, hardly felt the comparison stimuli to be equal to the standard at all. The subject's acute discriminative ability is evident in the very steep slopes of the Psychological Equal Function.

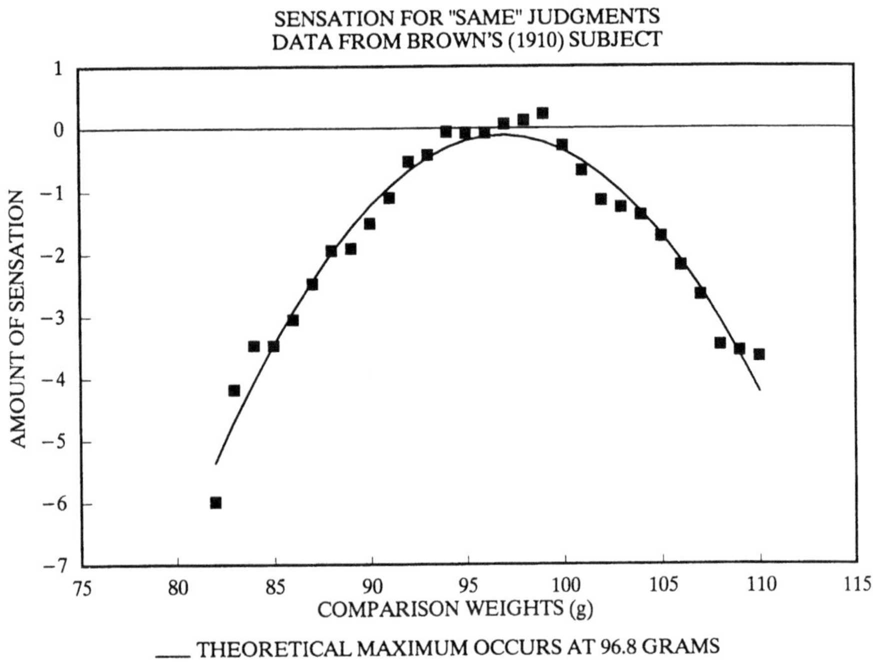

FIG. 17.7. Estimated values of $\theta{*}A$ for Brown's (1910) study of equal weight judgments.

THEORY OF IDEALS

A second form of equal judgment occurs in the many studies of Ideals described in Part I. Usually two stimuli are compared against a person's Ideal. For example, "If you were beginning a family would you rather have no children or 8 children?" The ideal number of children is a value in the mind of an individual. For stimuli ordered on a single dimension the experimenter's stimuli either bracket the Ideal or diverge in the same direction. Subjects are hypothesized to choose the stimulus that is closest to their Ideal. Thus, a person desiring a large family will choose the 8 children over no children. However, if the cost of children weighs heavily in this judgment then no children may be the best alternative. The Theory of Ideals provides methods for extracting the subject's known or unknown Ideal. The theory and methods are directly applicable to clinical assessment and diagnosis.

The Ideal Age for Marriage

The empirical facts about these judgments are easily demonstrated by an experiment on the ideal age for marriage (Garbas, 1985). The subjects, all unmarried

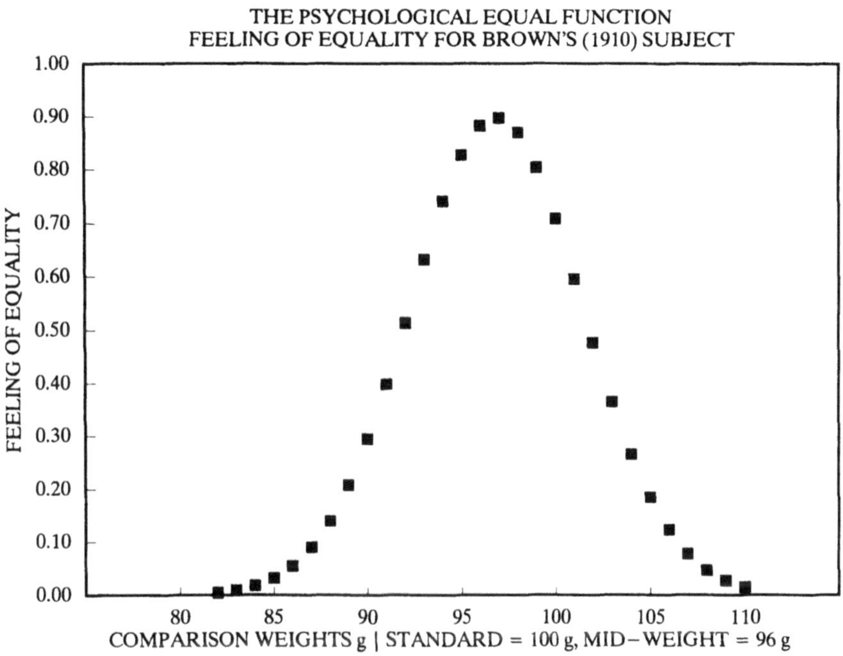

FIG. 17.8. The Psychological Equal Function determines the feeling of equality generated by comparison stimuli ranging from 84 g to 110 g.

college juniors 2 or 3 years apart in age, selected from two alternatives the age at which they preferred to marry. Table 17.1 presents summary data for the 30 student participants. Each pair of ages defines a row and column that intersect to form a cell. Within each cell the table shows the proportion for choosing the age at the heading of the column rather than the age for the row as the better age for marriage. A sample of all ages from 18 to 32 appears in Table 17.1. Ages 18, 21, 24, 25, 26, 27, 29, and 32 provide a good view of changes in response proportions when both males and females are asked to choose which of two ages seems better for marriage.

The first column of proportions reveals that 18 is not preferred to any greater age up to the maximum of 32 years. Examining the column of proportions for age 26 shows that 26 is preferred to every other age. The column for age 25 also reveals that only age 26 is preferred to age 25. For this group of males and females the Ideal age for marriage appears to be age 26. Perhaps more can be mined from these data than just this obvious conclusion.

The Theory of Ideals proposes that the difference between stimuli S_A and S_B and the Ideal are critical components of such comparative judgments. Assuming the same mechanism to apply to comparisons against an Ideal as applies in the case of *equality* judgments suggests that two comparison stimuli S_A and S_B form the boundaries on an interval that may contain the Ideal. If this analogy holds then, from Equation 17.2, the wave difference between the two stimuli and the Ideal takes the form,

$$\boldsymbol{D}_{\mathrm{AB}}(\Delta t) = [\boldsymbol{I}(\Delta t) - \boldsymbol{W}_{\mathrm{A}}(\Delta t)][\boldsymbol{W}_{\mathrm{B}}(\Delta t) - \boldsymbol{I}(\Delta t)]. \tag{17.7}$$

In this case stimuli on opposite sides of the Ideal generate a value of $\boldsymbol{D}_{\mathrm{AB}}$ with the same sign. Values of $\boldsymbol{W}_{\mathrm{A}} > \boldsymbol{I}$ and values of $\boldsymbol{W}_{\mathrm{B}} < \boldsymbol{I}$ are on opposite sides of the Ideal. They generate a positive value for $\boldsymbol{D}_{\mathrm{AB}}$. If $\boldsymbol{W}_{\mathrm{A}} > \boldsymbol{I}$ and $\boldsymbol{W}_{\mathrm{B}} > \boldsymbol{I}$ then the value of $\boldsymbol{D}_{\mathrm{AB}}$ is negative. Since the roles of $\boldsymbol{W}_{\mathrm{A}}$ and $\boldsymbol{W}_{\mathrm{B}}$ are arbitrary the stimuli may be reversed without the signs of the outcomes changing. Thus the sign of $\boldsymbol{D}_{\mathrm{AB}}$ denotes a qualitative difference between the stimuli and the Ideal.

The obvious problem with this formulation is that although the qualitative information correctly identifies the positioning of the stimuli with respect to the Ideal, there is no indication of which stimulus lies nearer the Ideal. Furthermore, stimuli equally distant, but on opposite sides of the Ideal, yield a measure of discrimination that is non-zero, but the data suggest that such stimuli are chosen only with probability 0.50, in which case $\boldsymbol{D}_{\mathrm{AB}}$ must be zero. Thus $\boldsymbol{D}_{\mathrm{AB}}$ is useful for determining whether two stimuli are on the same side of the Ideal, such as right- and left-wing political candidates. The degree of their opposition is measured by the increasingly negative value for $\boldsymbol{D}_{\mathrm{AB}}$ as either $\boldsymbol{W}_{\mathrm{A}}$ or $\boldsymbol{W}_{\mathrm{B}}$ deviates more from the Ideal, which represents the imagined dividing point between the two opposing political camps. But Equation 17.7 does not provide a foundation for choosing which stimulus is nearer the Ideal.

Some slight reformulation of Equation 17.7 may at first appear to capture the necessary information. Suppose that both stimuli are compared to the Ideal and that their product, which as an expected value is a cross-correlation or cross-covariation, defines the discrimination measure. That is,

$$\boldsymbol{D}_{\mathrm{AB}}(\Delta t) = [\boldsymbol{W}_{\mathrm{A}}(\Delta t) - \boldsymbol{I}(\Delta t)][\boldsymbol{W}_{\mathrm{B}}(\Delta t) - \mathbf{I}(\Delta t)]. \tag{17.8}$$

When $\boldsymbol{W}_{\mathrm{A}}$ and $\boldsymbol{W}_{\mathrm{B}}$ are both on the positive side of the Ideal, $\boldsymbol{D}_{\mathrm{AB}}$ must be positive. When both occur on the negative side, $\boldsymbol{D}_{\mathrm{AB}}$ is negative. This measure varies with the distance from the Ideal but does not determine which stimulus is nearer the Ideal. Although the subject must choose the stimulus nearest the Ideal this representation only distinguishes between the cases of stimuli found to be on either the opposite or the same side of the Ideal. Somehow both order and directionality need to be inserted into the measure. A different approach is required.

As an additional consideration the accumulation of values of $\boldsymbol{D}_{\mathrm{AB}}$ should tend to a choice of $\mathbf{S}_{\mathrm{A}}$ whenever $\boldsymbol{W}_{\mathrm{A}}$ is nearer $\boldsymbol{I}$ than $\boldsymbol{W}_{\mathrm{B}}$, regardless of whether $\boldsymbol{W}_{\mathrm{A}}$ is larger or smaller than $\boldsymbol{W}_{\mathrm{B}}$. If the two stimuli occur on the left side of the Ideal then $\boldsymbol{W}_{\mathrm{A}} > \boldsymbol{W}_{\mathrm{B}}$ should generally lead to the choice of $\mathbf{S}_{\mathrm{A}}$. If the two stimuli occur on the right side of the Ideal then $\boldsymbol{W}_{\mathrm{A}} < \boldsymbol{W}_{\mathrm{B}}$ should again produce a higher probability of choosing $\mathbf{S}_A$ over $\mathbf{S}_B$. That is, if $|\boldsymbol{W}_A - \boldsymbol{I}| < |\boldsymbol{W}_B - \boldsymbol{I}|$ then $\mathbf{S}_A$ is selected as nearer the Ideal.

But, the absolute value of a wave difference strips away information concerning the position of the stimuli with respect to each other, whereas, in judgments involving Ideals it is exactly this information that is critical. Simultaneously capturing the ordering of the stimuli with respect to the Ideal and their values with respect to each other requires some form of comparison of differences between stimuli and the Ideal. That is, a difference such as

$$\boldsymbol{D}_{AB}(\Delta t) = [\boldsymbol{W}_B(\Delta t) - \boldsymbol{I}(\Delta t)]^2 - [\boldsymbol{W}_A(\Delta t) - \boldsymbol{I}(\Delta t)]^2 \qquad (17.9)$$

This discriminative value compares the difference between $\boldsymbol{W}_A$ and $\boldsymbol{I}$ against the difference between $\boldsymbol{W}_B$ and $\boldsymbol{I}$, but ignores the information about which side of the Ideal $\boldsymbol{W}_A$ or $\boldsymbol{W}_B$ are on. When $\boldsymbol{W}_A$ is nearer $\boldsymbol{I}$ the result is positive. An accumulation of generally positive $\boldsymbol{D}_{AB}$ values creates a higher probability of choosing $\mathbf{S}_A$ rather than $\mathbf{S}_B$. If $\boldsymbol{W}_B$ is nearer $\boldsymbol{I}$ the difference is negative leading to an accumulation of generally negative values that produce the choice $\mathbf{S}_B$. If $\boldsymbol{W}_A$ and $\boldsymbol{W}_B$ are equally distant from $\boldsymbol{I}$ then $\boldsymbol{D}_{AB}$ is zero and chance performance results. The features of this comparison process fit the qualitative discoveries.

One of the most troublesome results from experiments involving Ideals, perhaps the motivating force behind the development of Coombs' ideas, derived from paired-comparison studies. In what might seem like a trivial experiment on choosing the best gray color-patch (chap. 6) subject's results filled paired-comparison response proportion tables with numerous choice intransitivities; intransitivities such as choosing $\mathbf{S}_A$ over $\mathbf{S}_B$, $\mathbf{S}_B$ over $\mathbf{S}_C$ but $\mathbf{S}_C$ over $\mathbf{S}_A$, that invalidated Thurstone's Law of Comparative Judgment. These intransitivities are a consequence of the form of judgment proposed by Equation 17.9.

The paired-comparison data also reveal that although chance performance results whenever two stimuli are indiscriminable from each other, chance responses also occur when two comparison stimuli are equally distant but on opposite sides of the Ideal. This feature is also captured by Equation 17.9. Expanding this equation shows there are two different ways that $\boldsymbol{D}_{AB}$ can equal zero:

$$\begin{aligned}
\boldsymbol{D}_{AB} &= \boldsymbol{W}_B^2 - 2\boldsymbol{W}_B\boldsymbol{I} + \boldsymbol{I}^2 - \boldsymbol{W}_A^2 + 2\boldsymbol{W}_A\boldsymbol{I} - \boldsymbol{I}^2 \\
&= (\boldsymbol{W}_B - \boldsymbol{W}_A)(\boldsymbol{W}_B + \boldsymbol{W}_A - 2\boldsymbol{I}) \\
&= 0 \begin{cases} \text{when } \boldsymbol{W}_A = \boldsymbol{W}_B \\ \text{when } \boldsymbol{W}_B + \boldsymbol{W}_A = 2\boldsymbol{I}. \end{cases}
\end{aligned} \qquad (17.10)$$

The last case occurs when both stimuli are equally distant but on opposite sides of the Ideal, that is $0.5(\boldsymbol{W}_B + \boldsymbol{W}_A) = \boldsymbol{I}$. Therefore, the expected value of $\boldsymbol{D}_{AB}$ equals zero when the stimuli are identical to each other or equally distant from the Ideal. When the expected value of $\boldsymbol{D}_{AB}$ is zero, the corresponding value of θ^* is zero and the sensation of difference is zero. Remembering that mean response time increases as the expected value of $\boldsymbol{D}_{AB}$ approaches zero makes evident the prediction that response times are large when the stimuli are identical or when they are opposite each other and equally distant from the Ideal.

As shown in Fig. 17.3, the discrimination measure for Ideals, $\boldsymbol{D}_{AB}$, accumulates until either the upper response threshold for choosing $\mathbf{S}_A$ or the lower threshold for choosing $\mathbf{S}_B$ is first exceeded. Thus the process of accumulation remains the same as in all previous examples, only the input to the accumulation process changes. The Wald Identity applies to this Martingale process and suggests that increases in the value of θ^* are related to increases in the value of $E\{\boldsymbol{D}_{AB}\}$. In fact, θ^* and $E\{\boldsymbol{D}_{AB}\}$ are of the same sign. When θ^* equals zero so does $E\{\boldsymbol{D}_{AB}\}$. Thus response proportions determine the shape of $\boldsymbol{D}_{AB}$ for a fixed Ideal but varying comparison stimuli.

THEORY OF PAIRED COMPARISONS FOR IDEALS

Generally the experimenter presents for comparison a large number of stimulus pairs drawn from the set $\mathbf{S}_1, \mathbf{S}_2, \ldots, \mathbf{S}_n$. These many paired comparisons are summarized by a paired-comparison choice matrix such as Table 17.1. When there is no response bias the analysis of paired comparisons in chap. 15 shows that for each stimulus pair $(\mathbf{S}_i, \mathbf{S}_j)$ a value of θ^*A derives from the relation

$$\theta^*_{ij}A = ln[p_{ij}/(1 - p_{ij})], \tag{17.11}$$

where p_{ij} is the probability of choosing stimulus $\mathbf{S}_i$ over $\mathbf{S}_j$. Such values are determined for each cell of the paired-comparison matrix.

TABLE 17.1
The Ideal Age for Marriage
Paired Comparisons From 15 Males and 15 Females

	Age								
Age	*18*	*21*	*23*	*24*	*25*	*26*	*27*	*29*	*32*
18	.00	.97	.97	.95	.94	.93	.89	.87	.83
21	.03	.00	.93	.90	.90	.84	.79	.72	.62
23	.03	.07	.00	.88	.80	.70	.64	.59	.43
24	.03	.10	.13	.00	.64	.59	.49	.48	.34
25	.06	.18	.24	.36	.00	.52	.43	.37	.24
26	.06	.16	.31	.41	.48	.00	.34	.27	.14
27	.12	.21	.36	.51	.57	.66	.00	.24	.14
29	.13	.28	.42	.53	.63	.73	.76	.00	.16
32	.17	.38	.58	.66	.76	.86	.86	.84	.00

Note. From Garbas, C. A. (1985) The ideal age for marriage.

The monotonic relation between $\theta^*_{ij}A$ and $E\{\boldsymbol{D}_{ij}\}$ suggests a distribution-free method of determining how values of $\theta^*_{ij}A$ may be averaged to reveal the effects of a single stimulus. Values of θ^*_{ij} follow the form of $E\{\boldsymbol{D}_{ij}\}$. Averaging values of $E\{\boldsymbol{D}_{ij}\}$ for a fixed value of i but across all values j = 1, 2, . . . , n gives,

$$\sum_{j=1}^{n} E\{\boldsymbol{D}_{ij}(\Delta t)\} = \sum_{j=1}^{n} E\{(\boldsymbol{W}_j(\Delta t) - \boldsymbol{I}(\Delta t))^2 - (\boldsymbol{W}_i(\Delta t) - \boldsymbol{I}(\Delta t))^2\}$$

$$= -nE\{(\boldsymbol{W}_i(\Delta t) - \boldsymbol{I}(\Delta t))^2\} + \sum_{j=1}^{n} E\{(\boldsymbol{W}_j(\Delta t) - \boldsymbol{I}(\Delta t))^2\}. \tag{17.12}$$

This is a quadratic function of the variable (stimulus) $\mathbf{S}_i$ with Wave representation $\boldsymbol{W}_i$. The right-most term is a constant. Thus, the shape of this function is parabolic, reaching a peak when a stimulus is presented that equals the Ideal. Because $\theta^*_{ij}A$ is monotonic with $E\{\boldsymbol{D}_{ij}\}$ so too is the average of $\theta^*_{ij}A$ values within a column of the paired-comparison matrix. Therefore, when the column averages of the $\theta^*_{ij}A$ are plotted as a function of the stimuli, the peak of the function identifies the stimulus that corresponds to the Ideal.

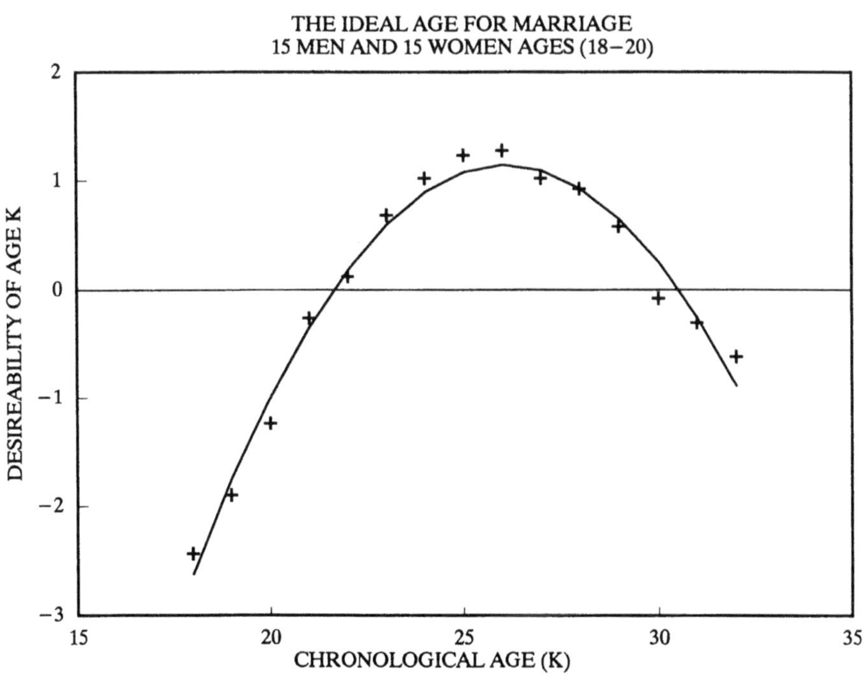

FIG. 17.9. Estimates of $\theta^*_i A$ for each age of marriage for 30 unmarried college students.

Figure 17.9 shows the averaged values of $\theta^*_{ij}A$ obtained from the Garbas study of the Ideal age of marriage. The graph consists of averaged values of $\theta^*_{ij}A$ obtained by converting all proportions in a choice matrix larger than, but inclusive of, Table 17.1 and averaging these within a column. As the age of marriage increases the average values of $\theta^*_{ij}A$ steadily increase toward a peak at an Ideal age for marriage, then the values decline.

Figure 17.10 shows values obtained for the 15 males and 15 females considered separately. For the males the curve is somewhat flatter, and the maximum occurs for an age greater than that for the females. Knowing that the theoretical function for the sensations θ^*A evoked by comparing the two stimuli must be parabolic provides a method of identifying the Ideal age for marriage for females and males. The best fitting theoretical function for each group yields parameter values that define the parabolic equation for each curve. Differentiating the equation, setting the derivative to zero, and solving for the age that gives the maximum value for θ^*A shows that for these females the Ideal age for marriage is 25.6 years, whereas the males' Ideal is 26.6 years.

The feelings of preference for an age of marriage are revealed by Fig. 17.11. The two curves (one for females and one for males) peak at the theoretical Ideal ages of 25.6 and 26.6 years, respectively. These measures show quite clearly that

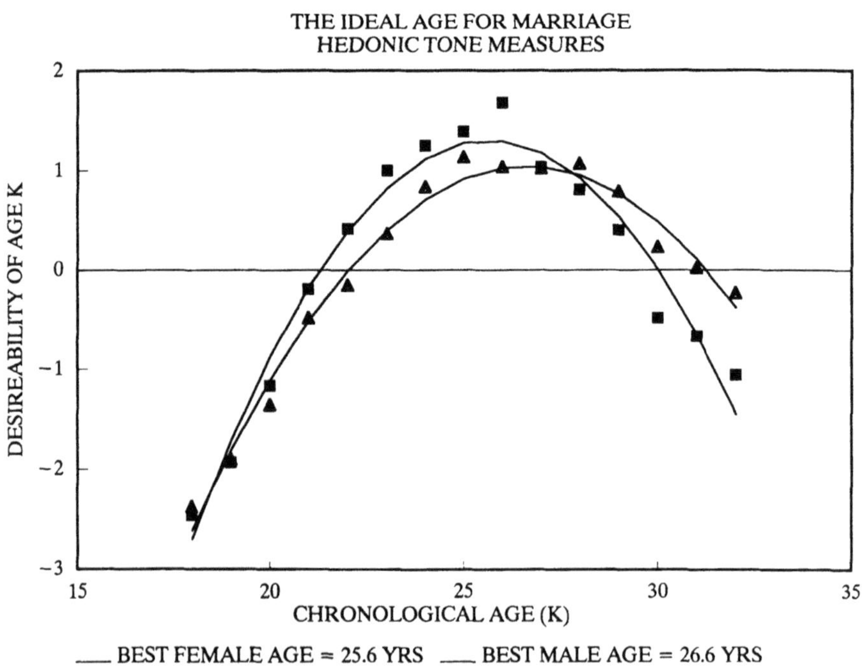

FIG. 17.10. Estimates of θ^*A for 15 male and 15 female unmarried college students. The solid line shows the best theoretical fit.

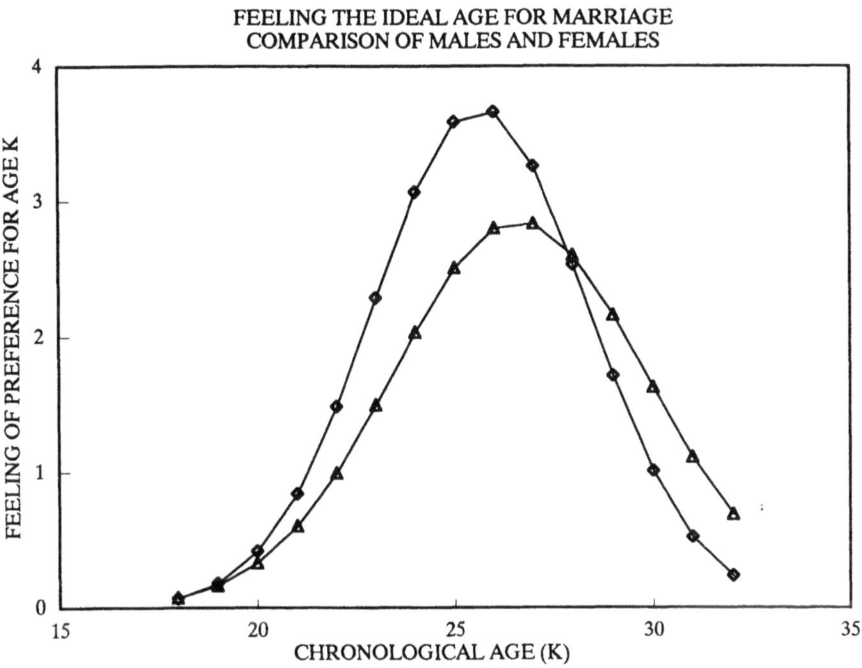

FIG. 17.11. Feelings of the preference for marriage for unmarried female (diamonds) and male (triangles) college students.

these women have feelings greater than a value of 1, which corresponds to $\theta^*A = 0$, for ages between 21 and 30. Males' feelings are above the value 1.0, that is they prefer ages of marriage between 22 and 31 years. Thus the 1 year shift of the Ideal age for marriage, from 25.6 for females to 26.6 for males, is a feature of the overall results. Males prefer to marry later than females.

The results point to a source of conflict between the male and female students in their view of the Ideal age for marriage. Given that the students are essentially the same age, their Ideal age for marriage differs by 1 year. When a male and female reach 26 years of age the female views the present time as ideal for marriage. The male, however, finds 26 somewhat premature. The woman finds 27 and 28 increasingly undesirable, with respect to her Ideal, whereas the male finds 26 or 27 about equally desirable. Based on obtaining maximum satisfaction in the choice of a year for marriage both should choose the age of 26.

The feelings of this couple are quite different from what is expected from examining the position of the Ideal. The female shows much more feeling about age 26 than does the male. If equal feelings are taken as a basis for selecting the best, but not Ideal, age for marriage, then age 28 is suitable because at age 28 each party to the decision has the same amount of feeling of preference. Feeling is not maximized but is equalized.

Perhaps neither party will compromise their Ideal. Then the couple must differ in chronological age in order for their ideals to be jointly satisfied in the same year. A male 1 year older than a female views 27 as an Ideal age for marriage. The female 1 year younger than the male views 26 as an Ideal age for marriage. A male 27 years old in the year 2000 and a female 26 years old in 2000 can both agree upon 2000 as an Ideal year for marriage.

The laboratory studies of the psychophysics of preferences for tones of gray or color dominated early experimental work. The Poisson signals for two different gray tones may be defined as $\boldsymbol{W}_{\mathrm{A}}$ and $\boldsymbol{W}_{\mathrm{B}}$. For Poisson signals $E\{\boldsymbol{W}_{\mathrm{A}}\} = \alpha$, $E\{\boldsymbol{W}_{\mathrm{A}}^2\} = \alpha(1 + \alpha)$ and $E\{\boldsymbol{W}_{\mathrm{B}}\} = \beta$, $E\{\boldsymbol{W}_{\mathrm{B}}^2\} = \beta(1 + \beta)$. Therefore,

$$
\begin{aligned}
E\{\boldsymbol{D}_{\mathrm{AB}}\} &= \beta(1 + \beta) - 2\boldsymbol{I}(\beta - \alpha) - \alpha(1 + \alpha) \\
&= (\beta - \alpha)(\beta + \alpha + 1 - 2\boldsymbol{I})\,. \qquad (17.13)
\end{aligned}
$$

When α and β are equal then, of course, neither is preferred to the other. When β and α are distinct but on opposite sides of the Ideal neither is preferred to the other. As in the case of the Ideal age for marriage, marginal estimates of θ^*A and of feeling e^{θ^*A} increase until a maximum is reached at the point of the Ideal.

To return to the introductory art history students, some may view the art of Titian and Caravaggio as similar because two obviously distinct paintings are equally distant from an Ideal. For some students the paintings are equally distant from the Ideal because the knowledge base places the paintings on the same side of the Ideal. For other students the paintings are distinct and on opposite sides of the Ideal. In spite of their quite different perceptions, both students view the paintings as similar. The Theory of Ideals provides a tool for distinguishing between these two unique perceptions and for measuring the feeling of pleasure derived from the aesthetic experience.

DÉNOUEMENT

Fechner's (1860/1966) extension of Gauss's (1809, 1821/1880) Theory of Error gave impetus to the movement toward a scientific psychology. Herbart (1824/1850) modelled mental processes and derived theoretical predictions, but Fechner brought science to psychology by creating the first bridge between theoretical predictions and their experimental verification. Fechner's theory, coupled with Weber's (1846) discovery of the *just noticeable difference*, provided the first measurement of mental abilities previously hidden from scientific scrutiny.

In spite of its broad success, Fechner's theory provoked counter-experiments that brought into question the role of the stimulus as the dominant force in the stimulus-response path. Although response proportions, times, and confidences changed as a function of stimulus-difference, disturbing "subjective" phenomena also appeared to influence discrimination. Furthermore, the so-called "Constant Error" revealed a ubiquitous deviation of performance from theoretical expectation. And, regardless of the success in measuring sensitivity to physical stimulation, Fechner's theory stumbled at the threshold to measuring more subjective experiences such as the appeal of a fragrance, a political candidate, a painting or a poem. Aesthetic judgments about stimuli with no physical measures lingered beyond the grasp of Fechner's psychophysics.

Extending Fechner's model, Thurstone (1927b) hypothesized that stimuli were defined by a distribution of psychological values distributed on a Psychological Continuum. By assuming that stimuli were compared directly against each other Thurstone created the Law of Comparative Judgment. In a further step, by treating the opinions of subjects as a result of societal variability, Thurstone (1927a) showed how to measure attitudes. In this way Thurstone produced a powerful theory capable of measuring such elusive psychological quantities as social value or preference for political candidates. Psychophysical methods no longer required a physical measure of the stimulus in order to measure differences between stimuli or the subject's sensitivity to them.

Yet, disturbing deviations from the predictions of the Law of Comparative Judgment soon appeared. Unexpected intransitivities among paired-comparison response proportions showed that a subject may prefer food item A to B, and B to C, but prefer C to A! Such studies of affective value continued to pose difficulties for Thurstone's theory until Coomb's suggested that instead of comparing stimuli against each other, stimuli were compared against an Ideal. Then the subject chose the stimulus nearest the Ideal. This theory predicted the existence of paired-comparison intransitivities and did much to resolve previous psychophysical paradoxes.

Other experiments showed that more than just the mechanism of comparison influenced performance. Both response proportions and response times were influenced by "subjective factors" that caused uncontrollable shifts in the Psychometric and Chronometric Functions, erratic changes in measurements of thresholds, and tradeoffs between the speed and accuracy of responses. Although Fernberger (1913, 1930) focussed attention on the possibility that subjective factors influenced discriminative judgments, and Thomson (1920) offered a quantitative theory to account for displaced and asymmetric Psychometric Functions, the rigorous analysis of subjective factors awaited Blackwell's (1953) creation of the idea of response bias.

Blackwell's (1953) experiments suggested that subjects exerted control over performance by biasing a response independent of the magnitude of stimulus-difference. Changes in bias proved to be ubiquitous in psychological experiments. Soon Peterson, Birdsall and Fox (1954) and Tanner and Swets (1954) proposed Signal Detectability Theory (SDT), a generalization of Fechner's theory, that allowed the response threshold hypothesized by Fechner to move at the will of the subject. Many experimenters produced Receiver Operating Characteristics (ROCs) predicted by SDT by using rewards and punishments to influence the subject. Soon other experimenters showed that ROCs quite different from the curves predicted by SDT also occurred in the context of very similar experimental procedures. Atkinson (1963), Blackwell (1963), and Luce (1963) each suggested theories substantially different from SDT and showed results compatible with their theories. The Receiver Operating Characteristic came to be viewed as the signature of operation of a particular theory.

Response times, as shown by Cattell (1886), became an especially valuable tool for the analysis of performance. Even when discrimination is nearly perfect, response times bear a somewhat reciprocal relation to stimulus-difference (Welford, 1980); faster response times occur to more discrepant stimuli. Yet response time, often thought of as the time for a mental reaction and therefore often called reaction time, also changes due to "subjective factors." Tradeoffs of speed for accuracy, clearly demonstrated by Johnson (1939), showed that subjects in some way exert influence over their own reaction times. Then Cartwright (1941b) discovered the peculiar shape of the Chronometric Function for *equal* judgments and, with Festinger, produced the first opponent-process theory of discrimination. The theory was unique in predicting both response probabilities and response times as due to a single discriminative mechanism.

The advent of Information Theory (Shannon, 1948) offered psychologists an opportunity to apply new ideas to old problems. Viewing the subject as a communication channel gave rise to the study of reaction time as a function of information. Bertelson (1961) demonstrated the strong influence of particular stimulus-response sequences on choice reaction time and response probability. The Wave Theory analysis of sequence effects shows the source of these variations to be response bias.

The measures of response proportion and response time provide different windows on discriminative abilities. The measurement of these abilities depends upon both objective and subjective factors that Fernberger forced to the attention of a skeptical psychological community. Once accepted these subjective factors change radically the scientist's view of the discriminating subject. Rather than simply responding to stimuli in a reactive sequence of thoughtless events, like a pinball machine reacting to the play of a pinball, the subject becomes an active participant in the experimental task. As many studies prove, a subject's response bias dramatically influences the various measures of discriminative performance.

Wave Theory reveals the existence of yet another subjective factor; the subject's resistance to respond. Resistance affects performance in discrimination experiments as well as in the measurement of sensation and feeling. Resistance determines how much information must accumulate before a response occurs. By controlling resistance the subject indirectly controls both response time and response probability in order to comply with the experimenter's performance demands. A high resistance to respond permits many sensory events that otherwise interfere with ongoing activities to slip beneath awareness. As the resistance to respond declines, responses become more likely and occur more quickly. By altering the resistance to respond subjects produce response proportions for Hits and False Alarms that mimic the Receiver Operating Characteristics of Signal Detection Theory, High Threshold Theory, Low Threshold Theory, and Variable Sensitivity Theory. Changes in response resistance also cause predictable changes in Psychometric Functions and introduce uncontrolled variation into the determination of psychophysical thresholds.

The two variables under the direct control of the subject prove to be controlling features of observable behavior. Response bias proves to be a ubiquitous property of psychophysical experiments and is a cause for the often observed Constant Error. The resistance to respond is the second factor under direct subjective control. Resistance to respond influences both response probability and response time. A high resistance to respond results in long response times and quite accurate discrimination. A subject hurried through an experimental session or a physician forced to a quick diagnosis can complete the task in a short time but at the cost of an increased error rate. This speed-accuracy tradeoff is a desirable, practical, feature of human performance and occurs under the subject's direct control.

Fechner and Thurstone relied on the Gaussian Theory of Error in order to measure psychological phenomena. Wave Theory presents an alternative view based on quantum units of stimulation modelled as Poisson variability, but not error. Poisson waves are the representations of mental events that are messages transmitted over time. The oscillations of a wave occur on a dimension like the Thurstonian Psychological Continuum, along which stimuli were once scaled in units of sensory or psychological error. Now, from this new view, measurement with respect to error is no longer necessary. Variability in the Poisson wave arises from the quantum nature of the mechanism that creates the waves themselves, not from error, sensory or otherwise. Error is no longer a basis for the measurement of sensation.

Wave Discrimination Theory postulates that judgments between stimuli occur through the direct comparison of two stimulus waves. The output from this wave analysis accrues over time. When a response threshold is first exceeded a response occurs. The application of powerful Martingale results permits the theory to unite response times with response probabilities and response confidences. Predictions derived from simple Martingale properties suggest the existence of empirical laws relating response time, response probability, and response confidence.

With respect to response probability the presentation of two stimuli, $\mathbf{S}_A$ and $\mathbf{S}_B$, generates waves $\boldsymbol{W}_A$ and $\boldsymbol{W}_B$ that produce a measure of discriminability equal to θ^*. For 73% correct responding the value θ^* proves to be Weber's Constant. This fact ties the inner state of mind together with external physical events. The characteristics of the external physical stimuli are passed through the surface shield of the body in a manner that presents the mind with an accurate image of the physical world.

Fechner believed that the Psychophysical Function relating stimulus magnitude to sensation magnitude was a logarithmic function. S. S. Stevens, among others, demurred and supported the idea of Hering that the Psychophysical Law was a power function of stimulus magnitude. A resolution to this controversy follows from the Wave Theory prediction that for Poisson Signals the amount of sensation equals the measure of discriminability, θ^*, times the subject's resistance to respond, A. That is, Weber's Constant times the resistance to respond equals the amount of sensation. A prediction of this theory is that in S. S. Stevens' extensive experiments on the matching of sensations, relative Power Law exponents should equal relative Weber's Constants. The confirmation of this prediction unites within a single theory two divergent paths of psychophysical analysis.

Stevens postulated a Power Law to account for linear relations between logarithmic functions. Wave Theory shows how the observed logarithmic relations derive from a mechanism for the comparison of sensory signals. Stevens' speculation, that some mechanism must give rise to a Power Law relation between the physical stimulus and sensation, is not a feature of sensation. Rather the Power Law appears in an important new context relating feeling, rather than sensation, to stimulus magnitude. The Theory of Feeling describes the generation of feeling from the comparison of Poisson waves. Because concepts themselves have wave representations comparisons may occur between concepts with feeling arising from these comparisons. The successful analysis of the feeling based on Guilford's (1931) study of lifted weights generalizes to the amount of feeling generated by Thurstone's comparisons of serious crimes. The new techniques for the rigorous measurement of feeling extend psychophysical analysis to the core of human experience.

The kind of judgment determines the form of analysis to which stimulus waves are subjected. In the case of comparing one stimulus against another, for example *larger* or *smaller* judgments, the waves characterizing the two stimuli were compared by subtraction of one wave from the other. Several other forms of judgment depended upon a cross-correlation of wave differences. Even when judging one stimulus against another the comparison can be conceived of as the cross-correlation of two wave differences.

Imagine the standard to set an upper limit of an Interval of Equality and the Absolute Threshold to set the lower limit. The difference,

$$\boldsymbol{D} = (\boldsymbol{W}_A - \boldsymbol{T}_0)(\boldsymbol{W}_A - \boldsymbol{W}_S)$$

where $\boldsymbol{W}_A$ is the comparison stimulus
$\boldsymbol{T}_0$ is the absolute threshold
$\boldsymbol{W}_S$ is the value of the standard

will be positive when the stimulus is greater than $\boldsymbol{W}_S$ and negative when $\boldsymbol{W}_A$ is smaller than $\boldsymbol{W}_S$ but larger than the Absolute Threshold. Because most discrimination experiments require supra-threshold stimuli the Absolute Threshold plays only a minor unnoticed role. Because there are no comparison values smaller than $\boldsymbol{T}_0$, the value of $\boldsymbol{D}$ is positive only when $\boldsymbol{X}_A$ exceeds the standard $\boldsymbol{X}_S$. When $\boldsymbol{X}_A$ is within the interval of equality the value of $\boldsymbol{D}$ is negative.

This consolidation of different forms of judgments into a single cross-correlation analysis suggests a new summary of the psychology of difference and similarity. The response discrimination variable:

$$\boldsymbol{D} = (\boldsymbol{W}_A - b_1)(\boldsymbol{W}_B - b_2),$$

creates the information from which all forms of discrimination arise:

Absolute discrimination: $\boldsymbol{W}_A = \boldsymbol{W}_B$ and $b_1 = b_2 = \boldsymbol{T}_0$

Comparative discrimination:

Greater-Lesser	$\boldsymbol{W}_A = \boldsymbol{W}_B$ and $b_1 = \boldsymbol{T}_0$, $b_2 = \boldsymbol{W}_S$
	b_1 lower limit of the Interval of Equality
Equal-Different	b_2 upper limit of the Interval of Equality

Beyond these discriminations lie more complicated mental judgments that require more extensive analyses. For matching against Ideals:

$$\boldsymbol{D} = (\boldsymbol{W}_A - \boldsymbol{I})^2 - (\boldsymbol{W}_B - \boldsymbol{I})^2,$$

where $\boldsymbol{I}$ is the wave produced by the Ideal.

Whatever the form of judgment, the stimulus-response path terminates with a single two-choice response mechanism. The subject controls the response mechanism through the direct control of response bias and resistance to respond. Wave Theory suggests techniques to measure the influence of these two "subjective factors."

The study of judgment processes may be limited by the choice of stimuli but the stimuli may range from paintings or social issues, or even ideas, to the more sensory stimuli of early experimental psychology. A subject's mental analysis of these stimuli depends upon the nature of the question put to the subject. Certain complicated judgments, not yet investigated, may require the accumulation of wave values drawn from several mechanisms. Some judgments may require parallel or even competing mental processes. The combination of primitive mental processes provides an avenue for developing new models of thought processes and feelings.

The study of such processes opens to analysis mental processes not in the repertoire of human mental abilities. Such theoretical processes may play an important role in the analysis and resolution of conflicts and in communicating with, or creating, other forms of intelligence. Thinking and feeling are major features of human existence, but human beings may be incapable of thinking certain thoughts or experiencing certain feelings. Some thought processes may be forever closed to direct human experience. The methods of Wave Theory analysis, shown to apply to human discrimination, suggest the existence of theoretical thoughts, sensations, and feelings that transcend human mental abilities. By creating new mental processes Wave Theory brings to human understanding thoughts we cannot think and emotions we cannot feel.

APPENDICES

Appendix A
Accumulation of Wave Differences

Martingales and random walks play an important part in determining features of the stochastic process of accumulating the amplitudes of stationary signals. Treating the process of accumulation as a Martingale process places the psychophysical theory in a well-developed mathematical workshop. Martingales possess a simple property characteristic of many natural phenomena. In gambling, for example, imagine that the value of winnings after game play N is V_N. If, on the next bet, the average value still equals V_N, and if this is true for any play of the game, then the sequential process generating these values is called a "Martingale." Thus, if you have \$10 and you place a bet with the assurance that your expected winnings will remain \$10, the game is a Martingale[1].

Figure A.1 illustrates these ideas. The accumulated value of winnings is shown by \$ symbols connected by a line to make the flow of the game easier to read. At time n the value is about \$18. At time n + 1 the accumulated value is expected to stay at the same value. Of course, the actual value after bet n + 1 will most likely deviate from the average.

In the context of signal analysis let a signal's value at time N be V_N and at the next moment in time the expected value of the signal still be V_N, that is $E\{V_{N+1} \mid V_N\} = V_N$. If this property is true for all N then the stochastic process characterized by V_N is said to be a Martingale. Because V_N and V_{N+1} may be dependent random variables, Martingale theory provides especially useful tools for the study of both sequential processes and dependent random variables.

The powerful Martingale theorems also apply to sub-Martingales, those pro-

[1]This usage differs from the original "Martingale" gamble, sometimes called the St. Petersberg game where, after each loss, the stakes are doubled.

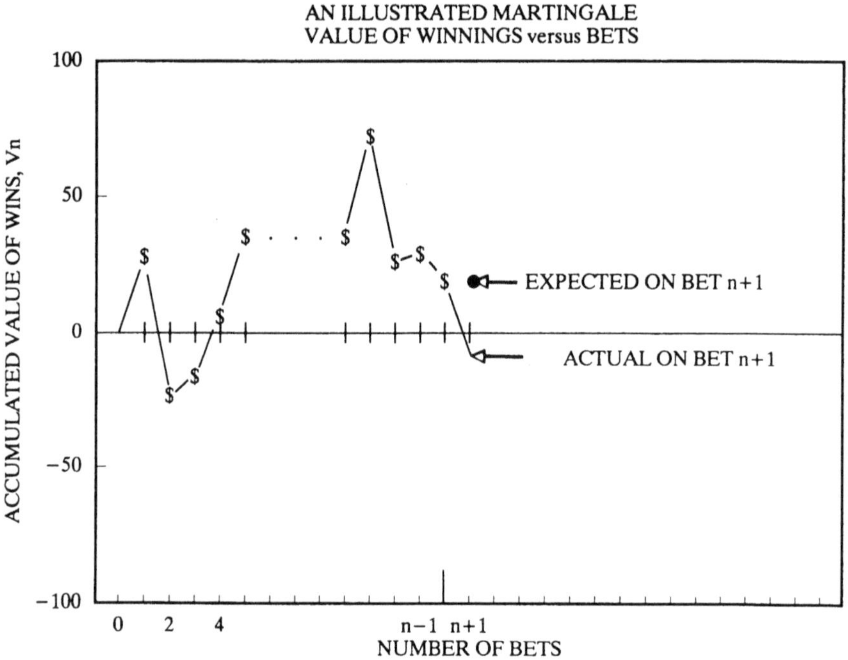

FIG. A.1. A Martingale is defined by expected values. The expected value at time N + 1 must equal the actual value at time N.

cesses for which $E\{V_{N+1} \mid V_N\} = V_N + v$, where v is a constant. A gambling game where the winnings continually increase by an average amount on each play equal to v forms a sequence of winnings that is a sub-Martingale. The sub-Martingale is of major interest in the analysis of signals and is directly applicable to Wave Theory. Rather than repeat Martingale derivations that are presented more rigorously elsewhere, only some properties of Martingale theory are used below. For those interested in the conclusions regarding response probability and response time the results are applied to data in chap. 10.

A convenient method of analyzing this judgment process is to analyze the expected value of the accumulated wave differences. For those wishing to continue with the exposition, consider $E\{\boldsymbol{W}_A(\Delta t)\}$ to be the expected wave value of stimulus $\boldsymbol{S}_A$ during time epoch Δt. Let $E\{\boldsymbol{W}_R(\Delta t)\}$ be the expected wave value for a referent stimulus during time epoch Δt. The expected difference between these waves is $E\{\boldsymbol{D}_A(\Delta t)\} = E\{\boldsymbol{W}_A(\Delta t) - \boldsymbol{W}_R(\Delta t)\}$. The judgment process continues the accumulation of wave differences until one or the other of the two thresholds is exceeded. Because the sum of differences is fixed, the random variables entering into the sum are dependent, leading to some difficulties in analysis. The analysis of this process of summation is greatly facilitated by the use of Martingale Theory.

To derive properties of the judgment process let sequential samples of W_{Ai} and W_{Ri} occur during epochs 1, 2, 3, . . . , i , ... The $D_{Ai} = W_{Ai} - W_{Ri}$ form the stochastic sum, $D_{A1} + D_{A2} + \ldots + D_{AN} = V_N$, with the property that

$$E\{V_{N+1} \mid V_N\} \geq V_N \ (N \geq 0) \tag{A.1}$$

when the expected difference between the two waves is greater than or equal to zero. That is, the ***expected value*** of the accumulation at time N + 1 will be greater than or equal to the ***actual value*** at time N. Clearly, the ***actual value*** V_{N+1} may, or may not, exceed the value at time N. This process of accumulating differences satisfies the requirements for a Martingale or sub-Martingale depending upon whether the expected value at time N + 1 equals or exceeds V_N.

Two important consequences of Martingale theory, known as the Wald Identities (first proved by using the methods of Wald, 1947) use the notion of a stopping time, a time when the accumulation stops. Let $D_{A1}, D_{A2}, \ldots, D_{Ai}$ be independent identically distributed random variables with expected values $E\{D_A\} = \mu < \infty$ and N be a stopping time with $E\{N\} < \infty$. Then, (Shiryayev, 1984)

$$E\{D_{A1} + D_{A2} + \ldots + D_{AN}\} = E\{D_A\}\, E\{N\}. \tag{A.2}$$

Moreover, if the variances of the D_A are also finite then,

$$E\{(D_{A1} + D_{A2} + \ldots + D_{AN}) - N \times E\{D_A\}\}^2 = Var\{D_A\}\, E\{N\}. \tag{A.3}$$

Result A.2 shows the expected value of the stopped Martingale; its average terminal value equals the expected difference between W_A and W_R times the expected number of differences accumulated in order for the accumulation to stop. In a sense, the average distance travelled on the axis of Informative Value equals the average step size per unit time multiplied by the average number of samples. Alternatively, if the average value when the process stops is known, then dividing by the average difference yields the average time when the accumulation stops. Or, dividing the average stopped value by the average time yields a measure of the average difference. Result A.3 ensures that a value of the average stopped process exists even when the expected difference is zero. In this case the variance of the differences plays a critical role.

With respect to sensory processes these two simple results show that when there are two threshold values, such as A and $-A$ in Fig. 9.4, then the average time to exceed a threshold, either threshold, is determined by the probabilities of exceeding either one, and the expected value or variance of the differences being accumulated. One of the most famous sequential sampling theorems determines the probability of exceeding these thresholds.

WALD'S FUNDAMENTAL IDENTITY

Let $\boldsymbol{D}_{A1}, \boldsymbol{D}_{A2}, \ldots, \boldsymbol{D}_{AN}$ be a sequence of independent identically distributed random variables summed to form $\boldsymbol{V}_N = \boldsymbol{D}_{A1} + \boldsymbol{D}_{A2} + \ldots + \boldsymbol{D}_{AN}$, $(N \geq 1)$. In the present case $\boldsymbol{V}_N$ is a sequence of summed differences between the standard and referent waves, that is, a sequence of summed amplitude differences.

The analysis of such sequential processes is simplified by using the moment-generating function (mgf) of a random variable. This function, like a Martingale, is defined in terms of expected values. If $\boldsymbol{X}$ is a random variable satisfying properties typical of the real world, then its moment-generating function is the expected value of $e^{-\theta X}$. This is a function, M, of the real variable θ. Thus, if $\boldsymbol{X}$ is a binomial random variable for which the value $-\boldsymbol{1}$ occurs with probability $1 - p$ and the value $\boldsymbol{1}$ occurs with probability p, the moment-generating function $M(\theta)$ is the expected value of $e^{-\theta X}$,

$$E\{e^{-\theta X}\} = (1 - p)e^{-\theta(-1)} + pe^{-\theta(1)}. \tag{A.4}$$

When plotted as a function of θ the moment-generating function appears as a convex function. Its properties aid in derivations of important predictions (Cox & Miller, 1965). Two of these properties are illustrated in Fig. A.2. First, $E\{e^{-\theta X}\} = 1$ is an equation with two roots. One is clearly $\theta = 0$, the other occurs at the value $\theta^* = ln[p/(1 - p)]$. Second, the slope of the line tangent to the mgf at $\theta = 0$ equals $-E\{X\}$. As $E\{X\}$ approaches zero the line tangent to the mgf at $\theta = 0$ must become horizontal because the slope of this line measures $-E\{X\}$, which is zero. Then $E\{X\} = 0$ and the convex properties of this mgf insure this is the only value of θ for which $M(\theta) = 1$.

Let the moment-generating function of $\boldsymbol{D}_{Ai}$ be $M(\theta) = E\{e^{-\theta D_{Ai}}\}$ and for some $\theta^* \neq 0$, suppose the moment-generating function exists and $M(\theta^*) = 1$. Random variables having both positive and negative values produce this shape. If N is a stopping time such that $|V_N| \leq$ a constant, and $E\{N\} < \infty$, then

$$E\left\{\frac{e^{-\theta^* V_N}}{M(\theta^*)^N}\right\} = 1. \tag{A.5}$$

To show how this truly remarkable identity generates the probabilities of reaching a response threshold, note that there exists a value of θ such that $M(\theta^*) = 1$. For this value of θ^* the denominator then equals 1 and the identity reduces to

$$E\{e^{-\theta^* V_N}\} = 1. \tag{A.6}$$

Expanding this expected value in terms of the response-threshold values and (unknown) probabilities of reaching these thresholds produces a solution to determining the unknown probabilities of reaching A or $-A$. As shown in chap. 10, this method determines the probabilities of responding *larger* and *smaller*. Once

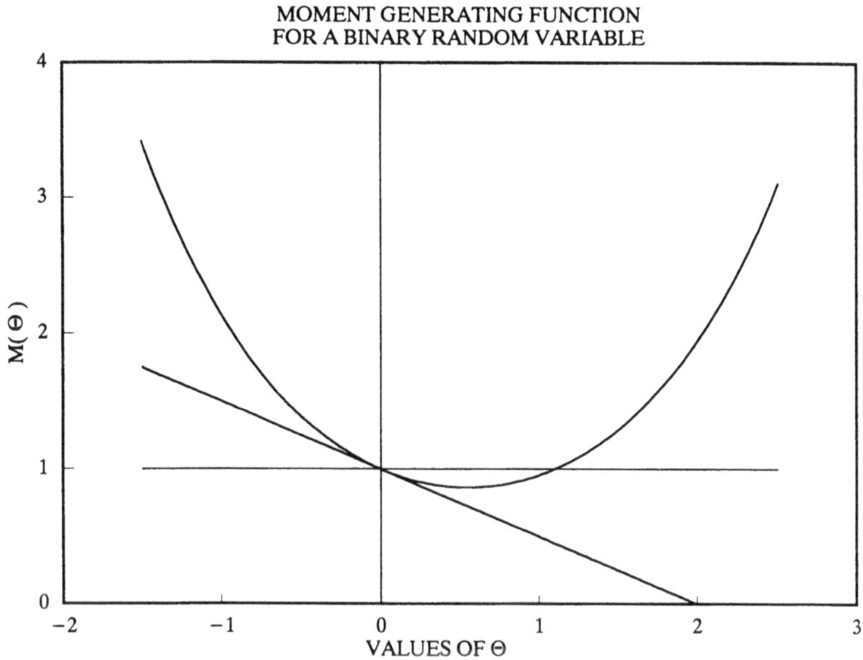

Fig. A.2. The moment-generating function is an expected value that changes with θ. Shown is the mgf for a binary random variable assuming values of -1 and $+1$ with probabilities $p = 0.25$ and $1 - p = 0.75$, respectively.

these values are determined Equations A.2 or A.3 determine the corresponding decision times. Because both response probabilities and decision times are the result of a single process, the theory predicts the existence of lawful relations between response probability and decision time. These relations are derived in chap. 10 and illustrated in the remaining chapters by applications to various psychological experiments.

Appendix B
Origins of the Logistic Response Function

Peripheral transceivers exist to receive peripheral stimulation, different transceivers listening to narrow energy bands. Within an energy band, the transceivers are considered theoretically similar when their Poisson intensities are equal. For these transceivers the difference between standard and comparison stimuli depends, then, only on the number of transceivers stimulated, as in von Békésy's (1930) Neural Quantum Theory. This Poisson basis for quantum stimulation provides an alternative to the Gaussian view of an otherwise chaste nature defiled by error.

Several features of the Poisson-quantum process are observable in the probability of responding *larger*, which in general equals

$$P_A = \frac{1}{1 + e^{-\theta^* A}}. \tag{B.1}$$

This is the often cited form for the Logistic Response Function in either the variable θ^* or A. Under the more specific Poisson assumptions (see chap. 12) Equation B.1 also yields,

$$P_A = \frac{1}{1 + e^{-[ln(L\alpha/M\beta)]A}}. \tag{B.2}$$

When the transceivers are similar with equal intensity parameters Equation B.1 simplifies to,

$$P_A = \frac{1}{1 + e^{-[ln(L/M)]A}}. \tag{B.3}$$

In the Method of Constant Stimuli the standard, with value M, remains constant whereas the comparison stimulus, with value L, varies. Changes in

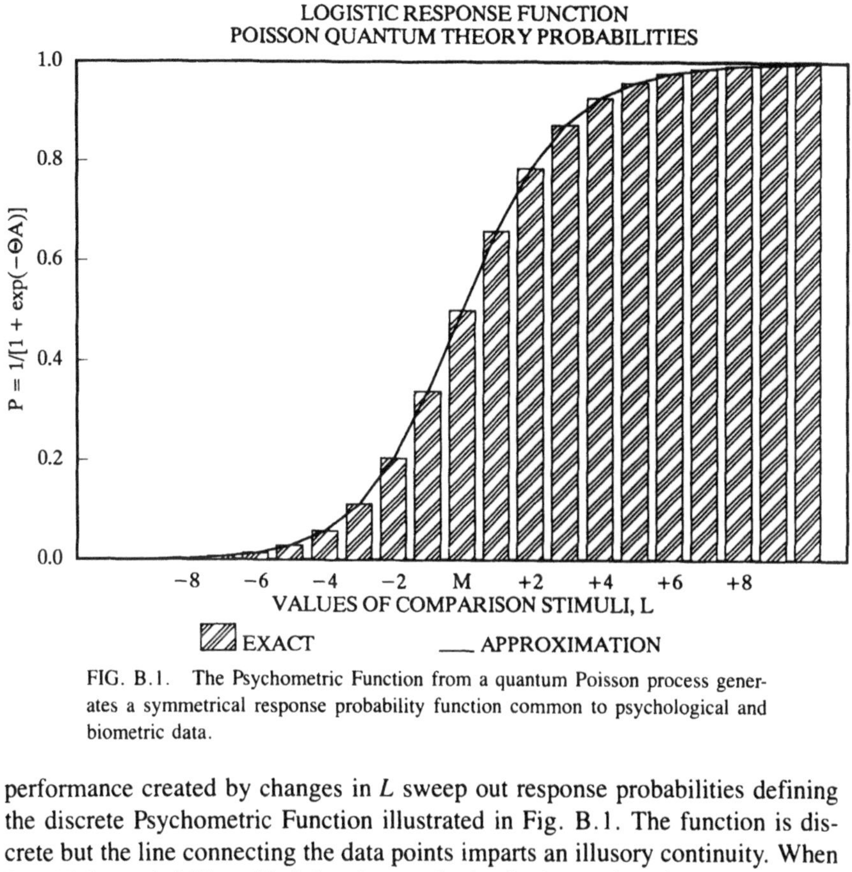

FIG. B.1. The Psychometric Function from a quantum Poisson process generates a symmetrical response probability function common to psychological and biometric data.

performance created by changes in L sweep out response probabilities defining the discrete Psychometric Function illustrated in Fig. B.1. The function is discrete but the line connecting the data points imparts an illusory continuity. When $L = M$ the probability of judging the standard to be larger than the comparison is 0.50. Approaching this value on the left, the function increases, like the Psychometric Function of Gaussian theories. As stimulus intensity recruits more transceivers into the Poisson process the number of potential transceivers increases and the value of P_A appears to increase in the manner suggested by Gaussian error theories.

Thurstone (1928b) in a classic study of the Phi-gamma Hypothesis argued that

> The phi-gamma hypothesis states that the psychometric curve is the function described by the integral of the normal probability curve. . . . The phi-gamma curve $\phi(\gamma)$ is symmetrical about its origin in the sense that the upper and lower quartile points of the curve are equidistant from the mean. The phi-gamma hypothesis simply states that the psychometric curve can be fitted by the integral of the normal probability curve. (pp. 296–297)

Thurstone was at pains to prove that in fact the psychometric curve (function) should not be symmetric at all because data from brightness discrimination by rats appeared to show a positively skewed Psychometric Function. That is, if the probability of responding *smaller* given a stimulus difference of $-\Delta s$ equals P_S and the probability of responding *larger* to the symmetric stimulus difference $+\Delta s$ equals P_L then $P_L < P_S$. To prove a lack of symmetry Thurstone assumed all the usual properties of Gaussian distributions as the basis of the Law of Comparative Judgment. Then, professed Thurstone (1928b):

> My first theorem merely asserts that if the absolute limen (threshold) increases with the stimulus, then the psychometric curve for two categories of judgment is positively skewed. (p. 297)

The reason for a lack of symmetry in the Psychometric Function arises from the idea that as stimulus magnitude increases and the variance of the comparison stimulus increases the variance of the Gaussian distributed difference between the standard and comparison stimuli must increase. For two comparisons equally but oppositely spaced from the standard, the variance for the smaller comparison stimulus is smaller. Thus the discrimination should be more accurate. The Psychometric Function should be skewed with sharper discrimination for values of the comparison smaller than the standard.

The response proportions from the Method of Constant Stimuli create a Psychometric Function with the abscissa scaled in units of stimulus difference. The standard $\mathbf{S}_B$ is constant and the comparison $\mathbf{S}_A$ varies to produce differences of the form,

$$\mathbf{S}_A - \mathbf{S}_B = \Delta s.$$

Response proportions or estimates of theoretical parameters are plotted as a function of Δs. From Equation 12.13, θ^* is a function of Δs,

$$\theta^* \cong \Delta s(1/\mathbf{S}_B). \tag{B.4}$$

For a fixed standard the value of the standard may be used as a unit of measurement by setting it equal to 1.0, then $\theta^* \cong \Delta s'$. Substituting into the *larger* response probability gives,

$$P_A = \frac{1}{1 + e^{-\Delta s' A}}. \tag{B.5}$$

A remarkable property of this probability is its easy conversion to values of $\theta^* A$, similar to Berkson's (1944) "logits." (These are not "logits" only because the probabilities involved do not derive from the logistic probability distribution.) The unknown value of $\theta^* A$ is obtained by computing

$$\begin{aligned} ln[P_A/(1 - P_A)] &= \Delta s' A \\ &= \theta^* A. \end{aligned} \tag{B.6}$$

A graph of Equation B.6 should produce a linear function with slope equal to the unknown value A.

Values of θ^*A estimated from the experiment by Peirce and Jastrow (1884, see chap.2, this volume) are shown in Fig. 10.2. Quite clearly the data conform well to the hypothesis that the physical stimuli produce linear increases in the values of the discrimination parameter θ^*. What remains unexplained is the Constant Error so often discovered in studies of the Psychometric Function. Quite clearly, the value of θ^* reaches zero before Δs reaches zero.

Kellogg (1931) generated a set of four increasing luminances ($\mathbf{S}_1$, $\mathbf{S}_2$, $\mathbf{S}_3$, $\mathbf{S}_4$) that were "felt" to give equal increments in brightness. The luminances appeared side-by-side on a split disk. Seven pairs of brightness differences formed a set of paired comparisons. The probabilities of choosing the brighter stimulus yielded values of θ^*A shown in Fig. 10.5 as a function of the number of units of brightness difference between the left- and right-hand sides of the disk (-3, -2, -1, 0, 1, 2, and 3). Again the values of θ^*A are linearly related to brightness differences. Again θ^* equals zero at a value of Δs unequal to zero.

Except for the obvious Constant Error (chap. 11 contains the relevant discussion) these results support the view that changes in stimulus intensity, selected by the experimenter for their linear effect, are responded to by the subject as if the discrimination parameter increased linearly. Thus the response probabilities plotted as a function of stimulus magnitude agree with those of a logistic distribution.

GAUSSIAN COMPARISONS

These results are not unusual. Statisticians (e.g., Berkson, 1944) claim that the logistic distribution fits Psychometric and Biometric Functions better than the Gaussian distribution. Figure B.2 shows two Psychometric Functions, one from a Gaussian distribution with mean zero and variance one, the other is a Logistic Response Function matched to the mean and variance of the Gaussian. Distinguishing between these two formulations is at best difficult, although there is an important distinction between them. The Gaussian distribution depends upon error to produce changes in probability, the logistic arises as a mathematical function, not a probability distribution function, relating response probability to stimulus difference.

In previous theories the response proportions from these experiments were treated as the consequence of Gaussian error. As the comparison stimulus deviated from the standard, the error played a less significant role in the judgment. **In Wave Theory, error is not a principle of discrimination.** Rather, the variability of a signal is an inherent part of the way signals are transmitted through the nervous system. The binary nature of transceivers, the Poisson representation of nervous signals, and the large number of transceivers available for activation produce a signal whose time average and power are messages about stimulus difference and intensity.

Why, then, does Gaussian theory work so well? One reason is the originality

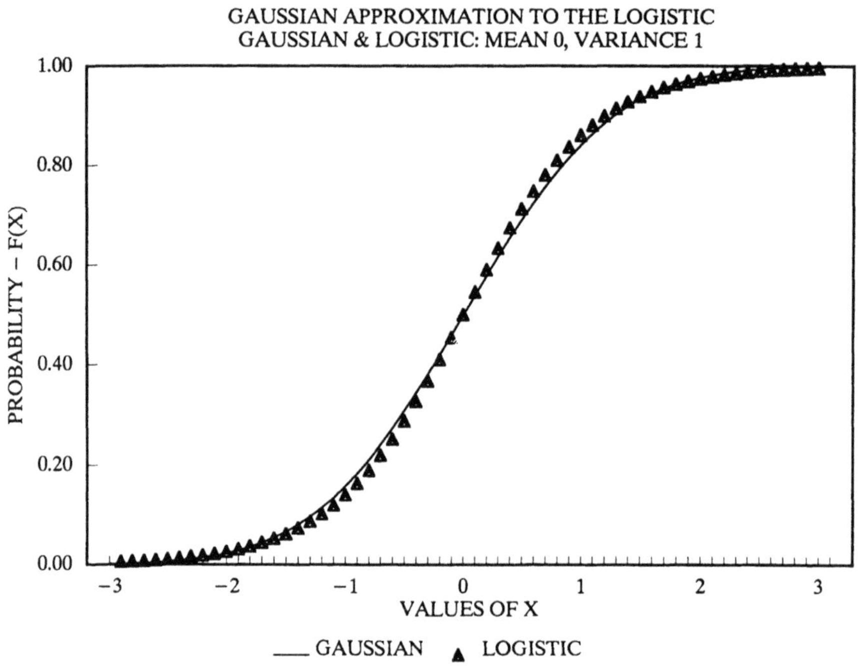

FIG. B.2. The cumulative Gaussian distribution advocated by 19th-century psychophysical theories closely matches the Logistic Response Function.

of Fechner's accomplishment. The theory blazed a trail into the wilderness of mental measurement and for many years the theory suggested exciting experimental and theoretical discoveries. It created a handy account of seemingly disparate results. These successes confirmed the value of Gaussian thinking and its application to psychophysical data.

A second reason is the good approximation of the Gaussian distribution to many probability distributions. The closeness of the approximation is shown in Fig. B.3. The continuous line forms the Gaussian density function. The logistic function is discrete because the data generating the function are from a necessarily finite number of transceivers. It is a mathematical function of Δs and not a probability density function. Yet, when there is no reason to suspect otherwise, the tendency to interpret this derivative as a probability density function is irresistible.

In an odd twist of fate the Logistic Response Function came to be considered an approximation to the Gaussian distribution. Bock and Jones (1968), for example, commented that:

> The logistic density is slightly leptokurtic to the normal curve, but the difference is so slight that the two functions are indistinguishable in practical work. The choice between them is largely a question of computational convenience, but even in this connection the difference is slight. . . . The logistic response function also has

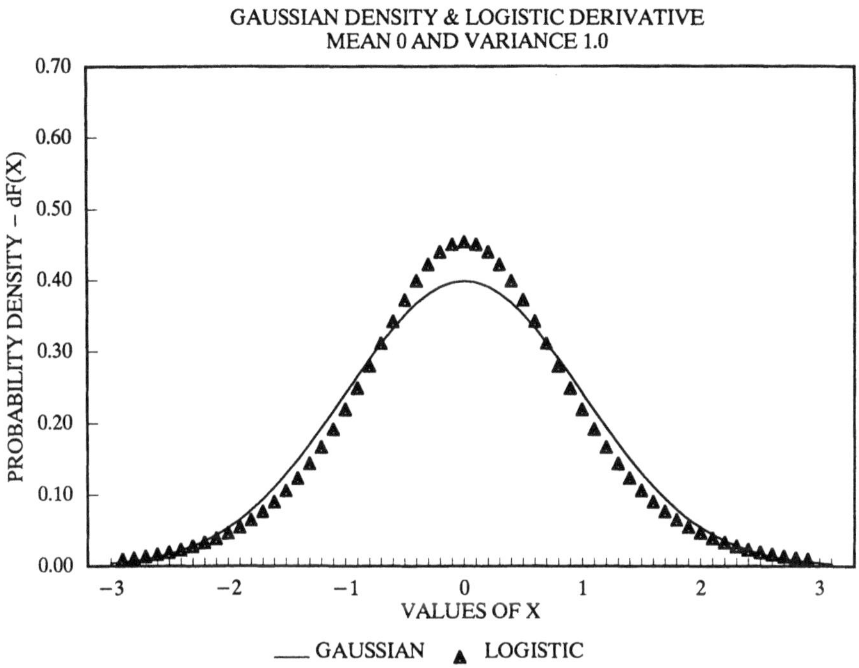

FIG. B.3. The Gaussian probability density function and the derivative of the logistic function have similar shapes. For most practical purposes they are indistinguishable.

interesting applications to the method of paired comparisons and prediction of choice. (p. 71)

The "logistic density" of Fig. B.3 is not a probability density function at all. This logistic function is only the derivative of a mathematical function that relates response probability to stimulus difference.

If the Gaussian is a good approximation to the logistic, then the logistic is an equally good approximation to the Gaussian. Instead of imagining stimulus difference in units of variability, the standard deviation, it is more to the point to treat experimental results as the consequence of stimulus magnitudes. In fact, the discriminability parameter θ^* equals the logarithm of the ratio of the comparison stimulus magnitude to the magnitude of the standard, $\theta^* = ln(\mathbf{S}_i/\mathbf{S})$. Experimental tradition sets $\theta^* = \Delta s/\mathbf{S}$ which is a very close approximation. The goodness of fit is excellent and the number of parameters needed to account for the data is reduced to only A.

On this basis the 19th-century view that error is the foundation for psychophysical performance can be replaced with the viewpoint that transceiver activation is the basis for psychophysical discrimination. Variability in neural firing may be as essential to the central representation of a stimulus as eye movements are to maintaining a visual image. Variability is not error.

The Poisson distribution characterizes both spatial and temporal dimensions. In space the Poisson characterizes the combined output of many transceivers fed to a common collector of sensory inputs. These collectors may in turn feed another step of collectors. The collected sensory input is still Poisson distributed and eventually transmitted to a new site for analysis. A remarkable property of the Poisson distribution is that the collected spatial inputs can be transmitted through a Poisson process with a pulse rate that reflects the collected input.

Thus a pattern of spatial transceivers can be gated to a neural structure responsible for the temporal production of pulses. The pulse rate maintains the Poisson parameter characterizing the intensity of the stimulation. The product of intensity and number of transceivers produces only a single parameter, thus some information about the source of stimulation is lost but the benefits of the transmission system are great. The digital properties of energy transfer from the world surrounding the sensory surface is maintained. At a terminal site the temporal sequence of pulses is decoded into a spatial representation of the original stimulus. Successive codings and decodings create a Poisson chain that preserves the sum of digital stimulation fed into a network of higher order sensory collectors.

Both theory and data support the view that stimulation of the body's sensory surface leads to a Poisson representation of the stimulus. The geometrical organization of such response units is not critical to this mathematical representation. Inputs from different sources, such as exterior body surfaces or internal regions of tissue, have a common mathematical expression that captures the major features of discrete events occurring within small units of space or volume or, as is equally possible in some sense modalities, events that occur over time within a small region of space.

Appendix C
Symmetric Stimuli

A fundamental relation between the probabilities of responding *larger* or *smaller* exists. The probabilities for negative values of θ are related to positive values of θ by a simple equation suggested by the symmetry of $L(\theta)$ and $S(\theta)$. Necessarily the probability of a *larger* response equals one minus the probability of a *smaller* response,

$$L(\theta) = 1 - S(\theta).$$

Simple algebra easily confirms the symmetry of these functions as shown in Fig. C.1. The probability of a *larger* response given θ equals the probability of a *smaller* response given $-\theta$. That is,

$$L(\theta) = S(-\theta). \qquad \text{(C.1)}$$

Replacing the *smaller* response probability by its equivalent in terms of *larger* response probabilities shows that the probability of a *larger* response for θ equals one minus the probability of a *smaller* response for $-\theta$,

$$L(\theta) = 1 - L(-\theta),$$

and,

$$S(\theta) = 1 - S(-\theta).$$

These relations play a particularly significant role in chaps. 11 and 16, which introduce the effects of response bias on response probability and response time. These relations are generated by stimuli with symmetric parameters, θ and $-\theta$. Such pairs of stimuli are referred to as *Symmetric Stimuli*.

In those experiments where two stimuli are presented directly for comparison

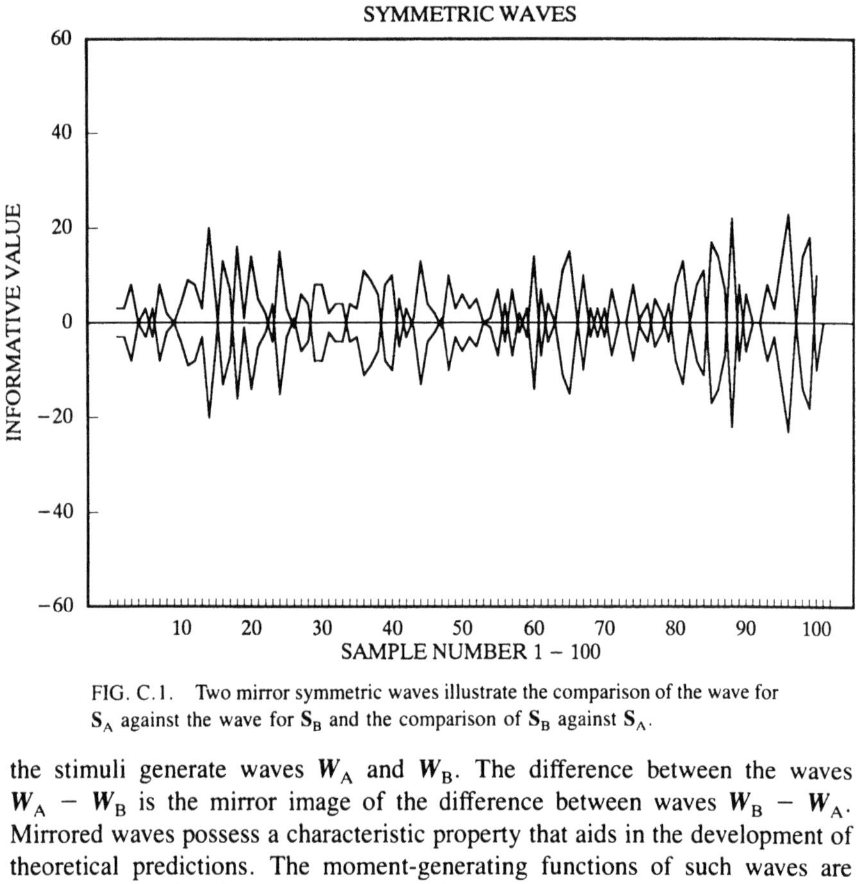

FIG. C.1. Two mirror symmetric waves illustrate the comparison of the wave for $\mathbf{S}_A$ against the wave for $\mathbf{S}_B$ and the comparison of $\mathbf{S}_B$ against $\mathbf{S}_A$.

the stimuli generate waves W_A and W_B. The difference between the waves $W_A - W_B$ is the mirror image of the difference between waves $W_B - W_A$. Mirrored waves possess a characteristic property that aids in the development of theoretical predictions. The moment-generating functions of such waves are mirror symmetric with respect to $\theta = 0$. For example, the mgf for the difference $W_A - W_B$ is $M_{WA-WB}(\theta)$. The reversed comparison, $W_B - W_A$ produces a mgf $M_{WB-WA}(\theta) = M_{WA-WB}(-\theta)$. This symmetry is a natural property of the two different orders for directly comparing two stimuli.

In experiments where only a single stimulus is presented during an experimental trial the standard is implicit but based upon the ensemble of stimuli experienced by the subject. Such experiments typify the mental comparisons of daily life, comparisons based on memories derived from previous experience. Often in these experiments only two stimuli are ever presented. The subject may adopt a referent based on previous experience, the geometry of the signals, or the experimenter's manipulation of the stimulus ensemble.

In this regard, symmetry proves to be a ubiquitous property of the mixture of waves. The trial-by-trial random presentation of two different stimuli provides the data for creating a mixture of stimulus waves that acts as a referent against which a comparison stimulus is judged. The mixture may depend upon the probabilities with which the components occur, but in the case of a binary mixture of equally probable stimuli, a case found often in studies of discrimination, symmetry exists regardless of the form of the stimulus waves. This symmetry provides a basis for many commonly observed results.

To demonstrate that the waves themselves need not be symmetric in order for the comparisons against a referent to be symmetric it is sufficient to show that the mgfs of the wave differences from the referent are themselves symmetric. Let $\mathbf{S}_A$ and $\mathbf{S}_B$ be the two stimuli that occur with probability p and $1 - p$, respectively. Let the waves characterizing these stimuli be $\boldsymbol{W}_A$ and $\boldsymbol{W}_B$, respectively. Let the mixture $p\mathbf{S}_A + (1 - p)\mathbf{S}_B$ define the referent $\mathbf{S}_R$ with wave,

$$\boldsymbol{W}_R = p\boldsymbol{W}_A + (1 - p)\boldsymbol{W}_B.$$

To prove that any two waves always produce symmetric differences when the waves are mixed with equal probability note that $\boldsymbol{W}_A - \boldsymbol{W}_R = (1 - p)(\boldsymbol{W}_A - \boldsymbol{W}_B)$. The moment-generating function $M_{AR}(\theta)$ of this difference equals $(1 - p)M_{WA-WB}(\theta)$. The other difference, $\boldsymbol{W}_B - \boldsymbol{W}_R$, has moment-generating function $M_{BR}(\theta) = pM_{WB-WA}(\theta)$. Because $\boldsymbol{W}_A - \boldsymbol{W}_B$ is symmetric with $\boldsymbol{W}_B - \boldsymbol{W}_A$, $M_{WA-WB}(\theta) = M_{WB-WA}(-\theta)$. Consequently,

$$\begin{aligned} M_{BR}(-\theta) &= pM_{WB-WA}(-\theta) \\ &= pM_{WA-WB}(\theta) \\ &= [p/(1 - p)]M_{AR}(\theta). \end{aligned}$$

When $p = 0.50$,

$$M_{BR}(-\theta) = M_{AR}(\theta).$$

This condition defines mirror symmetry with respect to $\theta = 0$. Most important, the stimulus waves for $\mathbf{S}_A$ and $\mathbf{S}_B$ need not be symmetric in order for the wave differences $\boldsymbol{W}_A - \boldsymbol{W}_R$ or $\boldsymbol{W}_B - \boldsymbol{W}_R$ to be symmetric.

Mirror symmetric waves correspond to physically realizable mental processes and lead easily to testable theoretical predictions. Also, the two stimuli undergoing comparison may be vastly different yet, if each is compared against a mental standard defined by an equal mixture of the two, the resulting difference waves are symmetric.

Symmetry leads to simple derivations concerning response probability and response time. For example, when $\mathbf{S}_A$ is compared against an internal referent, the value of θ^* satisfying Wald's Identity $M_{AR}(\theta^*) = 1$ is equal but opposite in sign to the value of θ^* satisfying Wald's Identity for the comparison of $\mathbf{S}_B$ against $\mathbf{S}_R$. That is,

$$M_{AR}(\theta^*) = M_{BR}(-\theta^*).$$

The waves appear to depend on the stimuli, whereas the referent, which is essential to the discrimination, is only a transparent partner. If the two stimuli are compared directly against each other the referent wave is the wave for the alternative stimulus. Then,

$$M_{AB}(\theta^*) = M_{BA}(-\theta^*),$$

so that symmetry is preserved.

For example, when the difference between $\boldsymbol{W}_A$ and $\boldsymbol{W}_B$ is characterized by a discriminability parameter θ then the difference between $\boldsymbol{W}_B$ and $\boldsymbol{W}_A$ is characterized by a discriminability parameter $-\theta$. The Poisson waves introduced in chap. 9 conveniently illustrate the point. When $\boldsymbol{W}_A$ and $\boldsymbol{W}_B$ are Poisson waves with parameters α and β respectively, then $\boldsymbol{W}_A - \boldsymbol{W}_B$ generates a value of $\theta^*_{AB} = ln(\alpha/\beta)$. The difference between $\boldsymbol{W}_B$ and $\boldsymbol{W}_A$ generates a value of $\theta^*_{BA} = ln(\beta/\alpha) = -ln\theta^*_{AB}$.

These various measures, $\boldsymbol{W}_A$, $\boldsymbol{W}_R$, and $\boldsymbol{D}_A = \boldsymbol{W}_A - \boldsymbol{W}_R$, are the outcomes of random processes. For stationary random processes the amplitude is easily considered as a value sampled from a sample space with an unknown probability distribution. Two examples of such probability distributions are shown in Fig. 12.2. For illustration these are Poisson distributions, such as those associated with the activation of sensory receptors (Hecht, Shlaer, & Pirenne, 1942; McGill, 1964). They differ in both mean and variance because they depend upon different sized sensory-transceiver areas.

The sum of Informative Values, itself an Informative Value, is the driving force toward a response. As is shown in chap. 11 if, after time epoch N, the average amplitude difference remains zero, then the probability that response R_A occurs equals $(A + \mathbf{V}_N)/2A$. Thus, when the forces acting to determine a response are equal, the accumulated difference becomes a disposition toward one response or the other, no longer propelled by a directional force.

References

Angell, F. (1901). Discrimination of clangs for different intervals of time. *American Journal of Psychology*, 12, 58–79.

Anscombe, F. J. (1956). On estimating binomial response relations. *Biometrika,* 43, 461–464.

Atkinson, R. C. (1963). A variable sensitivity theory of signal detection. *Psychological Review*, 70, 91–106.

Atkinson, R. C., & Kinchla, R. A. (1965). A learning model for forced-choice detection experiments. *British Journal of Mathematical and Statistical Psychology*, 18, Part 2, 183–206.

Atkinson, R. C., Carterette, E. C., & Kinchla, R. A. (1962). *Sequential phenomena in psychophysical judgments: A theoretical analysis*. (Tech. Rep. No. 46). Stanford: Stanford University, Institute for Mathematical Studies in the Social Sciences.

Atkinson, R. C., Carterette, E. C., & Kinchla, R. A. (1964). The effect of information feedback upon psychophysical judgments. *Psychonomic Science*, 1, 83–84.

Attneave, F. (1959). *Applications of information theory to psychology*. New York: Henry Holt & Company.

Ayer, A. J. (1968). *The origins of pragmatism*. London: Macmillian and Co.

Azzarello, M. A. (1985). *Numerical comparisons: An application of relative judgment theory*. Honours B.Sc. Thesis. McMaster University, Department of Psychology, Hamilton, Canada.

Beebe-Center, J. G. (1932). *The psychology of pleasantness and unpleasantness*. New York: Van Nostrand.

Berkson, J. (1944). Application of the logistic function to bio-assay. *Journal of the American Statistical Association*, 39, 357–365.

Bertelson, P. (1961). Sequential redundancy and speed in a serial two-choice responding task. *Quarterly Journal of Experimental Psychology*, 13, 90–102.

Bindra, D., Donderi, D. C., & Nishisato, S. (1968). Decision latency of "same"-"different" judgments. *Perception & Psychophysics*, 3, 121–130.

Blackwell, H. R. (1953). Psychophysical thresholds: Experimental studies of methods of measurement. *University of Michigan Engineering Research Institute Bulletin*, 36, Ann Arbor: University of Michigan Press.

Blackwell, H. R. (1963). Neural theories of simple visual discriminations. *Journal of the Optical Society of America*, 53(1), 129–160.

Bloxom, B. (1985). A constrained spline estimator of a hazard function. *Psychometrika*, 50(3), 301–321.

Bock, R. D., & Jones, L. V. (1968). *The measurement and prediction of judgment and choice*. San Francisco: Holden-Day.

Boring, E. G. (1950). *A history of experimental psychology*. (2nd ed.) New York: Appleton-Century-Crofts.

Brown, W. (1910). The judgment of difference with special reference to the doctrine of the threshold, in the case of lifted weights. *University of California Publications in Psychology*, 1(1), 1–71.

Brown, W. (1914). The judgment of very weak sensory stimuli with special reference to the absolute threshold of sensation for common salt. *University of California Publications in Psychology*, 1(3), 199–268.

Brunswik, E. (1956). *Perception and the representative design of psychological experiments*. Berkeley: University of California Press.

Bush, R. R. (1963). Estimation and Evaluation. In R. D. Luce, R. R. Bush, & E. Galanter (Eds.), *Handbook of mathematical psychology. Vol. 1*. (pp. 421–469). New York: Wiley.

Bush, R. R., & Mosteller, F. (1955). *Stochastic models for learning*. New York: Wiley.

Carlson, W. R., Driver, R. C., & Preston, M. G. (1934). Judgment times for the method of constant stimuli. *American Journal of Psychology*, 17, 113–118.

Carterette, E. C., Friedman, M. P., & Cosmides, R. (1965). Reaction-time distributions in the detection of weak signals in noise. *Journal of the Acoustical Society of America*, 38(4), 531–542.

Cartwright, D. (1941a). Decision-time in relation to the differentiation of the phenomenal field. *Psychological Review*, 48, 425–442.

Cartwright, D. (1941b). Relation of decision-time to the categories of response. *American Journal of Psychology*, 54, 174–196.

Cartwright, D., & Festinger L. (1943). A quantitative theory of decision. *Psychological Review*, 50, 596–621.

Castellan, N. J., Jr., & Restle, F. (1978). *Cognitive theory. Vol. 3*. Hillsdale: Lawrence Erlbaum Associates.

Cattell, J. M. (1886). The influence of the intensity of the stimulus on the length of the reaction time. *Brain*, 8, 512–515.

Cattell, J. M. (1887). Experiments on the association of ideas. *Mind*, 2, 69–74.

Cattell, J. M. (1893). On errors of observation. *American Journal of Psychology*, 5, 284–293.

Cattell, J. M. (1902). The time of perception as a measure of differences in intensity. *Philosophische Studien*, 19, 63–68.

Clarke, F. R. (1960). Confidence ratings, second-choice responses and confusion matrices in intelligibility tests. *Journal of the Acoustical Society of America*, 32, 35–46.

Cochran, W. G. (1976). Early development of techniques in comparative experimentation. In Owen, D. B. (Ed.) *On the history of probability and statistics* (pp. 1–25). New York: Marcel Dekker, Inc.

Cohn, J. (1894). Experimentelle untersuchungen über die gefuhlsbetonung der farben, helligkeiten und ihrer combinationen. *Philosophische Studien*, 10, 562.

Coombs, C. H. (1950). Psychological scaling without a unit of measurement. *Psychological Review*, 57, 145–158.

Coombs, C. H. (1954). A method for the study of interstimulus similarity. *Psychometrika*, 19, 183–194.

Coombs, C. H. (1958). On the use of inconsistency of preferences in psychological measurement. *Journal of Experimental Psychology*, 55, 1–7.

Coombs, C. H. (1960). A theory of data. *Psychological Review*, 67, 143–159.

Coombs, C. H. (1964). Some symmetries and dualities among measurement data matrices. In N. Frederiksen & H. Gulliksen (Eds.). *Contributions to mathematical psychology*. (pp. 41–59). New York: Holt, Rinehart & Winston.

Cooper, G. R., & McGillem, C. D. (1986). *Probabilistic methods of signal and system analysis*. New York: Holt, Rinehart & Winston.

Cornsweet, T. N. (1971). *Visual perception*. New York: Academic Press.

Cox, D. R., & Miller, H. D. (1965). *The Theory of Stochastic Processes*. London: Methuen.

Crossman, E. R. F. W. (1953). Entropy and choice time: the effect of frequency unbalance on choice-response. *Quarterly Journal of Experimental Psychology*, 5(2), 41–51.

Crossman, E. R. F. W. (1956). The measurement of discriminability. *Quarterly Journal of Experimental Psychology*, 7, 176–195.

Culler, E. (1926a). Studies in psychometric theory. *Psychological Monographs*, 35, 56–137.

Culler, E. (1926b). Studies in Psychometric Theory. XIII. The normal law: Its place in quantitative psychology. *Journal of Experimental Psychology*, 9(4), 271–298.

Culler, E. (1927). The accuracy of the method of constant stimuli: a reply to Dr. Urban. *American Journal of Psychology*, 38, 307–312.

Dashiell, J. F. (1937). Affective value-distances as a determinant of esthetic judgment-times. *American Journal of Psychology*, 50, 57–67.

David, H. A. (1969). *The method of paired comparisons*. London: Charles Griffin & Co.

deJaager, J. J. (1970). Reaction time and mental processes. In J. Brozek, & M. S. Sibinga, (Eds. and Trans.). *Origins of psychometry: Johan Jacob Jaager*. Nieuwkoop: De Graaf. (Original work published 1865)

Doob, J. L. (1940). Regular properties of certain families of chance variables. *Transactions of the American Mathematical Society*, 47, 455–485.

Donders, F. C. (1868). Die schnelligkeit psychischer processe. *Archiv für Anatomie und Physiologie und Wissenschaftlich Medizin*, 657–681.

Egan, J. P. (1976). *ROC analysis*. New York: Academic Press.

Egan, J. P., & Clarke, F. R. (1956). Source and receiver behavior in the use of a criterion. *Journal of the Acoustical Society of America*, 28, 1267–1269.

Emerson, P. L. (1970). Simple reaction time with Markovian evolution of Gaussian discriminal processes. *Psychometrika*, 35(1), 99–109.

Estes, W. K. (1950). Toward a statistical theory of learning. *Psychological Review*, 57, 94–107.

Estes, W. K. (1959). A random walk model for choice behavior. In K. J. Arrow, S. Karlin, & P. Suppes (Eds.), *Mathematical methods in the social sciences, 1959*. Stanford: Stanford University Press.

Falmagne, J. C. (1961). *Un modèle linéaire pour les T.R. de choix. Application a des resultats experimentaux*. Université de Bruxelles. Unpublished manuscript.

Falmagne, J. C. (1965). Stochastic models for choice reaction time with applications to experimental results. *Journal of Mathematical Psychology*, 2, 77–124.

Falmagne, J. C. (1968). Note on a simple fixed-point property of binary mixtures. *British Journal of Mathematical and Statistical Psychology*, 21(1), 131–132.

Falmagne, J. C., Cohen, S. P., & Dividi, A. (1975). Two-choice reactions as an ordered memory scanning process. In P. M. A. Rabbitt & S. Dornic (Eds.), *Attention and performance V*. New York: Academic Press.

Falmagne, J. C., & Iverson, G. J. (1979). Conjoint weber's law and additivity. *Journal of Mathematical Psychology*, 20(2), 164–183.

Fechner, G. T. (1860). *Elements of psychophysics*. Leipzig: Breitkopf & Härtel.

Fechner, G. T. (1966). *Elements of psychophysics. Vol. 1*. (E. G. Boring & D. H. Howes, Eds.; H. E. Adler, Trans.). New York: Holt, Rinehart & Winston. (Original work published 1860)

Fechner, G. T. (1871). *Zur experimentellen aesthetik*. Leipzig: Hirzel.

Fechner, G. T. (1876). *Vorschule der aesthetik. Vols. 1 & 2*. Leipzig: Breitkopf & Härtel. (reprinted 1925)

Feller, W. (1939). On the logistic law of growth and its empirical verifications in biology. *Acta Biotheoretica*, 5, 51–66.

Feller, W. (1957). *An introduction to probability theory and its applications*. (2nd ed.). New York: Wiley.

Feller, W. (1966). *An introduction to probability theory and its applications. Vol. II*. New York: Wiley.

Feller, W. (1968). *An introduction to probability theory and its applications. Vol. 1*. (3rd ed.). New York: Wiley.

Fernberger, S. W. (1913). On the relation of the methods of just perceptible differences and constant stimuli. *Psychological Monographs*, 19(4), 1–81.

Fernberger, S. W. (1914). The effect of the attitude of the subject upon the measure of sensitivity. *American Journal of Psychology*, 25, 538–543.

Fernberger, S. W. (1918). An analysis of the process of comparing. *Psychological Monographs*, 26, 1–161.

Fernberger, S. W. (1920). Interdependence of judgments within the series for the method of constant stimuli. *Journal of Experimental Psychology*, 3, 126–150.

Fernberger, S. W. (1930). The use of equality judgments in psychophysical procedures. *Psychological Review*, 37, 107–112.

Fernberger, S. W. (1931). On absolute and relative judgments in lifted weight experiments. *American Journal of Psychology*, 43, 560–578.

Fernberger, S. W., & Irwin, F. W. (1932). Time relations for the different categories of judgment in the absolute method in psychophysics. *American Journal of Psychology*, 44, 505–525.

Fernberger, S. W., Glass, I., Hoffman, I., & Willig, M. (1934). Judgment times of different psychophysical categories. *Journal of Experimental Psychology*, 17, 286–293.

Festinger, L. (1943). Studies in decision: I. Decision-time, relative frequency of judgment and subjective confidence as related to physical stimulus difference. *Psychological Review*, 50(3), 291–306.

Festinger, L. (1947). The treatment of qualitative data by "scale analysis." *Psychological Bulletin*, 44, 149–161.

Fullerton, G. S., & Cattell, J. M. (1892). *On the perception of small differences: with special reference to the extent, force and time of movement*. Publications of the University of Pennsylvania. Philosophical Series, 2. Philadelphia.

Garbas, C. A. (1985). *The ideal age of marriage*. B.A. Honours Thesis. McMaster University, Department of Psychology, Hamilton, Canada.

Garner, W. R. (1954). An informational analysis of absolute judgments of loudness. *Journal of Experimental Psychology*, 46, 373–380.

Garrett, H. E. (1922). A study of the relation of accuracy to speed. *Archives of Psychology*, 56, 1–104.

Gauss, C. F. (1880). Theoria combinationis observationum erroribus minimis obnoxiae, pars prior. In *Carl Friedrich Gauss Werke. Vol. 4*. (pp. 1–26). Göttingen: Königlichen Gesellschaft der Wissenschaften zu Göttingen. (Original work published 1821)

Geissler, H. G. (1990). Foundations of quantized processing. In H. G. Geissler, (Ed.), *Psychophysical explorations of mental structures*. (pp. 193–210). Toronto: Hogrefe & Huber.

George, S. S. (1917). Attitude in relation to the psychophysical judgment. *American Journal of Psychology*, 28, 1–37.

Gikham, I. I., & Skorokhod, A. V. (1969). *Introduction to the theory of random processes*. Philadelphia: Saunders.

Ghosh, B. K. (1970). *Sequential tests of statistical hypotheses*. Reading MA: Addison-Wesley.

Green, D. M. (1960). Psychoacoustics and detection theory. *Journal of the Acoustical Society of America*, 32, 1189–1203.

Green, D. M. (1961). Detection of auditory signals of uncertain frequency. *Journal of the Acoustical Society of America*, 33, 897–903.

Green, D. M., & Swets, J. A. (1966). *Signal detection theory and psychophysics*. New York: Wiley.

Guilford, J. P. (1928). The method of paired comparisons as a psychometric method. *Psychological Review*, 35, 494–506.
Guilford, J. P. (1931). Some empirical tests of the method of paired comparisons. *Journal of General Psychology*, 5, 64–77.
Guilford, J. P. (1936). *Psychometric methods*. New York: McGraw-Hill.
Guilford, J. P. (1954). *Psychometric methods*. (2nd ed.), New York: McGraw-Hill.
Guttmann, L. (1947). On Festinger's evaluation of scale analysis. *Psychological Bulletin*, 44, 451–465.
Guttmann, N. (1965). Effects of discrimination formation on generalization measured from a positive-rate baseline. In D. I. Mostovsky (Ed.), *Stimulus generalization*. (pp. 210–217). Stanford: Stanford University Press.
Hanson, H. M. (1959). Effects of discrimination training on stimulus generalization. *Journal of Experimental Psychology*, 58, 321–334.
Hecht, S., Shlaer, S., & Pirenne, M. H. (1942). Energy, quanta, and vision. *Journal of General Physiology*, 25, 819–846.
Helmholtz, H. (1950). Über die methoden, kleinsts zeittheile zu messen, und ihre anwendung für physiologische zwecke. Trans. On the methods of measuring very small portions of time, and their application to physiological processes. *Philosophical Magazine*, 1853, Series 4, 6, 313–325.
Helson, H. (1959). Adaptation level theory. In S. Koch (Ed.), *Psychology: A study of a science, Vol. 1*. (pp. 565–621). New York: McGraw-Hill.
Helson, H. (1964). *Adaptation-level theory*. New York: Harper and Row.
Henmon, V. A. C. (1906). The time of perception as a measure of differences in sensation. *Archives Philosophy Psychology and Scientific Method*, 8, 5–75.
Henmon, V. A. C. (1911). The relation of the time of a judgment to its accuracy. *Psychological Review*, 18, 186–201.
Herbart, J. F. (1850). Psychologie als wissenschaft, neu gegründet auf erfahrung, metaphysik, und mathematik. In G. Hartenstein (Ed.) *John Friedrich Herbart's Sämmtliche Werke. Vol. 5*. Leipzig: Leopold Voss. (Original published 1824)
Hick, W. E. (1952). On the rate of gain of information. *Quarterly Journal of Experimental Psychology*, 4, 11–26.
Hinrichs, J. V., & Krainz, P. L. (1970). Expectancy in choice reaction time: anticipation of stimulus or response? *Journal of Experimental Psychology*, 85, 330–334.
Hollingworth, H. L. (1939). Psycho-dynamics of chewing. *Archives of Psychology*, 34, No. 239, 1–90.
Hunt, W. A., & Volkmann, J. (1937). The anchoring of an affective scale. *American Journal of Psychology*, 49, 88–92.
Hurvich, L. M., & Jameson, D. (1989). Leo M. Hurvich and Dorothea Jameson. In G. Lindzey (Ed.), *A history of psychology in autobiography VIII*, (pp. 157–206). Stanford: Stanford University Press.
Hyman, R. (1953). Stimulus information as a determinant of reaction time. *Journal of Experimental Psychology*, 45, 188–196.
James, W. (1893). *Psychology*. New York: Henry Holt and Co.
James, W. (1890). *The principles of psychology. Vol. 1*. New York: Henry Holt & Co.
Jastrow, J. (1888). A critique of psycho-physic methods. *American Journal of Psychology*, 6, 271–309.
Johnson, D. M. (1939). Confidence and speed in the two-category judgment. *Archives of Psychology*, No. 241, 1–52.
Kahn, D. (1967). *The code breakers*. New York: Macmillan.
Kellogg, W. N. (1931). The time of judgment in psychometric measures. *American Journal of Psychology*, 43, 65–86.
Kinchla, R. A. (1966). *A comparison of sequential effects in detection and recognition*. (Experi-

mental Psychology Series, Tech. Rep. No. 1). New York: New York University, Department of Psychology.

Kirby, N. (1980). Sequential effects in choice reaction time. In A. T. Welford (Ed.). *Reaction times*. (pp. 129–172). New York: Academic Press

Kolmogoroff, A. (1933). Grundbegriffe der wahrscheinlichkeits-rechnung. *Ergebnisse der Mathematik und Ihrer Grenzgebiete*. Herausgegeben von der Schriftleitung des Zentralblatt für Mathematik. Bd. 2, H. 3. Berlin: Julius Springer.

Kowalewski, A. (1904). Studien zur psychologie des pessimismus. *Grenzfragen des Nerven und Seelenlebens*. Heft 24.

Laberge, D. L. (1959). A model with neutral elements. In R. R. Bush and W. K. Estes (Eds.), *Studies in mathematical learning theory*. (pp. 53–64). Stanford: Stanford University Press.

Laming, D. (1986). *Sensory analysis*. London: Academic Press.

Levy, P. (1937). *Theorie de l'addition des variables aleatoires*. Paris: Gauthier-Villars.

Link, S. W. (1971). Applying RT deadlines to discrimination reaction time. *Psychonomic Science*, 25(6), 355–358.

Link, S. W. (1975). The relative judgment theory of two choice response time. *Journal of Mathematical Psychology*, 12, 114–135.

Link, S. W. (1978a). The relative judgment theory of the psychometric function. In J. Requin (Ed.), *Attention & performance VII*. (pp. 619–630). Hillsdale, NJ: Lawrence Erlbaum Associates.

Link, S. W. (1978b). The relative judgment theory of response time deadline experiments. In N. J. Castellan, Jr. & F. Restle (Eds.), *Cognitive theory. Vol. 3*. Hillsdale, NJ: Lawrence Erlbaum Associates.

Link, S. W. (1980). Crime vs. punishment. Paper presented at the 41st Annual Convention of the Canadian Psychological Association. Calgary, Alberta.

Link, S. W. (1982). Correcting response measures for guessing and partial information. *Psychological Bulletin*, 92 (2), 469–486.

Link, S. W. (1985, November). *How numerical comparisons occur*. Paper presented at the 26th Annual Meeting of the Psychonomic Society, Boston, MA.

Link, S. W. (1990a). Modeling imageless thought: the relative judgment theory of numerical comparisons. *Journal of Mathematical Psychology*, 34, 2–41.

Link, S. W. (1990b). The psychological design of psychological experiments. In H. G. Geissler, (Ed.), Psychophysical explorations of mental structures. Toronto: Hogrefe and Huber.

Link, S. W. (in press) Imitatio Estes. Stimulus sampling origins of Weber's law. In Healy, A., Kosslyn, S., & Shiffrin, S. (Eds.), *From learning theory to connectionist theory: Essays in honor of W. K Estes. Volume 1*. Hillsdale, NJ: Lawrence Erlbaum Associates.

Link, S. W., & Tindall, A. B. (1970). *Reaction times to comparative judgements of line length*. (Tech. Rep. No. 32). Hamilton, Ontario: McMaster University, Department of Psychology.

Link, S. W., & Tindall, A. B. (1971). Speed and accuracy in comparative judgments of line length. *Perception and Psychophysics*, 9, 284–288.

Link, S. W., & Heath, R. A. (1975). A sequential theory of psychological discrimination. *Psychometrika*, 40, 77–105.

Luce, R. D. (1959). *Individual choice behavior*. New York: Wiley.

Luce, R. D. (1960). The theory of selective information and some of its behavioral applications. In R.D. Luce (Ed.), *Developments in mathematical psychology*. Glencoe, Il: Freepress.

Luce, R. D. (1963). A threshold theory for simple detection experiments. *Psychological Review*, 70, 61–79.

Luce, R. D. (1986). *Response times: Their role in inferring elementary mental organization*. Belfast: Oxford University Press.

Luce, R. D. & Edwards, W. (1958). The derivation of subjective scales form just noticeable differences. *Psychological Review*, 65, 222–237.

Lupker, S. J., & Theios, J. (1977). Further tests of a two-state model for choice reaction times. *Journal of Experimental Psychology*, 3(3), 496–504.

Martin, L. J., & Müller, G. E. (1899). *Zur analyse der unterschiedsempfindlichkeit: experimentalle beiträge*. Leipzig: Barth.
Mason, S. J., & Zimmerman, H. J. (1960). *Electronic circuits, signals and systems*. New York: Wiley.
McGill, W. J. (1953). Multivariate information transmission. *Transactions of The Institute of Radio Engineers: 1954 Symposium on Information Theory*, 4, 93–111.
McGill, W. J. (1954). Multivariate information transmission. *Psychometrika*, 19, 97–116.
McGill, W. J. (1964). *Introduction to counter theory in psychophysics*. Unpublished manuscript. Institute for Mathematical Studies in the Social Sciences. Stanford University. Stanford, California.
Miller, G. A., & Frick, F. C. (1949). Statistical behavioristics and sequence of responses. *Psychological Review*, 56, 311–324.
Miller, G. A. (1953). What is information measurement? *American Psychologist*, 8, 3–11.
Mowrer, O. H. (1940). Preparatory set (expectancy): Some methods of measurement. *Psychological Monographs*, 52(2), 1–43.
Müller, G. E. (1878). *Zur grundlegung der psychophysik*. Grieben: Berlin.
Munsterberg, H. (1894). Studies from the Harvard psychological laboratory. *Psychological Review*, 1(1), 34–60.
Murray, H. (1938). *Explorations in personality*. New York: Oxford University Press.
Neyman, J. (1976). The emergence of mathematical statistics. In D. B. Owen, (Ed.), *On the history of statistics and probability*. (pp. 147–193). New York: Marcel Dekker, Inc.
Nickerson, R. S. (1972). Binary-classification reaction time: A review of some studies of human information-processing capabilities. *Psychonomic Monograph Supplements*, 4(17), 275–318.
Noreen, D. L. (1979). *Analysis of some mathematical models for choice and decision*. Doctoral dissertation. Rockefeller University. New York, New York.
Oberlin, K. W. (1936). Variation in intensitive sensitivity to lifted weights. *Journal of Experimental Psychology*, 19, 438–455.
Ogilvie, J. C., & Creelman, C. D. (1968). Maximum-likelihood estimation of ROC curve parameters. *Journal of Mathematical Psychology*, 5, 377–391.
Ollman, R. T. (1966). Fast guesses in choice-reaction time. *Psychonomic Science*, 6, 155–156.
Parducci, A. (1991). Elaborations upon psychophysical contents for judgment: implications for cognitive modes. In H. G., Geissler, S. W., Link, & J. T., Townsend, (Eds.), *Cognition Information Processing and Psychophysics: Basic Issues*. Hillsdale, N.J.: Lawrence Erlbaum Assoc.
Parzen, E. (1960). *Modern probability theory and its applications*. New York: Wiley.
Parzen, E. (1962). *Stochastic processes*. San Francisco: Holden-Day.
Peirce, C. S. (1877). Illustrations of the logic of science. *The Popular Science Monthly*, 12, 1–15, 286–302, 604–615, 705–718.
Peirce, C. S., & Jastrow, J. (1884). On small differences in sensation. *Memoirs of the National Academy of Sciences*, 3, 73–83.
Peterson, W. W., & Birdsall, T. G. (1953). *The theory of signal detectability*. (Tech. Rep. No. 13). Ann Arbor: University of Michigan, Electronic Defense Group.
Peterson, W. W., Birdsall, T. G., & Fox, W. C. (1954). The theory of signal detectability. *Transactions of The Institute of Radio Engineers: 1954 Symposium on Information Theory*, 4, 171–212.
Petrusic, W. M. (1966a). Latency, stochastic dominance, and laterality effects. *Psychonomic Science*, 5(10), 375–376.
Pieron, H. (1914). Recherches sur les lois de variation des temps de latence sensorielle en fonction des intensités excitatrices. *L'Année Psychologique*, 20, 17–96.
Pieron, H. (1920). Nouvelles recherches sur l'analyse du temps de latence sensorialle et sur la loi qui relie de temps a l'intensité d'excitation. *L'Année Psychologique*, 22, 58–142.
Pirenne, M. H., & Marriott, F. H. C. (1959). The quantum theory of light and the psychophysiology of vision. In S. Koch, (Ed.), *Psychology: A study of a science. Vol. 1*. (pp. 288–361) New York: McGraw-Hill.

Poisson, S. D. (1837). *Recherches sur la probabilité des jugements en matière criminelle et en matière civile, précedées des règles générales du calcul des probabilitiés*. Paris.

Quine, W. V. (1985). Four hot questions in philosophy. *The New York review of books*. 32(2), 32.

Rabbitt, P. M. A. (1966). Errors and error-corrections in choice response tasks. *Journal of Experimental Psychology*, 71, 264–272.

Rabbitt, P. M. A. (1968). Three kinds of error-signalling responses in a serial choice task. *Quarterly Journal of Experimental Psychology*, 20, 179–188.

Rabbitt, P. M. A. (1969). Psychological refractory delay and response-stimulus interval duration in serial, choice-response tasks. In W. G. Kostner (Ed.), *Attention and performance II*. (pp. 195–219). Amsterdam: North Holland.

Rashevsky, N. (1938). *Mathematical biophysics*. Chicago: University of Chicago Press.

Rashevsky, N. (1960). *Mathematical biophysics. Vol II*. (3rd revised edition). New York: Dover Publications.

Rasch, G. (1960). *Probabilistic models for some intelligence and attainment tests*. Copenhagen: Danmarks Paedogogishe Institut.

Remington, R. J. (1969). Analysis of sequential effects in choice reaction time. *Journal of Experimental Psychology*, 82, 250–257.

Restle, F., & Greeno, J. G. (1970). *An introduction to mathematical psychology*. Reading, Mass.: Addison-Wesley.

Schlosberg, H., & Solomon, R. L. (1943). Latency of response in a choice discrimination. *Journal of Experimental Psychology*, 33, 22–39.

Schoenberg, I. J. (1946). Contributions to the problem of approximation of equidistant data by analytic functions, Part A: On the problem of smoothing a graduation, a first class of analytic approximation formulae. *Quarterly Journal of Applied Mathematics*, 4, 45–99.

Schumaker, L. (1981). *Spline functions*. New York: Wiley.

Schumann, F. (1898). Zur psychologie der zeitanschauung. *Zietshrift fur Psychologie*, 17, 106–148.

Schumann, F. (1902). Beiträge zur analyse der gesichtwahrnehmungen. *Zietshrift fur Psychologie*, 30, 241–291, 321–339.

Shannon, C. E., & Weaver, W. (1963). *The mathematical theory of communication*. Chicago: University of Illinois Press. (Shannon's original work published 1948)

Shipley, W. C., Coffin, J. I., & Hadsell, K. C. (1945). Affective distance and other factors determining reaction time in judgments of color preference. *Journal of Experimental Psychology*, 35, 206–215.

Shipley, W. C., Norris, E. D., & Roberts, M. L. (1946). The effect of changed polarity of set on decision time of affective judgments. *Journal of Experimental Psychology*, 36, 237–243.

Shiryayev, A. N. (1984). *Probability*. New York: Springer-Verlag.

Skinner, B. F. *The behavior of organisms: An experimental analysis*. New York: Appleton-Century.

Steinman, A. R. E. (1944). Reaction time to change compared with other psychophysical methods. *Archives of Psychology*, 292, 5–60.

Stevens, J. C., Mack, J. D., & Stevens, S. S. (1960). Growth of sensation on seven continua as measured by force of handgrip. *Journal of Experimental Psychology*, 59, 60–67.

Stevens, S. S. (1961). To honor Fechner and repeal his law. *Science*, 133, 80–86.

Stevens, S. S. (1966). A metric for the social consensus. *Science*, 151, 530–541.

Stevens, S. S. (1975). *Psychophysics: Introduction to its perceptual, neural and social prospects*. (G. Stevens, Ed.). New York: Wiley.

Stevens, S. S., Morgan, C. T., & Volkmann, J. (1941). Theory of the Neural Quantum in the Discrimination of Loudness and Pitch. *American Journal of Psychology*, 54(3), 315–335.

Stone, M. (1960). Models for choice reaction time. *Psychometrika*, 25(3), 251-260.

Stroud, J. M. (1955). The fine structure of time. In H. Quastler (Ed.), Information theory in psychology. (pp. 174–207). Glencoe, Ill.: Free Press.

Sully, J. (1874). *Sensation and intuition: studies in psychology and esthetics*. London: Longmans.

Swets, J. A. (1961). Is there a sensory threshold? *Science*, 134, 168–177.

Swets, J. A. (1963). Central factors in auditory frequency selectivity. *Psychological Bulletin*, 60, 429–440.

Swets, J. A., Tanner, W. P., Jr., & Birdsall, T. G. (1955). *The evidence for a decision-making theory of visual detection*. (Tech. Rep. No. 40). University of Michigan: Electronic Defense Group.

Swets, J. A., Green, D. M., & Tanner, W. P., Jr. (1962). On the width of critical bands. *Journal of the Acoustical Society of America*, 34, 108–113.

Tanner, W. P., Jr. (1961). Application of the theory of signal detectability to amplitude discrimination. *Journal of the Accoustical Society of America*, 33, 1233–1244.

Tanner, W. P., Jr., & Swets, J. A. (1954). A decision-making theory of visual detection. *Psychological Review*, 61(6), 401–409.

Tanner, W. P., Jr., & Swets, J. A. (1954). The human use of information. I. Signal detection for the case of the signal known exactly. *Transactions of The Institute of Radio Engineers: 1954 Symposium on Information Theory*, 4, 213–221.

Tanner, W. P., Jr., Swets, J. A., & Green, D. M. (1956). *Some general properties of the hearing mechanism*. (Tech. Rep. No. 30). University of Michigan: Electronics Defense Group.

Teghtsoonian, R. (1971). On the exponents in Stevens' law and the constants in Ekman's law. *Psychological Review*, 78, 71–80.

Theios, J. & Smith, P. G. (1972). Can a two-state model account for two-choice reaction time data? *Psychological Review*, 79, 172–177.

Theios, J., & Walter, D. G. (1974). Stimulus and response frequency and sequential effects in memory scanning reaction times. *Journal of Experimental Psychology*, 102, 1092–1099.

Thomson, G. H. (1913). Note on the probable error of Urban's formula for the method of just perceptible differences. *British Journal of Psychology*, 6(2), 217–222.

Thomson, G. H. (1920). A new point of view in the interpretation of threshold measurements in psychophysics. *Psychological Review*, 27, 300–307.

Thurstone, L. L. (1927a). The method of paired comparisons for social values. *Journal of Abnormal Psychology*, 21, 384–400.

Thurstone, L. L. (1927b). A law of comparative judgment. *Psychological Review*, 34, 273–286.

Thurstone, L. L. (1928a). The measurement of opinion. *Journal of Abnormal Psychology*, 22, 415–430.

Thurstone, L. L. (1928b). The phi gamma hypothesis. *Journal of Experimental Psychology*, 11, 293–305.

Thurstone, L. L. (1931). Rank order as a psychophysical method. *Journal of Experimental Psychology*, 14, 187–201.

Thurstone, L. L. (1937). Psychology as a quantitative rational science. *Science*, 85, 228–232.

Thurstone, L. L. (1952). Thurstone, L. L. In E. G. Boring & R. M. Yerkes (Eds.), *A history of psychology in autobiography. Vol. 4*. Worcester, Mass: Clark University Press.

Thurstone, L. L. (1954). The measurement of values. *Psychological Review*, 61, 47–58.

Thurstone, L. L. (1955). *The measurement of values*. Chicago: University of Chicago Press.

Titchener, E. B. (1901a). *Experimental Psychology. A manual of laboratory practice. Vol. I, Part I*. New York: Macmillan Company.

Titchener, E. B. (1901b). *Experimental Psychology. A manual of laboratory practice. Vol. 1. Part II*. New York: Macmillan Company.

Titchener, E. B. (1902). Ein versuch die methode der paarweisen vergleichung auf die vershiedenen gefühlsrichtungen anzuwenden. *Philosophische Studien*, 20, 382.

Titchener, E. B. (1905a). *Experimental Psychology. Vol. II, Part I*. New York: Macmillan Company.

Titchener, E. B. (1905b). *Experimental Psychology. Vol. II, Part II*. New York: Macmillan Company.

Titchener, E. B. (1929). *Systematic Psychology Prologamena*. Ithaca: Cornell University Press.

Townsend, J. T., & Ashby, F. G. (1983). *Stochastic modeling of elementary psychological processes*. New York: Cambridge University Press.

Urban, F. M. (1907). On the method of just perceptible differences. *Psychological Review*, 14, 244–253.

Urban, F. M. (1908). *The Application of Statistical Methods to the Problems of Psychophysics*. Philadelphia: Psychological Clinic Press.

Urban, F. M. (1910). The method of constant stimuli and its generalizations. *The Psychological Review*, 17(4), 229–259.

Uspensky, J. V. (1937). *Introduction to mathematical probability*. New York: McGraw-Hill.

Uttal, W. R. (1975). *An autocorrelation theory of form detection*. Hillsdale, N.J.: Erlbaum.

Volkmann, J. (1934). The relation of the time of judgment to the certainty of judgment. *Psychological Bulletin*, 31, 672–673.

von Békésy, G. (1929). Zur theorie des hörens; Über die eben merkbare amplituden- und frequenzänderung eines tones; Die theorie der schwebungen. *Physik. Zeits.* 30, 721–745.

von Békésy, G. (1930). Uber das Fechnersche gesetz und seine bedeutung für die theorie der akustischen beobachiungsfehler und die theorie des horens. *Annalen Der Physik*, S-5, 7, 329–359.

von Békésy, G. (1960). *Experiments in hearing*. (E. G. Wever, Ed. and Trans). New York: McGraw-Hill.

von Neumann, J. (1955). *Mathematical foundations of quantum mechanics*. (R. T. Beyer, Trans.). Princeton: Princeton University Press.

Wald, A. (1947). *Sequential Analysis*. New York: Wiley.

Ward, J. (1876). An attempt to interpret Fechner's Law. *Mind*, 1, 452–466.

Warren, H. C. (1897). The reaction time of counting. *Psychological Review*, 4, 569–591.

Watson, J. B. (1919). *Psychology from the standpoint of a behaviorist*. Philadelphia: Lippincott.

Weber, E. H. (1934). *De pulsu, resportione, auditu et tactu*. Leipzig: Koehler.

Weber, E. H. (1846). Der tastsinn und das gemeingefühl. In R. Wagner (Ed.), *Handwörterbuch der physiologie. Vol. 3.* (pp. 481–588). Brunswick: Vieweg.

Weber, E. H. (1978). De subtilitate tactus. In H. E. Ross & D. J. Murray (H. E. Ross, Trans.). *E. H. Weber: The sense of touch*. London: Academic Press. (Original published 1834)

Weber, E. H. (1978). Der tastsinn und das gemeingefühl. In H. E. Ross & D. J. Murray (D. J. Murray, Trans.). *E. H. Weber: The sense of touch*. London: Academic Press. (Original published 1846)

Welford, A. T. (Ed.), (1980). *Reaction times*. New York: Academic Press.

Wever, E. G., & Zener, K. E. (1928). The method of absolute judgment in psychophysics. *Psychological Review*, 35, 466–493.

Whipple, G. M. (1901). An analytic study of the memory image and the process of judgment in the discrimination of clangs and tones. *American Journal of Psychology*, 12(4), 409–457.

Wickelgren, W. A., & Norman, D. (1966). Strength models and serial position in short-term recognition memory. *Journal of Mathematical Psychology*, 3, 316–347.

Wiener, N. (1948). Cybernetics. New York: Wiley.

Wolfowitz, J. (1947). The efficiency of sequential estimates and Wald's equation for sequential processes. *Annals of Mathematical Statistics*, 18, 215–230.

Wundt, W. (1865). *Lehrbuch der physiologie des menschen*. Stuttgart: Ferdinand Enke.

Wundt, W. (1878). *Lehrbuch der physiologie des menschen*. Stuttgart: Ferdinand Enke.

Wundt, W. (1896). *Lectures on human and animal psychology*, (2nd ed.) (Trans. J. Creighton & E. Titchener). New York: Swan Sonnenschein.

Wundt, W. (1904). *Principles of physiological psychology*, (E. B. Titchener, Trans.). New York: MacMillan. (Original work published 1902)

Yellott, J. I. Jr. (1967). Correction for guessing in choice reaction time. *Psychonomic Science*, 8, 321–322.

Yellott, J. I. Jr. (1971). Correction for fast guessing and the speed-accuracy tradeoff in choice reaction time. *Journal of Mathematical Psychology*, 8, 159–199.

SUPPLEMENTAL REFERENCES

Anderson, N. H. (1981). *Foundations of information integration theory*. New York: Academic Press.

Anderson, N. H. (1982). *Methods of information integration theory*. New York: Academic Press.

Arthur, A. Z. (1965). 'Cross-over' effect and subjective intensity. *British Journal of Psychology*, 56(4), 477–478.

Attneave, F. (1954). Some informational aspects of visual perception. *Psychological Review*, 61, 183–193.

Audley, R. J. (1973). Some observations on theories of choice reaction time. In S. Kornblum (Ed.), *Attention and performance IV*. New York: Academic Press. 509–545.

Audley, R. J., & Mercer, A. (1968). The relation between decision time and relative response frequency in a blue-green discrimination. *British Journal of Mathematical and Statistical Psychology*, 21, 183–192.

Audley, R. J., & Pike, A. R. (1965). Some alternative stochastic models of choice. *British Journal of Mathematical and Statistical Psychology*, 18(2), 207–225.

Audley, R. J., & Wallis, C. P. (1964). Response instructions and the speed of relative judgements. 1. Some experiments on brightness discrimination. *British Journal of Psychology*, 55(1), 59–73.

Audley, R. J., & Wallis, C. P. (1965). 'Cross-over' effect: A reply to Arthur. *British Journal of Psychology*, 56(4), 479–482.

Banks, W. P., White, Hedy, Sturgill, W., & Mermelstein, R. (1983). Semantic congruity and expectancy in symbolic judgments. *Journal of Experimental Psychology: Human Perception and Performance*, 9, 560–582.

Bamber, D. (1969). Reaction times and error rates for "same"-"different" judgments of multidimensional stimuli. *Perception & Psychophysics*, 6(3), 169–174.

Bennett, W. F., Fitts, P. M., & Noble, M. (1954). The learning of sequential dependencies. *Journal of Experimental Psychology*, 48(4), 303–312.

Berkson, J. (1972). Minimum discrimination information, the 'NO INTERACTION' problem and the logistic function. *Biometrics*, 28, 443–468.

Berlyne, D. E. (1957a). Conflict and choice time. *British Journal of Psychology*, 48, 106–118.

Berlyne, D. E. (1957b). Conflict and information-theory variables as determinants of human perceptual curiosity. *Journal of Experimental Psychology*, 53(6), 399–405.

Berlyne, D. E. (1957c). Uncertainty and conflict: A point of contact between information-theory and behavior-theory concepts. *Psychological Review*, 64(6), 329–339.

Bernstein, I. H., & Reese, C. (1965). Behavioral hypotheses and choice reaction time. *Psychonomic Science*, 3, 259–260.

Bertelson, P. (1963). S-R relationships and reaction times to new versus repeated signals in a serial task. *Journal of Experimental Psychology*, 65, 478–484.

Bertelson, P. (1965a). Central intermittency twenty years later. Lecture delivered on May 4, 1965. St. John's College, Cambridge, England.

Bertelson, P. (1965b). Serial choice reaction-time as a function of response versus signal-and-response repetition. *Nature*, 206, 217–218.

Bertelson, P., & Barzeele, J. (1965). Interaction of time-uncertainty and relative signal frequency in determining choice reaction time. *Journal of Experimental Psychology*, 70, 448–451.

Bevan, W., & Dukes, W. F. (1953). Preparatory set (expectancy)—an experimental reconsideration of its 'central' locus. *British Journal of Psychology*, 44, 52–56.

Birnbaum, M. H. (1978). Differences and ratios in psychological measurement. In N. J. Castellan, Jr. & F. Restle (Eds.), *Cognitive theory. Vol 3*. Hillsdale, NJ: Lawrence Erlbaum Associates.

Birren, J. E., & Botwinick, J. (1955). Speed of response as a function of perceptual difficulty and age. *Journal of Gerontology*, 10, 433–436.

Blackman, A. R. (1972a). Influence of stimulus and response probability on decision and movement latency in a discrete choice reaction task. *Journal of Experimental Psychology*, 92(1), 128–133.

Blackman, A. R. (1972b). Stimulus probability and choice reaction time. *Perception & Psychophysics*, 12(2), 146–150.

Boeving, R. S., & Randolph, J. J. (1975). The "gambler effect": Choice and the probability of reinforcement of small response outputs. *Psychological Reports*, 37, 55–62.

Borger, R. (1963). The refractory period and serial choice-reactions. *Quarterly Journal of Experimental Psychology*, 15(1), 1–12.

Boring, E. G. (1924). Is there a generalized psychometric function? *American Journal of Psychology*, 35, 75–78.

Botwinick, J., & Brinley, J. F. (1962). An analysis of set in relation to reaction time. *Journal of Experimental Psychology*, 63(6), 568–574.

Bradshaw, J. L. (1969). Background light intensity and the pupillary response in a reaction time task. *Psychonomic Science*, 14(6), 271–272.

Brainard, R. W., Irby, T. S., Fitts, P. M., & Alluisi, E. A. (1962). Some variables influencing the rate of gain of information. *Journal of Experimental Psychology*, 63, 105–110.

Brebner, J., & Gordon, I. (1962). Ensemble size and selective response times with a constant signal rate. *Quarterly Journal of Experimental Psychology*, 14, 113–116.

Brebner, J., & Gordon, I. (1964a). The influence of signal probability and the number of non-signal categories on selective response times. *Quarterly Journal of Experimental Psychology*, 16, 56–60.

Brebner, J., & Gordon, I. (1964b). Some variables affecting selective response latencies. *Nature*, 201, 1109–1110.

Bricker, P. D. (1955). The identification of redundant stimulus patterns. *Journal of Experimental Psychology*, 49(2), 73–81.

Castaneda, A., & Worrell, L. (1961). Differential relation of latency and response vigour to stimulus similarity in brightness discrimination. *Journal of Experimental Psychology*, 61, 309–314.

Cowan, T. M., & Monroe, B. D. (1970). Choice reactions with a null stimulus event. *Perception and Psychophysics*, 7(5), 267–270.

Cross, D. V. (1973). Sequential dependencies and regression in psychophysical judgments. *Perception and Psychophysics*, 14, 547–552.

Davis, H. (1964). Enhancement of evoked cortical potentials in humans related to a task requiring a decision. *Science*, 145, 182–183.

Dillon, P. J. (1966). Stimulus versus response decisions as determinants of the relative frequency effect in disjunctive reaction-time performance. *Journal of Experimental Psychology*, 71(3), 321–330.

Donchin, E. (1979). Event-related brain potentials: A tool in the study of human information processing. In H. Begleiter (Ed.), *Evoked brain potentials and behavior*, 13–75. New York: Plenum Press.

Donchin, E. (1981). Surprise!. . . . Surprise?. *Psychophysiology*, 18(5), 493–513.

Donchin, E., & Lindsley, D. B. (1966). Average evoked potentials and reaction times to visual stimuli. *Electroencephalography and Clinical Neurophysiology*, 20, 217–223.

Drazin, D. H. (1961). Effects of foreperiod, foreperiod variability and probability of stimulus occurrence on simple reaction time. *Journal of Experimental Psychology*, 62, 43–50.

Edwards, W. (1965). Optimal strategies for seeking information: Models for statistics, choice reaction times and human information processing. *Journal of Mathematical Psychology*, 2, 312–329.

Egan, J. P., Lindner, W. A., & McFadden, D. (1969). Masking-level differences and the form of the psychometric function. *Perception and Psychophysics*, 6, 209–215.

Egeth, E., & Smith, E. E. (1967). On the nature of errors in a choice reaction task. *Psychonomic Science*, 8, 345–346.

Eisler, H. (1974). The derivation of Stevens' psychophysical power law. In H. R. Moskowitz, B. Scharf, & J. C. Stevens (Eds.), *Sensation and measurement*. (pp. 61–64). Dordrecht, Holland: Reidel.

Ekman, G. (1959). Weber's law and related functions. *Journal of Psychology*, 47, 343–351.

Ellis, H. D. (1972). Adaptation-level theory and context effects on sensory judgments: Perception or response? *Perception*, 1, 101–109.

Ellis, H. D. (1973a). An examination of the adaptation-level theory account of time errors. *Journal of Experimental Psychology*, 98(1), 160–163.

Ellis, H. D. (1973b). Proactive effects of interpolated anchors. *Journal of Experimental Psychology*, 98(2), 233–238.

Ells, J. G., & Gotts, G. H. (1977). Serial reaction time as a function of the nature of repeated events. *Journal of Experimental Psychology: Human Perception and Performance*, 3(2), 234–232.

Engen, T., & Levy, N. (1955). The influence of standards on psychophysical judgments. *Perceptual and Motor Skills*, 5, 193–197.

Eriksen, C. W., O'Hara, W. P., & Eriksen, B. (1982). Response competition effects in *same-different* judgments. *Perception and Psychophysics*, 32(3), 261–270.

Estes, W. K., & Wessel, D. L. (1966). Reaction time in relation to display size and correctness of response in forced-choice visual signal detection. *Perception and Psychophysics*, 1, 369–373.

Fagot, R. F. (1963). On the psychophysical law of estimation procedures in psychophysical scaling. *Psychometrika*, 28, 145–160.

Falmagne, J. C. (1971). The generalized Fechner problem and discrimination. *Journal of Mathematical Psychology*, 8, 22–43.

Falmagne, J.C. (1982). Psychometric functions theory. *Journal of Mathematical Psychology*, 25, 1–50.

Falmagne, J. C. (1985). *Elements of psychophysical theory*. New York: Oxford University Press.

Fechner, G. T. (1871). Zur experimentellen aesthetik, Abh. d. Math. - Phys. Cl. d. k. Sachs. *Ges. d. Wissensch*, 9, 555–635.

Fitts, P. M. (1966). Cognitive aspects of information processing III, set for speed versus accuracy. *Journal of Experimental Psychology*, 71, 847–849.

Fitts, P. M., & Deininger, R. L. (1954). S-R compatibility: Correspondence among paired elements within stimulus and response codes. *Journal of Experimental Psychology*, 48, 483–492.

Fitts, P. M., Peterson, J., & Wolpe, G. (1963). Cognitive aspects of information processing: II. Adjustments to stimulus redundancy. *Journal of Experimental Psychology*, 65, 507–514.

Fitts, P. M., & Seeger, C. M. (1953). Stimulus-response compatibility: Spatial characteristics of stimulus response codes. *Journal of Experimental Psychology*, 46, 199–210.

Fitts, P. M., & Switzer, G. (1962). Cognitive aspects of information processing: I. The familiarity of S-R sets and subsets. *Journal of Experimental Psychology*, 63, 321–329.

Foley, P. J., & Humphries, M. (1962). Blocking in serial simple reaction tasks. *Canadian Journal of Psychology*, 16, 237–238.

Forrin, B., Kumler, M. L., & Morin, R. E. (1966). The effects of response code and signal probability in a numerical-naming task. *Canadian Journal of Psychology*, 20(2), 115–124.

Forrin, B., & Morin, R. E. (1966). Effect of contextual associations upon selective reaction time in a numerical-naming task. *Journal of Experimental Psychology*, 71(1), 40–46.

Forrin, B., & Morin, R. E. (1967). Effects of context on reaction time to optimally coded signals. *Acta Psychologica*, 27, 188–196.

Fraisse, P., & Blancheteau, M. (1962). The influence of the number of alternatives on the perceptual recognition threshold. *Quarterly Journal of Experimental Psychology*, 14, 52–55.

Gahagan, L. (1929). On the absolute judgment of lifted weights. *Journal of Experimental Psychology*, 12, 490–501.

Gaillard, A. W., & Näätänen, R. (1973). Slow potential changes and choice reaction time as a function of interstimulus interval. *Acta Psychologica*, 37, 173–186.

Gauss, C. F. (1809). Theoria motus corporum coelestium. Hamburg: Certhes and Besser.

Gibbon, J. (1977). Scalar expectancy theory and Weber's Law in animal timing. *Psychological Review*, 84(3), 279–325.

Granjon, M., Requin, J., Durup, H., & Reynard, G. (1973). Effects of timing signal on simple reaction time with 'non-aging' foreperiods. *Journal of Experimental Psychology*, 101, 139–145.

Granjon, M., & Reynard, G. (1977). Effect of the length of the runs of repetitions on the simple RT-isi relationship. *Quarterly Journal of Experimental Psychology*, 29, 283–295.

Green, D. M., and McGill, W. J. (1970). On the equivalence of detection probabilities and well-known statistical quantities. *Psychological Review*, 77(4), 294–301.

Greenberg, G. Z. (1962). *Cueing signals and frequency uncertainty in auditory detection*. (Tech. Rep. No. ESD=TDR=62=38). Bloomington, Indiana: Indiana University, Hearing and Communication Laboratory.

Greenwald, A. G., & Rosenberg, K. E. (1976). Sequential effects of distracting stimuli in a selective attention reaction time task. In J. Requin (Ed.), *Attention and performance, VII*. (pp. 487–504). Hillsdale, NJ: Lawrence Erlbaum Associates.

Grice, G. R. (1972). Application of a variable criterion model to auditory reaction time as a function of the type of catch trial. *Perception and Psychophysics*, 12, 103–107.

Grossberg, S. (1975). Neural model of attention, reinforcement and discrimination-learning. *International Review of Neurobiology*, 8, 263–327.

Guilford, J. P. (1932). A generalized psychophysical law. *Psychological Review*, 39, 72–86.

Guilford, J. P., & Park, D. G. (1931). The effect of interpolated weights upon comparative judgments. *American Journal of Psychology*, 43, 589–599.

Harris, W. P. (1957). A revised law of comparative judgment. *Psychometrika*, 22(2), 189–198.

Hegelmeyer, F. (1852). Ueber das gedächtniss für linearanschauungen. *Archiv für Physiologisch Heilkunde*, 11, 844–853.

Hellyer, S. (1963). Stimulus-response coding and amount of information as determinants of reaction time. *Journal of Experimental Psychology*, 65, 521–522.

Helson, H., Michels, W. C., & Sturgeon, A. (1931). The use of comparative rating scales for the evaluation of psychophysical data. *American Journal of Psychology*, 43, 321–326.

Herrick, R. M. (1973). Psychophysical methodology: VI. Random method of limits. *Perception and Psychophysics*, 13(3), 548–554.

Hillyard, S. A., & Galambos, R. (1967). Effects of stimulus and response contingencies on a surface negative slow potential shift in man. *Electroencephalography and Clinical Neurophysiology*, 22, 297–304.

Hoisington, L. B. (1917). An example of the fractionation of data from the method of constant stimuli for the two-point limen. *American Journal of Psychology*, 28, 588–596.

Holt, E. B. (1904). The classification of psycho-physic methods. *Psychological Review*, 11(6), 343–369.

Hopkins, G. W., & Kristofferson, A. B. (1980). Ultrastable stimulus-response latencies: Acquisition and stimulus control. *Perception and Psychophysics*, 27(3), 241–250.

Hovland, C.I. (1938). A note on Guilford's generalized psychophysical law. *Psychological Review*, 45, 430–434.

Hufford, L. E. (1964). Reaction time and the retinal area-stimulus intensity relationship. *Journal of the Optical Society of America*, 54(11), 1368–1373.

Hulin, W. S., & Frazier, L. R. (1928). Two comparison stimuli with one initial standard in lifted weights. *American Journal of Psychology*, 40, 626–627.

Indow, T. (1980). Errors in simple and monotone tasks: A proposal of testing on site. *Behaviormetrika*, 7, 1–12.

Iverson, G. J. (1979). Conditions under which Thurstone Case III representations for binary choice probabilities are also Fechnerian. *Journal of Mathematical Psychology*, 20(3), 263–271.

Jamieson, D. G., & Petrusic, W. M. (1975). Relational judgments with remembered stimuli. *Perception and Psychophysics*, 18(6), 373–378.

Jamieson, D. G., & Petrusic, W. M. (1976). On a bias induced by the provision of feedback in psychophysical experiments. *Acta Psychologica*, 40, 199–206.

Jamieson, D. G. & Petrusic, W. M. (1977). Preference and the time to choose. *Organizational Behavior and Human Performance*, 19, 56–67.

Jastrow, J. (1893). On the judgment of angles and positions of lines. *American Journal of Psychology*, 5, 214–248.

Johnson, R., Jr., & Donchin, E. (1980). P300 and stimulus categorization: Two plus one is not so different from one plus one. *Psychophysiology*, 2, 167–178.

Jones, L. E., & Parducci, A. (1967). Stimulus uncertainty and the latency of category judgment. *Psychonomic Science*, 7(2), 71–72.

Kahneman, D., & Tversky, A. (1982). The psychology of preferences. *Scientific American*, 246(1), 160–173.

Kamlet, A. S. (1969). *Reaction time: A bibliography with abstracts*. Aberdeen Research & Development Center, Aberdeen Proving Ground, Maryland.

Kantowitz, B. H. (1978). On the accuracy of speed-accuracy tradeoff. *Acta Psychologica*, 42, 79–80.

Karlin, L. (1959). Reaction time as a function of foreperiod duration and variability. *Journal of Experimental Psychology*, 58, 185–191.

Karlin, L. (1966). Development of readiness to respond during short foreperiods. *Journal of Experimental Psychology*, 72, 505–509.

Kaswan, J., & Young, S. (1965). Effect of luminance, exposure duration, and task complexity on reaction time. *Journal of Experimental Psychology*, 69(4), 393–400.

Kellogg, W. N. (1929). An improved photometer for measuring brightness thresholds in hundredths of a meter candle. *Journal of General Psychology*, 2, 369–377.

Klemmer, E. T. (1956). Time uncertainty in simple reaction time. *Journal of Experimental Psychology*, 51, 179–184.

Knight, J. L. Jr., & Kantowitz, B. H. (1974). Speed-accuracy tradeoff in double stimulation: Effects on the first response. *Memory and Cognition*, 2(3), 522–532.

Knight, J. L. Jr., & Kantowitz, B. H. (1976). Speed-accuracy tradeoff in double stimulation: II. Effects on the second response. *Memory and Cognition*, 4(6), 690–700.

Kornblum, S. (1973). Simple reaction time as a race between signal detection and time estimation: A paradigm and model. *Perception and Psychophysics*, 13, 108–112.

Kunst-Wilson, W. R., & Zajonc, R. B. (1980). Affective discrimination of stimuli that cannot be recognized. *Science*, 207, 557–558.

Lachenmeyer, C. W. (1975). Measurement and investigation: Five demands for social science measurement. *Journal of General Psychology*, 93, 173–182.

Laming, D. R. J. (1973). *Mathematical psychology*. New York: Academic Press.

Laming, D. (1979a). Autocorrelation of choice-reaction times. *Acta Psychologica*, 43, 381–412.

Laming, D. (1979b). Choice reaction performance following an error. *Acta Psychologica*, 43, 199–224.

Lappin, J. S., & Disch, K. (1972). The latency operating characteristic: I. Effects of stimulus probability on choice reaction time. *Journal of Experimental Psychology*, 92, 419–427.

Larimer, J., & Thompson, K. (1973). A sufficient condition for conditional expectation to be monotonically related to distribution parameters. *Journal of Mathematical Psychology*, 10, 223–231.

Lemmon, V. W. (1927). The relation of reaction time to measures of intelligence, memory and learning. *Archives of Psychology*, 15(94), 1–38.

Loveless, N. E. (1957). Signal detection with simultaneous visual and auditory presentation. *Flying personnel research committee*, 1–14, Nuffield Department of Industrial Health, University of Durham, Dec. 1957.

Luce, R. D. (1977). Thurstone's discriminal processes fifty years later. *Psychometrika*, 42(4), 461–489.

Lufkin, H. (1928). The best fitting frequency function for Urban's lifted-weight results. *American Journal of Psychology*, 40, 75–82.

Lufkin, H. M. (1927). The accuracy of the method of constant stimuli. *American Journal of Psychology*, 38, 666–667.

Marks, D. F. (1972). Relative Judgment: A phenomenon and a theory. *Perception and Psychophysics*, 11, 156–160.

Marley, A. A. J. (1976). A revision of the response ratio hypothesis for magnitude estimation. *Journal of Mathematical Psychology*, 14, 252–254.

McAdam, D. W., Knott, J. R., & Rebert, C. S. (1969). Cortical slow potential changes in man related to interstimulus interval and to pretrial prediction of interstimulus interval. *Psychophysiology*, 5, 349–358.

McCarthy, G., & Donchin, E. (1981). A metric for thought: A comparison of P300 latency and reaction time. *Science*, 211, 22–80.

McFadden, D., & Skinner, R. B. (1974). On the form of psychometric functions for taste. *Perception and Psychophysics*, 15(2), 379–382.

McGill, W. J. (1963). The general-gamma distribution and reaction times.(Available from W. J. McGill, Center for Human Information Processing. University of California, San Diego. La Jolla, California.)

McGill, W. J. (1967). Neural counting mechanism and energy detection in audition. *Journal of Mathematical Psychology*, 4, 351–376.

Meischner, W., & Eschler, E. (1979). Wilhelm Wundt. Leipzig: Urania.

Michels, W. C., & Helson, H. (1954). A quantitative theory of time-order effects. *American Journal of Psychology*, 67, 327–334.

Michie, P. T., Clarke, A. M., Sinden, J. D., & Glue, L. C. T. (1975). Lateral facilitation of Hoffmann-reflexes prior to voluntary movement in choice reaction time task. *Applied Neurophysiology*, 38, 191–196.

Mo, S. S., & George, E. J. (1977). Foreperiod effect on time estimation and simple reaction time. *Acta Psychologica*, 41, 47–59.

Morin, R. E., & Forrin, B. (1962). Mixing of two types of S-R associations in a choice reaction time task. *Journal of Experimental Psychology*, 64, 137–141.

Morin, R. E., Konick, A., Troxell, N., & McPherson, S. (1965). Information and reaction time for "naming" responses. *Journal of Experimental Psychology*, 70, 309–314.

Mowbray, G. H. (1960). Choice reaction times for skilled responses. *Quarterly Journal of Experimental Psychology*, 12, 193–202.

Mowbray, G. H. (1964). Subjective expectancy and choice reaction times. *Quarterly Journal of Experimental Psychology*, 16, 216–223.

Mowbray, G. H., & Rhoades, M. V. (1959). On the reduction of choice reaction times with practice. *Quarterly Journal of Experimental Psychology*, 11, 16–23.

Moyer, R. S., & Bayer, R. H. (1976). Mental comparison of symbolic distance effect. *Cognitive Psychology*, 8(2), 228–246.

Näätänen, R. (1970). The diminishing time uncertainty with the lapse of time after the warning signal in reaction time experiments with varying foreperiods. *Acta Psychologica*, 34, 399–419.

Näätänen, R. (1971). Non-aging foreperiods and simple reaction time. *Acta Psychologica*, 35, 316–327.

Näätänen, R., & Gaillard, A. W. (1974). The relationship between the contingent negative variation and the reaction time under prolonged experimental conditions. *Biological Psychology*, 1, 277–291.

Noreen, D. L. (1981). Optimal decision rules for some common psychophysical paradigms. *Mathematical Psychology and Psychophysiology*, 13, 227–279.

Olson, C. L., & Ogilvie, J. C. (1972). Method of constant stimuli with 2 or more categories of response. *Journal of Mathematical Psychology*, 9, 320–338.

Olton, D. S., & Samuelson, R. (1974). Decision making in the rat: Response-choice & response-time measures of discrimination reversal learning. *Journal of Comparative and Physiological Psychology*, 87(6), 1134–1147.

Osaka, N. (1976). Visual reaction time as a function of target size and retinal eccentricity in the peripheral visual field. *Japanese Psychological Research*, 18(4), 183–190.

Pachella, R. G., & Fisher, D. (1972). Hick's law and the speed-accuracy trade-off in absolute judgment. *Journal of Experimental Psychology*, 92(3), 378–384.

Pachella, R. G., & Pew, R. W. (1968). Speed-accuracy tradeoff in reaction time: Effect of discrete criterion times. *Journal of Experimental Psychology*, 76(1), 19–24.

Pease, V. (1964). The intensity-time relation of a stimulus in simple visual reaction time. *The Psychological Record*, 14, 157–164.

Patton, R. A. (1972). A parametrization of the logistic function which admits uncorrelated maximum likelihood estimators. *Biometrics*, 28, 469–474.

Pease, V. P., & Sticht, T. G. (1965). Reaction time as a function of onset and offset stimulation of the fovea and periphery. *Perceptual and Motor Skills*, 20, 549–554.

Petrusic, W. M., & Baranski, J. V. (1989). Semantic congruity effects in perceptual comparisons. *Perception and Psychophysics*, 45(4), 439–452.

Petrusic, W. M., & Jamieson, D. G. (1976). Studying preferential choice without stable probability estimates: Temporal transitivity analyses. *Acta Psychologica*, 40, 375–383.

Petrusic, W. M., & Jamieson, D. G. (1978). The relation between probability of preferential choice and time to choose changes with practice. *Journal of Experimental Psychology*, 4(3), 471–482.

Petrusic, W. M., & Jamieson, D. G. (1979). Resolution time and the coding of arithmetic relations on supraliminally different visual extents. *Journal of Mathematical Psychology*, 19(2), 89–107.

Pew, R. W. (1969). The speed accuracy operating characteristic. In W. G. Kostner (Ed.), *Attention and performance II*. (pp. 16–26). Amsterdam: North-Holland.

Pierrel, R., & Savino-Murray, C. (1963). Some relationships between comparative judgment, confidence, and decision-time in weight-lifting. *American Journal of Psychology*, 76, 28–38.

Pike, A. R. (1968). Latency and relative frequency of response in psychological discrimination. *British Journal of Mathematical and Statistical Psychology*, 21, 161–182.

Pike, A. R. (1971). The latencies of correct and incorrect responses in discrimination and detection tasks: Their interpretation in terms of a model based on simple counting. *Perception and Psychophysics*, 9(6), 455–460.

Pike, A. R. (1973). Response latency models for signal detection. *Psychological Review*, 80, 53–68.

Pike, A. R., McFarland, K., & Dalgleish, L. (1974). Speed-accuracy tradeoff models for auditory detection with deadlines. *Acta Psychologica*, 38, 379–399.

Pike, R., & Ryder, P. (1973). Response latencies in the yes/no detection task: An assessment of two basic models. *Perception and Psychophysics*, 13(2), 224–232.

Possamai, C. A., Granjon, M., Requin, J., & Reynard, G. (1973). Sequential effects related to foreperiod duration in simple reaction time. *Perceptual and Motor Skills*, 36, 1185–1186.

Pratt, C. C. (1933). The time-error in psychophysical judgments. *American Journal of Psychology*, 45, 292–297.

Rabbitt, P. M. A., & Rodgers, B. (1977). What does a man do after he makes an error? An analysis of response programming. *Quarterly Journal of Experimental Psychology*, 29, 727–743.

Rebert, C. S. (1972). The effect of reaction time feedback on reaction time and contingent negative variation. *Psychophysiology*, 9, 334–339.

Reed, C. (1975). RT's for a same-different discrimination of vowel-consonant syllables. *Perception and Psychophysics*, 18(2), 65–70.

Requin, J., & Granjon, M. (1969). The effect of conditional probability of the response signal on simple reaction time. *Acta Psychologica*, 31, 129–144.

Rich, G. J. (1925). The measure of approximation of data to the phi-gamma hyothesis. *American Journal of Psychology*, 36, 615–620.

Ross, H. E. (1964). Constant errors in weight judgments as a function of the size of the differential threshold. *British Journal of Psychology*, 55(2), 133–141.

Schouten, J. F., & Bekker, J. A. M. (1967). Reaction time and accuracy. *Acta Psychologica*, 27, 143–153.

Seward, G. H. (1928). Recognition times as a measure of confidence. *Archives of Psychology*, 99, 1–52.

Shallice, T., & Vickers, D. (1964). Theories and experiments on discrimination times. *Ergonomics*, 7, 37–49.

Shepard, R. N. (1978). On the status of "Direct" psychophysical measurement. In Savage, C.W. (Ed.), *Perception and cognition: Issues in the foundations of psychology*. Minneapolis: University of Minnesota Press.

Smith, E. E. (1968). Choice reaction time: An analysis of the major theoretical positions. *Psychological Bulletin*, 69, 77–110.

Smith, P. T. (1968). Cost, discriminability, and response bias. *British Journal of Mathematical and Statistical Psychology*, 21, 35–60.

Sorkin, R. (1962). Extension of the theory of signal detection to matching procedures in psychoacoustics. *Journal of the Acoustical Society of America*, 34, 1745–1751.

Snodgrass, J. G. (1969). Foreperiod effects in simple reaction time: Anticipation or expectancy? *Journal of Experimental Psychology Monographs*, 79(3), 1–19.

Snodgrass, J. G. (1972). Matching patterns vs. matching digits. The effect of memory dependence and complexity on "same"-"different" reaction times. *Perception and Psychophysics*, 11, 341–349.

Snodgrass, J. G., & Kass, R. (1972). Time and consistency as measures of psychological distance. *Perceptual and Motor Skills*, 35, 635–645.

Spence, K. W. (1954). The relation of response latency and speed to the intervening variables and N in S-R theory. *Psychological Review*, 61, 209–216.

Squires, L., Petuchowski, S., Wickens, G., & Donchin, E. (1977). The effects on stimulus sequence on event related potentials: A comparison of visual & auditory sequences. *Perception and Psychophysics*, 22(1), 31–40.

Sternberg, S. (1969). The discovery of processing stages: extensions of Donder's method. In W. G. Kostner (Ed.), *Attention and performance II*. Amsterdam: North Holland.

Stevens, S. S. (1957). On the psychophysical law. *Psychological Review*, 64, 153–181.

Stevens, S. S. (1958). Adaptation-level vs. the relativity of judgment. *American Journal of Psychology*, 71, 633–64.

Stevens, S. S., & Galanter, E. H. (1957). Ratio scales and category scales for a dozen perceptual continua. *Journal of Experimental Psychology*, 54(6), 377–411.

Stilitz, I. (1972). Conditional probability and components of RT in the variable foreperiod experiment. *Quarterly Journal of Experimental Psychology*, 24, 159–168.

Still, A. W., & MacMillan, A. S. (1977). Response bias and the measurement of choice alternation. *Quarterly Journal of Experimental Psychology*, 29, 319–325.

Suci, G. J., Davidoff, M. D., & Surwillo, W. W. (1960). Reaction time as a function of stimulus information and age. *Journal of Experimental Psychology*, 60, 242–244.

Sully, J. (1884). *Outlines of psychology*. London: Longmans, Green.

Swensson, R. G.(1972) The elusive trade-off speed vs. accuracy in visual discrimination tasks. *Perception and Psychophysics*, 12, 16–32.

Swensson, R. G., & Green, D. M. (1977). On the relation between random walk models for two-choice response times. *Journal of Mathematical Psychology*, 15(3), 282–291.

Swets, J. A., Tanner, W. P., Jr., & Birdsall, T. G. (1961). Decision processes in perception. *Psychological Review*, 68, 301–340.

Symington, L. E. (1971). *Reaction time: a bibliography with abstracts supplement I with index for entire bibliography*. Human Engineering Laboratories, Aberdeen Proving Ground, Maryland.

Takane, Y., & Sergent, J. (1983). Multidimensional scaling models for reaction times and same-different judgments. *Psychometrika*, 48(3), 393–423.

Tanaka, Y. (1972). Muller-Urban weights. *Japanese Psychological Research*, 14, 43–46.

Tanner, W. P., Jr. (1956). Theory of recognition. *Journal of the Accoustical Society of America*, 28, 882–888.

Tanner, W. P., Jr., & Norman, R. Z. (1954). The human use of information. II. Signal detection for the case an unknown signal parameter. *Transactions of The Institute of Radio Engineers: 1954 Symposium on Information Theory*, 4, 222–227.

Teich, M. C., Prucnal, P. R., Vannucci, G., Breton, M. E., & McGill, W. J. (1982). Multiplication noise in the human visual system at threshold: I. Quantum fluctuations and minimum detectable energy. *Journal of the Optical Society of America*, 72(4), 419–431.

Theios, J. (1975). The components of response latency in simple human information processing tasks. In Rabbitt, P., & Dornic, S. (Eds.), *Attention and performance V*. (pp. 418–440). London: Academic Press.

Thomas, E. A. C. (1967). Reaction time studies: The anticipation and interaction of responses. *British Journal of Mathematical and Statistical Psychology*, 20, 1–29.

Thomas, E. A. C. (1971). Sufficient conditions for monotone hazard rate and application to latency-probability curves. *Journal of Mathematical Psychology*, 8, 303–332.

Thomas, E. A. C. (1973). On expectancy and the speed of accuracy of responses. In S. Kornblum, (Ed.), *Attention and performance IV*, (pp. 613–626). New York: Academic Press.

Thomas, E. A. C. (1975). A note on the sequential probability ratio test. *Psychometrika*, 40(1), 107–111.

Thomson, G. H. (1914). The accuracy of the phi-gamma process. *British Journal of Psychology*, 7, 44–55.

Thurstone, L. L. (1927c). Equally often noticed differences. *Journal of Educational Psychology*, 18, 289–293

Todhunter, I. (1949). *A history of the mathematical theory of probability*. New York: Chelsea. (Original work published 1865).

Torgerson, W. S. (1958). *Theory and methods of scaling*. New York: Wiley.

Townsend, J. T. (1971). Theoretical analysis of an alphabetic confusion matrix. *Perception and Psychophysics*, 9, 40–50.

Turner, W. D. (1931). Intra-serial effects with lifted weights. *American Journal of Psychology*, 43, 1–25.

Urban, F. M. (1927). The accuracy of the method of constant stimuli. *American Journal of Psychology*, 38, 252–256.

Vickers, D. (1979). *Decision processes in visual perception*. New York: Academic Press.

Vickers, D., Nettelbeck, T., & Willson, R. J. (1972). Perceptual indices of performance: the measurement of 'inspection time' and 'noise' in the visual system. *Perception*, 1, 263–295.

Vierordt, K. (1870). Die abhängigkeit der ausbildung des raumsinnes der haut von der beweglichkeit der körpertheile. *Zeitschrift für Biologie*, 6, 53–73.

Wandell, B. A. (1977). Speed-accuracy tradeoff in visual detection. Applications of neural counting and timing. *Vision Research*, 17, 217–225.

Wilding, J. M. (1974). Effects of stimulus discriminability on the latency distribution of identification responses. *Acta Psychologica*, 38, 483–500.

Wright, A. A. (1974). Psychometric and psychophysical theory within a framework of response bias. *Psychological Review*, 81(4), 322–347.

Wundt, W. (1913). Grundriss der Psychologie. Leipzig: Kröner.

Yellott, J. I. Jr. (1975). The relationship between Luce's choice axiom, Thurstone's theory of comparative judgment, and the double exponential distribution. *Social Sciences working paper No. 74*. University of California, Irvine.

Zahn, T. P., & Rosenthal, D. (1966). Simple reaction time as a function of the relative frequency of the preparatory interval. *Journal of Experimental Psychology*, 72, 15–19.

Zajonc, R. B., & Cross, D. V. (1965). Stimulus generalization as a function of drive shift. *Journal of Experimental Psychology*, 69, 363–368.

Author Index

C

D

E

F

G

Subject Index

For Product Safety Concerns and Information please contact our EU
representative GPSR@taylorandfrancis.com
Taylor & Francis Verlag GmbH, Kaufingerstraße 24, 80331 München, Germany

www.ingramcontent.com/pod-product-compliance
Ingram Content Group UK Ltd.
Pitfield, Milton Keynes, MK11 3LW, UK
UKHW022207190926
475702UK00003B/241
* 9 7 8 0 3 6 7 1 5 0 4 6 4 *